Media Business and Innovation

Series editor

M. Friedrichsen, Stuttgart Media University, Germany and Humboldt School Stuttgart/Berlin, Germany

More information about this series at http://www.springer.com/series/11520

Artur Lugmayr • Cinzia Dal Zotto
Editors

Media Convergence Handbook - Vol. 1

Journalism, Broadcasting, and Social Media Aspects of Convergence

Springer

Editors
Artur Lugmayr
Visualisation and Interactive Media
School of Media, Culture,
and Creative Arts (MCCA)
Curtin University
Bentley/Perth, Western Australia
Australia

Cinzia Dal Zotto
Academy of Journalism and Media
Faculty of Economics and Business
University of Neuchâtel
Neuchâtel,
Switzerland

Media Business and Innovation
ISBN 978-3-642-54483-5 ISBN 978-3-642-54484-2 (eBook)
DOI 10.1007/978-3-642-54484-2

Library of Congress Control Number: 2015956388

Springer Heidelberg New York Dordrecht London

Printed on acid-free paper

Springer-Verlag GmbH Berlin Heidelberg is part of Springer Science+Business Media (www.springer.com)

Preface

We are very happy to present our edited book covering the thematic area of convergence from a cross-disciplinary viewpoint. The edited book constitutes an attempt to investigate several aspects of convergence such as the strategic, technological, consumer driven, and application-oriented convergence. As these aspects of convergence are interrelated, the book shall shed light into the complexity of the topic and act as reference book and educational resource in the field. Convergence is very often discussed at a rhetorical and political level, without practical convergence process issues or concerns about potential divergence effects being part of the discourse. Today media managers are faced with the need to satisfy users' expectations and at the same time undertake convergence as an industrial strategy to achieve economies of scope and scale. Media technology developers are faced with rapidly emerging new technologies and information systems to quickly develop additional media services. Media scholars are bound to understand technological developments and industry convergence processes in order for their research to be relevant. With this book, we thus attempt to give answers to media managers, media scholars, students, researchers, as well as to technology developers and help them get a clearer understanding of this thematic area.

When we launched our edited book initiative, our primary goal was to investigate potential interest in this thematic area through a call for abstracts. We were overwhelmed when we received 87 abstracts, which have been evaluated by us as editors. 73 abstracts were invited to contribute a full book chapter, while 15 abstracts were rejected and not invited for a final manuscript version. To ensure high scientific standards, we implemented a strict blind peer review process for each book chapter. After this first review, we accepted the chapters for which review statements supported the publication. The final version of all chapters has been reviewed one more time in order to check that all review comments were adequately integrated. We ended up with 39 accepted book chapters, including 2 chapters contributed by us as editors which provide a red line through the book. Thus, the acceptance rate was 45 % with respect to the submitted abstracts and 54 % with respect to the invited abstracts. The chapters were divided between two book volumes: the first volume includes 19 chapters on journalism, broadcasting, and social media aspects of convergence, while the second volume presents 20 chapters covering firm and user perspectives on convergence. Most of the thematic areas that we wanted to include in the book (see Fig. 1) as well as many different media genres are represented within the contributions (see Fig. 2).

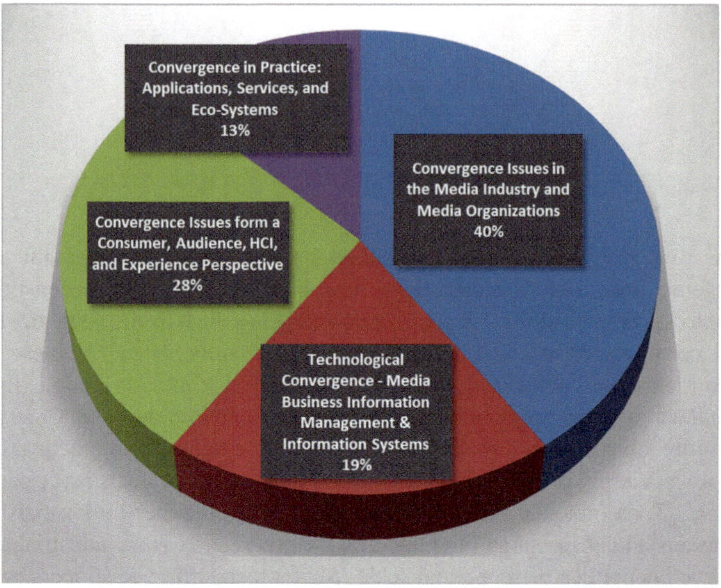

Fig. 1 Contributions to the thematic areas

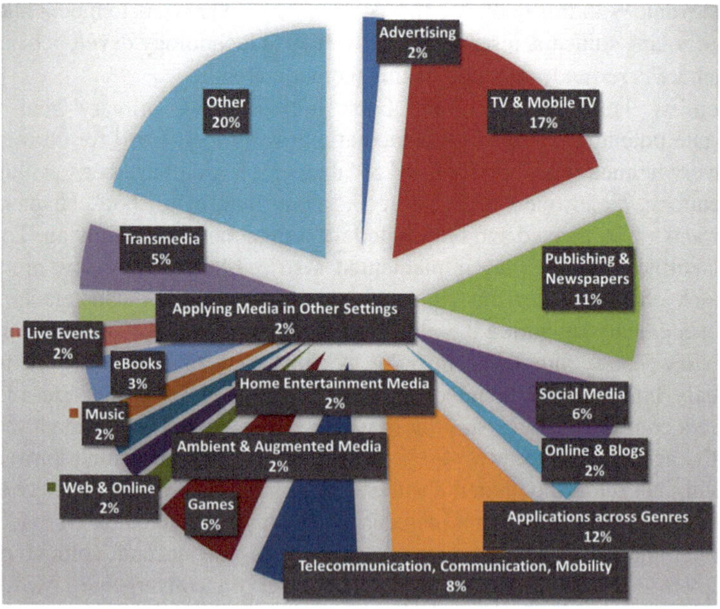

Fig. 2 Overview of the media genres that have been covered by contributions to this edited book

Fig. 3 Tag cloud of contributions

We started to work on the media convergence idea back in 2011, with a common contribution to the European Media Management Conference (EMMA) (Dal Zotto, Galichet, & Lugmayr, 2011). In 2013, we compiled another conference paper about this thematic area (Lugmayr & Dal Zotto, 2013). On both occasions, the audience was rather fascinated by our cross-disciplinary viewpoint on convergence. After many discussions, and in particular in acknowledgement to Prof. Greg Lowe, we started to work on this book project. By the end of 2012, we collected the extensive set of book abstracts which was rather surprising for us as editors. To cope with the high amount of submissions across a wide and thematic area (see Fig. 3 for a tag cloud of the chapters' keywords), Springer-Verlag gave us permission to divide the book into two volumes.

In particular, we would like to give our greatest appreciation to and warmly thank Prof. Dr. Greg Lowe, who supported our idea and initially helped us in setting up the book project. He gave us clarifying insights on convergence and very useful hints in which thematic area to develop the book project. We also would like to thank Springer-Verlag and in particular Barbara Fess who was extremely helpful, patient, and helped us though the book editing process with many questions. Last but not least, we definitely would like to thank all numerous reviewers of the book chapters. Without their hard work, we would have not been able to complete this extensive book. The first editor, Prof. Dr. Artur Lugmayr, would especially like to thank the Akerlund Foundation for funding his scientific activities and his professorship.

We would like to make you also aware about our Web site, which shall act as a collaborative platform for people interested in the thematic area. You can find our group email address, Facebook group, as well as other resources online on:

http://www.ambientmediaassociation.org/convergence. Further, we would like to pinpoint to other activities that we are currently conducting within the Association for Information Systems (AIS) in the Special Interest Group (SIG) eMedia: http://aisnet.org/group/SIG-eMedia or within the International Ambient Media Association (iAMEA): http://www.ambientmediaassociation.org. And finally, we wish you as reader of the book that you will have some useful and clarifying insights into such a complex thematic area as media convergence and gain a deeper understanding into this exciting topic.

We also would like to refer to our book website, which contains additional information, email lists, errata, and most importantly course and lecture slides: www.artur-lugmayr.com.

Bentley/Perth, Western Australia, Australia Artur Lugmayr
Neuchâtel, Switzerland Cinzia Dal Zotto

References

Dal Zotto, C., Galichet, G., & Lugmayr, A. (2011). *Managing media convergence: An economic and editorial project*. European Media Management Conference 2011, Moscow, Russia.

Lugmayr, A., & Dal Zotto, C. (2013). *Convergent divergence – cross-disciplinary viewpoint on media convergence*. European Media Management Education Association Conference 2013 – Digital Transformations and Transactions in Media Industries, Bournemouth, UK.

Contents

Part I

Media Convergence: Some Introductory Perspectives

Media Convergence as Evolutionary Process

Cinzia Dal Zotto and Artur Lugmayr

1 What Do We Understand Under Media Convergence?

One of the challenges of studying media convergence is that the concept is so broad that it has multiple meanings. As a result the academic literature in this area is diverse and underdeveloped from both a theoretical and an empirical perspective (Wirth, 2006). In this introductory chapter we will try to take the different interpretations of media convergence—which will be evident in the various book chapters—into account, but attempt to make them converge into some common ground, which we already investigated in (Lugmayr & Zotto, 2013; Dal Zotto, Galichet, & Lugmayr, 2011), AIS SIG eMedia (SIG-eMedia n.d.), and the International Ambient Media Association (iAMEA) (Anon n.d.).

According to the Merriam-Webster's Dictionary the general concept of convergence refers to "the act of converging and especially moving toward union or uniformity" (Mish, 1993). Within the media field, which is the research area that we are addressing in this book, convergence can be identified with the "ability to deliver different media channels via one digital platform" (McPhillips & Merlo, 2008). Previously broadcast media such as radio, television (e.g. Lugmayr, Niiranen, & Kalli, 2004) and the Internet, as well as print media have been distributed via different and well-distinguished platforms. Today content is becoming more and more digitalized: No matter the type of signal, any content can be transformed into undifferentiated bits of data that converge onto the same platform

C.D. Zotto (✉)
Academy of Journalism and Media, Faculty of Economics and Business, University of Neuchâtel, Neuchâtel, Switzerland
e-mail: cinzia.dalzotto@unine.ch

A. Lugmayr
Visualisation and Interactive Media, School of Media, Culture, and Creative Arts (MCCA), Curtin University, Bentley/Perth, Western Australia, Australia
e-mail: lartur@acm.org; artur.lugmayr@lugymedia.com

A. Lugmayr, C. Dal Zotto (eds.), *Media Convergence Handbook - Vol. 1*, Media Business and Innovation, DOI 10.1007/978-3-642-54484-2_1

(McPhillips & Merlo, 2008). Media convergence is therefore rather a process more than an outcome. However, digitalized content can be distributed on different digital platforms with little or no impact on costs. Content convergence can lead to distribution and thus consumption divergence. As such, the media convergence concept does not only refer to a technological shift but it includes changes within the industrial, cultural and social paradigms of our environment reflecting both media convergence and divergence processes.

Indeed media convergence alters relationships between technologies, industries, audiences, genres and markets. According to Rheingold (2000) advances in technology enable technological convergence, which in turn has "social-side effects" in that "the virtual, social and physical world are colliding, merging and coordinating". Jenkins (2006) states that media convergence is "the flow of content across multiple media platforms, the cooperation between multiple media industries, and the migratory behavior of media audiences" suggesting that media audiences nowadays play a crucial role in creating and distributing content, and convergence therefore has to be examined in terms of social, as well as technological changes within the society. Thanks to new technologies consumers are encouraged to actively seek out new information and make personalized connections among dispersed media content. Thus, media convergence requires media companies to rethink existing assumptions about media from a consumer's point of view, as these affect not only marketing but also production decisions (Krumsvik, 2013; Urban & Bodoky, 2013). With the proliferation of different media channels and an increasing portability of new telecommunications and computing technologies, we have entered into an era where media are constantly surrounding us. Furthermore, technology has empowered consumers who not only consume content from many different sources but also participate and contribute to content production activities. The subsequent oversupply and commoditization of media content have led to an efficiency decline of the traditional media business model (Dal Zotto, Dichamp, & Sommer, 2012; McPhillips & Merlo, 2008; Picard, 2000; Picard & Dal Zotto, 2006). As direct consequence the initial revolutionary digital paradigm shift presumed old media to be increasingly replaced by new media. However, the currently emerging convergence paradigm suggests that new and old media would interact in more complex ways than previously predicted (Wirtz, Schilke, & Ullrich, 2010).

Already in 2006 Jenkins saw media convergence as an ongoing process that should not be considered as a displacement of the old media, but rather as interaction between different media forms and platforms (Jenkins, 2006). Supporting this argument, Deuze (2011) suggests that media convergence should be viewed as 'cooperation and collaboration' between previously unconnected media forms and platforms. This holds even more true when we consider that, due to the multitude of available content on and off-line, the newly empowered consumers search for trustable media products and services. This has pushed media firms to develop into reputable brands that can offer content in a number of different forms and platforms. Newspapers for instance produce and sell books, or market a radio or web TV under their brand umbrella; the movie industry exploits the name of successful movies to produce and sell books, video games, or cartoons (Tameling & Broersma, 2013).

Branding encourages the expansion of a successful media concept into multiple platforms and therefore stimulates media industry convergence (Matteo, Spolaor, & Dal Zotto, 2013). However, this process does not enhance the creation of new ideas and further leads to technology divergence in the hardware industry: hardware needs to be specific to each function, so it diversifies in order to accommodate media convergence requirements. It thus appears that, in order to regain efficiency and success, media business models should not only react and adapt to the convergence trend, but also—and first of all—proactively factor in causes and effects of the media convergence process: this means to understand the multifaceted meaning of media convergence, including divergence and coexistence effects, develop a more customer oriented vision as well as to innovate by merging traditional with new elements, too.

2 Evolutionary Paths of Media Convergence

The development of media convergence processes is mainly due to the following three factors: (a) digitalization, (b) deregulation of media and communication markets, and (c) changes in user preferences. Digitalization offers new opportunities for the creation, presentation, storage and distribution of media products (Rayport & Jaworski, 2001). The standardization of storage media that were previously separated enhances the exploitation of economies of scope in preceding and subsequent stages of the value-creation chain (Wirtz, 2011). Further, with the conversion from analog to digital data communication, differing communication networks can be used for data transmission and become substitutable among each other. Internet access is for instance now possible via telephone, cable, satellite, mobile networks as well as wireless LANs.

As to the deregulation, since the middle of the 1990s measures have been introduced for liberalizing the telecommunication sectors both in the US and Europe. This has led to the emergence of competitive structures in the information, media and communication industries (Dal Zotto & Dowling, 2003; Dal Zotto & van Kranenburg, 2008). As a result the supply of media services has increased while audiences and media consumption were fragmentating. During the last decade consumers have been using an increasing number of different services to satisfy their information and entertainment needs (Aitamurto, Sirkkunen, & Lehtonen, 2011; Rayport & Jaworski, 2001). A clear change in user preferences towards personal information and communication tools can indeed be identified (Wirtz, 2011).

Media convergence has empowered users to produce and distribute their own content, viewing and listening schedules can be personalized via on-demand or search content and mobile services, users can engage with one another via peer to peer technology. Even if consumers can be distinguished between early adopters—the media literate and technology savvy ones—and late adopters—those who are happy with their current media experience—there is no doubt that such changes in user preferences together with the digitalization and deregulation of media and communication markets represent factors with a considerable impact on the competitive environment that media firms are facing (Picard, 2010). Depending on this impact we can distinguish different types and evolutionary paths of media convergence, as we illustrate below.

Technological convergence, that is the tendency of different technological systems to evolve towards performing similar tasks due to digitalization, leads on one hand to the convergence of different functionalities into a new product—*product convergence*—and on the other hand to an aggregation of markets that were previously distinguished from one another (Yoffie, 1997). In the market for communication services cable network and telecommunication companies, as well as mobile and satellite providers compete with each other. In order to differentiate themselves from each other they often pursue integration strategies in the area of content production: by enriching telecommunications products—such as Internet access or voice communication—with content, these operators provide consumers with accessory uses (Wirtz, 2011) and become a competitive threat to media companies. Once users consider products either substitutes or complements and as a consequence such products end up converging into a new product, we talk about *competitive or complementary convergence* (Dowling, Lechner, & Thilmann, 1998; Greenstein & Kanna, 1997; Picard, 2000, 2009).

Product level convergence can lead to *business level convergence* when a company realizes that it would be more advantageous to integrate convergent areas of business either within the company or through outside cooperation. This can stimulate economies of scale and innovations can emerge from a recombination of individual business units' performances (Wirtz, 2011). If the convergence of previously separated services concerns not only a business unit but the whole enterprise, this can lead to cooperations or mergers between previously separated companies. In this case we talk about *strategic or industrial structure convergence* (Chan-Olmsted & Kang, 2003; Thilmann & Dowling, 1999; Wirtz, 1999). Convergence is transforming the media and telecommunication industries from vertical businesses—telephone, television, computers—into horizontal segments—content, packaging, transmission network, manipulation infrastructure, and terminals. Established media firms have for long dominated the entire value chain, from creative inception to production, packaging, marketing and finally distribution. Thanks to technological progress production costs have substantially diminished and the Internet is providing an open-source route to market. Thus content producers are now able to easily bypass mainstream media and distribute their content independently (McPhillips & Merlo, 2008). In this case strategic convergence may lead to industrial structure divergence.

As a consequence of this evolution we have already been witnessing a consolidation phase among technology, media and telecommunication firms. Current structural changes point to a few major media players dominating the mass market and a few niche operators serving specialist areas. Middle size operators seem to be struggling to find scale enough to compete with larger firms. Media convergence seem thus to lead to a polarization of the industry structure with global conglomerates and small independent media owners co-existing (Picard, 2010).

Media ownership concentration, that is the control of media concentrated in the hands of a few private owners, is one of the major concerns when examining positive and negative consequences of media convergence (Jenkins, 2006). On the one hand, media ownership concentration—and thus industrial structure convergence—may cause the decline in the diversity of media products and service offerings and result in a tendency not to take into account voices of those lacking economic power (Horwitz, 2005; Valcke, Picard, Süksöd, & Sanders, 2009). On the other hand, it is argued that market driven media owned and controlled by big media corporations, thanks to their disposal of financial resources, can actually improve the value of the service, the plurality of topics and the competence of the contributors as well as enable technological developments, change the elitism of media professionals and create new general awareness (Barwise & Picard, 2012; Dal Zotto & Picard, 2015; Picard & van Weezel, 2008; Valcke, 2011).

3 The Convergence Process Between Advantages and Disadvantages

As mentioned above media convergence seems to represent an evolutionary process that accompanies the development of information technologies, telecommunications, media products and services, media industry and firm structures as well as media professionals. According to Stöber (2004) there are three stages of media evolution: invention, innovation and institutionalization. In the media evolution case of convergence, the **invention** stage consisted in the ability to convert differentiated bits of analogue data (such as voice, text, audio or visual) to undifferentiated bits of digital data. The **innovation** stage refers to the possibility to distribute and consume different media types, whether radio, TV or internet through the same platform and this phase is clearly completed. Indeed, the media industry is now finally moving towards the **institutionalisation** phase of convergence, which is defined as the adaptation to the environment to create a new « system entity » (McPhillips & Merlo, 2008). Such phase acquires more a revolutionary than an evolutionary character, as divergence rather than convergence processes can emerge as a result.

Stöber (2004) argues that four factors of change need to simultaneously occur to shape a new system entity and a viable business model: technological, cultural, political and economic factors. The first three are already in motion. *Advancements in technology* usually correspond to reductions in cost and significant increases in functionality. In the media industry this pattern can be identified with improvements in speed of data transfer for broadband, in quality of mobile devices as well as in costs of data storage. The high rate of adoption of broadband and other digital services confirm improvements of the perceived value of technological advancements. Furthermore, the time spent on consuming media services based on converged technologies such as mobile phone calls and Internet surfing, text messages, traditional and web radio listening as well as TV has rapidly increased and most of these consumption activities take place simultaneously. This is a clear sign of a *cultural shift* away from single source consumption patterns (McPhillips & Merlo, 2008) and denotes a device diverging trend. From a *political* point of view media technology convergence is enhanced by governments' encouragement if not enforcement of digital broadcast adoption.

A fourth critical factor of change towards the institutionalization phase of media convergence is the *economic factor*, which is still evolving. The emergence of an appropriate business model will determine the future of the media industry (McPhillips & Merlo, 2008). For the moment we have just seen media convergence leading to an over-supply and commoditization of media content, the decline of advertising effectiveness due to media fragmentation as well as the over-exposure of audiences to an infinite number of advertising touch points leading to the decline of the current media business model (Dal Zotto et al., 2012; Matteo & Dal Zotto, 2015; Picard & Dal Zotto, 2006). Increasing the number of vehicles, touch points and the volume of advertising may increase revenues in the short term but it eventually leads to a downward pressure on margins in the medium-long terms (McPhillips & Merlo, 2008).

Similarly, as the attention that consumers can dedicate to media is limited, the direct consequence of media fragmentation, or consumption divergence, is not only the erosion of advertising revenues but also the decline of subscription revenues. Information media have reacted to this effect by cutting budgets and costs. Technological and product convergence can thus have negative effects as it is the case for journalism and journalistic products (Deuze, 2004; Dupagne & Garrison, 2006; Fioretti & Russ-Mohl, 2009; Huang et al., 2006; Huang, Rademakers, Fayemiwo, & Dunlap, 2004; Lawson-Borders, 2006; Meier, 2007). Online information media have a 24-h constantly updated news cycle and require staff to be able to work multiplatform. Further, not only content is constantly updated—thus reducing staff's time to research, report or even think about their work (Jarvis, 2009; Klinenberg, 2005)—but it is also developed across multiple platforms and channels, a phenomenon called multimedia storytelling (Gray et al., 2012; Jenkins, 2001). This inevitably calls for a cultural and organizational change within media firms (Boczkowski, 2005) and increases the level of uncertainty among employees. Within web-driven newsrooms journalists have for instance less time to talk with sources and check facts (Porlezza & Russ-Mohl, 2013; Thurman & Myllylahti, 2009), they will therefore tend to report news that have already been or are being reported by other colleagues in other media in order to reduce uncertainty. As a result information becomes standardized in its format and uniform in its content making it difficult for media firms to distinguish themselves from each other and gain a competitive advantage. Furthermore, multiplatform content production and distribution reflects a process of editorial convergence: a converged newsroom provides content for multiple distribution platforms thanks to the contribution of different, complementary and multiskilled media professionals. In order to accommodate and take advantage of cross-platform media strategies journalistic and editorial work within newsrooms needs to be reorganized while staff recruiting, training and development measures become fundamental.

4 Managing the Media Convergence Process: Outline of the Book

So far we have learnt that factors such as digitalization, deregulation and the subsequent changes in user preferences enhance the media convergence process. Depending on the impact of those factors different types of media convergence can be distinguished, even if they may be complementary and share evolution paths.

Indeed, media industry convergence is the result of technological convergence and the subsequent need for media firms—which are characterized by high fixed costs and low marginal costs—to reach economies of scale. As we have seen, technological development has led to an increasing fragmentation of media and audiences, denoting in this case a divergence process, and subsequently to a products and services' price deflation. A viable way to address this inevitable price deflation seems to be the application of competition policy, allowing industries to become more concentrated and so reduce competitive pressures. Considering that the loss of pricing power quickly leads to a decline in industry profitability, the only possible reaction for the media industry has been a horizontal consolidation or vertical integration to gain gatekeeping power (Noam, 2006, 2009).

Furthermore, both technological and industry convergence leads to experience and social convergence at consumer level, as well as to cultural and editorial convergence within media organizations (See Fig. 1). Due to multimedia and multiplatform content production and distribution environments different professional cultures and tasks are thus melting together. It is therefore clear that, in order to successfully manage media convergence processes, not only technological and

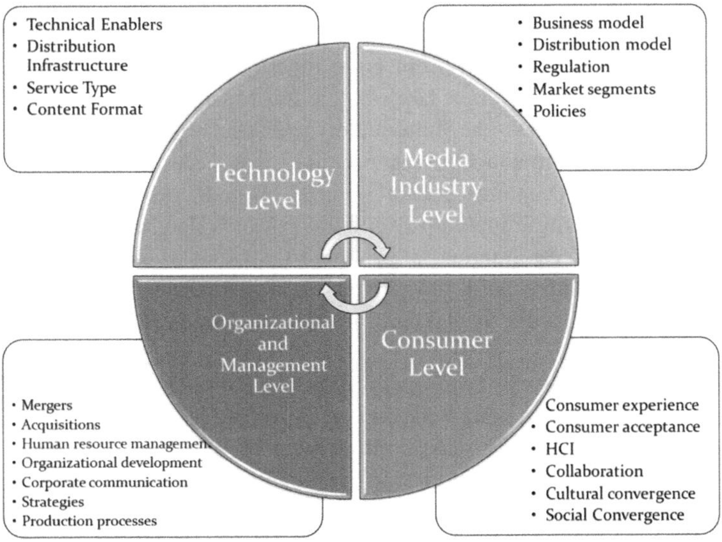

Fig. 1 Impact of media convergence at different levels

media economics aspects but also human resource and organizational development related aspects need to be taken into account.

In this handbook we have tried to touch upon all four main levels of media convergence—technology (e.g. information systems level (Lugmayr, 2013)), industry, consumer and organizational level—in order to show its multidisciplinary and multifaceted character (see Fig. 1). As contributions were many and converging around two main areas, we decided to organize them into two volumes, one focusing on the journalism, broadcasting and social media aspects of convergence, and a second one highlighting the perspectives of media firms and users.

4.1 Volume One

The *first volume* of this handbook starts with a closer look at the meaning of media convergence by presenting at the same time three different perspectives on the concept. As mentioned at the beginning of this introductory chapter, in order to analyze cause and effects of media convergence and thus conceive solutions to current media industry challenges, it is necessary to first understand the multifaceted meaning of media convergence. In the **first part** Arango-Forero, Roncallo-Dow and Uribe-Jongbloed address the idea of convergence both as an abstract concept and as a word ascribed to specific phenomena. They point out a divergent conceptualization of convergence in the literature so far and the necessity to disentangle the abstract concept from the working definition of convergence which refers to current geo-economic and culturally bound phenomena. The evolution of media convergence from its first stage of merging technologies to its second catalyzing stage is problematized by Marie-José Montpetit in the second chapter. Through a thorough discussion of network and device convergence, she explains how the advent of social connectivity has disrupted the media consumption experience and pushed convergence away from basic functional effects into a richer set of interdependent elements such as hardware, software, content and user interaction. Shantanu Dey concludes the introductory perspectives part on convergence by highlighting that media convergence does not necessarily lead to "power convergence"—i.e. "economic power" in terms of media ownership concentration is not necessarily being imposed on "cultural power". He claims that the existing open architecture of the Internet allows the articulation of divergent powers and proposes an "affirmative public ownership" policy at network infrastructure and applications level in order to maintain the structural openness of the system and enable free speech.

The volume continues by specifically addressing three media fields: journalism, broadcasting and social media. Focusing on convergence challenges in journalism and publishing, **part two** discusses the impact of news media online shift on journalistic quality. Christian Wellbrock argues that this shift is preventing mass media do adequately fulfill their function within democratic societies and suggests that public service broadcasters should be allowed to expand their activities online. On the same line, by discussing the implications of convergence on human resources and labor conditions, according to Spyridou and Veglis emerging convergence skills might reduce professional autonomy and journalistic performance. Kay Hamacher compares blogs and traditional media in order to see if digitalization increases the degree of content diversification, while Veglis, Dimoulas and Kalliris investigate technological issues when implementing cross-media publishing. Pointing out that media publishing channels differ in terms of production and consumption, they propose a modular model to appropriately manage cross-media publishing automation while respecting the peculiarities of the different publishing areas. Moving towards the consumer perspective, Mustic and Plenkovic highlight that the graphic representation of information affect the perception of content and this information credibility. Ebner, Schön and Alimucaj present a new print-on-demand service that can improve the perceived quality of open access online journals by allowing readers to assemble different articles and print their "personal book". Finally, the last paper in this chapter problematizes the fact that, in order for new business models to emerge and for the online shift to have success, media professionals need to reconsider their belief system.

Part three concentrates on broadcasting. In particular, Matteucci discusses intellectual property rights within the context of digital TV convergence in the European Union. He indicates how patent portfolio strategies and control of crucial copyrights become increasingly central for competing in a converging TV sector. Deliyannis looks at presentation requirements for a converged interactive television through some real-life case studies; Tuomi approaches mobile devices as second screen idea to enhance consumers' television experience in our present century, while Ala-Fossi problematizes existing divergences within converging radio and television newsrooms by highlighting the different approach followed by Europe and the USA on one hand and Japan on the other hand.

Social media aspects of convergence are the topic of **part four**. Villi, Matikainen and Khaldarova discuss technology-enabled user distributed content by exploring how the convergence enables a participatory audience to disseminate news media online content within social networks. Their study draws on the analysis of 15 news media within seven different countries. On the same path Duh, Meznaric and Korosak examine the possibility to use different social media channels for the automatic generation of interactive social media news streams. Zlatanov and Koleva argue that the raising information quantity determines the superiority of technology savvy and communication experienced individuals, as well as their subsequent major influence on information spread within online social networks. The authors point out that technological convergence might thus lead to social divergence. The tension between technological convergence and social divergence is problematized by Damasio, Henriques, Teixeira-Botelho and Dias, too. Their paper concludes the first volume of this handbook and focuses on mobile technology. According to the authors mobile Internet access facilitates more interactive and participative activities on one hand, but it results in network-based individualism on the other hand.

4.2 Volume Two

If the first volume focuses on the impact of digitalization on the media industry and thus on technology convergence, the *second volume* of the handbook considers structural/economic convergence issues as well as cultural convergence. **Part one** deals with the impact of convergence on media organizations and media markets. In the first paper Gershon examines media convergence effects on special design and operations within transnational organizations. Georgiades continues by addressing the critical issue of employees' involvement and highlighting how divergent understandings of concepts such as communication and employee involvement itself may hamper organizational convergence processes. Karmasin, Diehl and Koinig argue that media convergence implies the need for media firms to develop new business models, appropriate organizational structures as well as new managerial competences. Vukanovic concludes this part by analyzing how media convergence can create opportunities and shape successful business models.

Part two mainly discusses convergence in relation to media production activities. Denward's contribution deals with participation enabling media production. By analyzing the design and implementation phases of a drama production in Sweden, the author approaches various types of convergence—industrial, technology, cultural and social—and the difficulties they created in the production process. Noronha e Sousa, de Lemos Martins and Zagalo investigate the practical implications of producing and consuming transmedia narratives from both the perspective of producers and consumers. Innocenti, Pescatore and Rosati approach narratives, too. By using conceptual tools borrowed from the media, economics and information architecture studies, the authors offer a cross-disciplinary perspective on serial narratives as audiovisual production niche within a converging media environment. Indrek Ibrus applies an evolutionary approach to media change to interpret recent Northern European examples of how audiovisual micro-companies are confronting the "convergence culture" and innovating their output via cross-media strategies and transmedia storytelling. Foster and Brostoff's paper conclude part 2 discussing advances in 3D computer graphics technology and possible convergence enabled synergies between video game and urban design—two fields that share the goal of constructing a cohesive, credible and readable visual narrative about an alternative reality.

Part three focuses on user perspectives and experiences. Alexandre Fleury opens the part with an investigation about how convergence between mobile and television technology is experienced in highly connected societies. He explores current practices in Denmark and Japan and identifies a number of future trends for converging media solutions. Reichl continues by analyzing experience from a quality point of view. He addresses issues such as the development of laws as well as models to charge for quality of experience in converging communication systems. Tavares and Schofield deal with multisensory interaction design as critical success factor for a correct user experience of multiple devices, media and platforms. The following paper further addresses the implementation of a multimodal interface system within various media platforms to enhance the effectiveness of man-machine interaction. McKay points out though that effective media convergence depends on knowledge engineers' understanding of the changeable nature of the human-computer interaction environment. Finally, user experience design is approached from a transmedia storytelling point of view, early in the creative writing process. Ghellal, Wiklund-Engblom, Obal and Morrison define transmedia storytelling as evolving storylines running across multiple platforms. This definition highlights the complexity and importance of transmedia design, including for instance interdisciplinarity, genres and emergent production models.

Last but not least, **part four** presents a few papers discussing future media convergence trends. Grüblbauer and Haric conceptualize how to design and build a convergence application to effectively showcase big amounts of information and data coming from diversified sources. By analyzing existing theory on converging media and media planning, Svahn, Wahlund, Denward, Rademaker and Nilsson present a model for evaluating converging media for advertising purposes. Giesecke analyses the business models of 16 event ecosystem organizations in four different countries. His conclusions indicate that, in order to achieve convergence, the news media industry needs to learn from the entertainment media industry, build and bridge networks, and ensure affinity between content and advertisements. Finally, Giesecke and Immonen argue that mid-sized mass media companies suffer from a conservative approach to convergence, and encourage mass media company leaders to diverge and focus on business models based on a system approach. The future will show, if convergence will let media environments evolve towards more ubiquitous media or ambient media environments (Lugmayr, 2012).

5 Conclusion

All in all we believe that this handbook provides practitioners and researchers, experts and amateurs with an overview of the most important issues as well as with precious insights concerning media convergence processes. Furthermore, thanks to an in depth content analysis of all contributed chapters at the very end, the handbook contributes to the debate about causes and effects of the media evolution process by highlighting not only general trends in research but also and especially research gaps in the media convergence field.

References

Aitamurto, T., Sirkkunen, E., & Lehtonen, P. (2011). Trends in data journalism, next media.

Anon. International Ambient Media Association (iAMEA). http://www.ambientmediaassociation.org

Barwise, P., & Picard, R. G. (2012). *The economics of television in a digital world: What economics tells us for future policy debates*. RISJ Report, Reuters Institute, University of Oxford, Oxford.

Boczkowski, P. (2005). *Digitizing the news, innovation in online newspapers*. Cambridge, MA: MIT Press.
Chan-Olmsted, S. M., & Kang, J. (2003). Theorizing the strategic architecture of a broadband television industry. *Journal of Media Economics, 16*, 3–21.
Dal Zotto, C., Dichamp, E., & Sommer, C. (2012). Alte Rezepte und neue Businessmodelle: Die Medienkrise aus der Sicht der Verlagsmanager. In W. Meier, H. Bonfadelli, & J. Trappel (Eds.), *Gehen in den Leuchttürmen die Lichter aus? Was aus den Schweizer Leitmedien wird*. Münster: Lit Verlag.
Dal Zotto, C., & Dowling, M. (2003). Venture Capital Investitionen in Neugründungen der Telekommunikationsbranche: Erfahrungen aus der geplatzten "Blase" in den USA. In F. Habann (Ed.), *Innovations management in Medienunternehmen*. Wiesbaden: Gabler Verlag.
Dal Zotto, C., & Picard, R. G. (2015). The dimensions of ownership and control of media. In P. Valcke, M. Sükösd, & R. G. Picard (Eds.), *Media Pluralism and Diversity: Concepts, Risks, and Global Trends*. London: Palgrave McMillan.
Dal Zotto, C., & van Kranenburg, H. (2008). *Management and innovation in the media industry*. Cheltenham: Edward Elgar.
Dal Zotto, C., Galichet, J., & Lugmayr, A. (2011). Managing media convergence: An economic and editorial project. In *European Media Management Association Conference (EMMA)*, Moscow.
Deuze, M. (2004). What is multimedia journalism? *Journalism Studies, 5*(2), 139–152.
Deuze, M. (2011). *Managing media work*. Thousand Oaks, CA: Sage.
Dowling, M., Lechner, C., & Thilmann, B. (1998). Convergence—Innovation and change of market structures between television and online services. *Electronic Markets, 8*(4), 31–35.
Dupagne, M., & Garrison, B. (2006). The meaning and influence of convergence: A qualitative case study of newsroom work at the Tampa News Center. *Journalism Studies, 7*(2), 237–255.
Fioretti, N., & Russ-Mohl, S. (Eds.). (2009). *Merging media, converging newsrooms*. Lugano: Giampiero Casagrande editore.
Gray, J., Bounegru, L., & Chambers, L. (2012). *The data journalism handbook*. Sebastopol, CA: O'Reilly.
Greenstein, S. M., & Kanna, T. (1997). What does industry convergence mean? In D. B. Yoffie (Ed.), *Competing in the age of digital convergence*. Boston: Harvard Business School Press.
Horwitz, R. B. (2005). On media concentration and the diversity question. *The Information Society, 21*, 181–204.
Huang, E., Rademakers, L., Fayemiwo, M. A., & Dunlap, L. (2004). Converged journalism and quality: A case study of the Tampa Tribune news stories. *Convergence, 10*(4), 73–91.
Huang, E., Rademakers, L., Shreve, S., David, T., Bettendorf, E., & Nair, A. (2006). Facing the challenges of convergence: Media professionals' concern of working across media platforms. *Convergence, 12*(1), 83–98.
Jarvis, J. (2009, June 7). *Product v. process journalism: The myth of perfection v. beta culture*. http://buzzmachine.com/2009/06/07/processjournalism/
Jenkins, H. (2001, June). Convergence? I diverge. *Technology Review*, p. 93.
Jenkins, H. (2006). *Convergence culture: Where old and new media collide*. New York: New York University Press.
Klinenberg, E. (2005). News production in a digital age. *Annals of the American Academy of Political and Social Science, 597*, 48–64.
Krumsvik, A. H. (2013). Towards a typology of strategies for user-involvement. In M. Friedrichsen & W. Mühl-Benninghaus (Eds.), *Handbook of social media management*. Berlin: Springer.
Lawson-Borders, G. (2006). *Media organizations and convergence: Case studies of media convergence pioneers*. Mahwah, NJ: Lawrence Erlbaum.
Lugmayr, A. (2012). Ambient media culture: What needs to be discussed when defining ambient media from a media cultural viewpoint? *International Journal of Ambient Computing and Intelligence (IJACI), 4*(4), 58–64.

Lugmayr, A. (2013). Brief introduction into information systems and management research in media industries. In *IEEE International Conference on Multimedia and Expo Workshops (ICME)*, 2013.
Lugmayr, A., Niiranen, S., & Kalli, S. (2004). *Digital interactive TV and metadata—Future broadcast multimedia*. New York: Springer.
Lugmayr, A., & Dal Zotto, C. (2013). *Convergent divergence—cross-disciplinary viewpoint on media convergence*. European media management education association conference 2013—Digital transformations and transactions in media industries, Bournemouth, UK.
Matteo, S., & Dal Zotto, C. (2015). Native advertising, or how to stretch editorial to sponsored content within a transmedia branding era. In G. Siegert, K. Förster, S. Chan-Olmsted, & M. Ots (Eds.), *Handbook of media branding*. Berlin: Springer.
Matteo, S., Spolaor, G., & Dal Zotto, C. (2013). Branding with social media at RTS. In M. Friedrichsen & W. Mühl-Benninghaus (Eds.), *Handbook of social media management*. Berlin: Springer.
McPhillips, S., & Merlo, O. (2008). Media convergence and the evolving media business model: An overview and strategic opportunities. *The Marketing Review, 8*(3), 237–253.
Meier, K. (2007). Innovations in Central European newsrooms: Overview and case study. *Journalism Practice, 1*(1), 4–19.
Mish, F. C. (Ed.). (1993). *Merriam-Webster's collegiate dictionary* (10th ed.). Springfield, MA: Merriam Webster.
Noam, E. (2006). Deregulation and market concentration: An analysis of post-1996 consolidations. *Federal Communications Law Journal, 58*(3), 539–549.
Noam, E. (2009). *Media ownership and concentration in America*. Oxford: Oxford University Press.
Picard, R. G. (2000). Changing business models of online content services: Their implications for multimedia and other content producers. *The International Journal on Media Management, 1*, 31–34.
Picard, R. G. (2009). Media clusters and regional development: Reflections of the significance of location in media production. In I. Bernhard (Ed.), *The geography of innovation and entrepreneurship* (pp. 877–886). Trollhätten, Sweden: University West.
Picard, R. G. (2010). The future of the news industry. In J. Curran (Ed.), *Media and society*. London: Bloomsbury Academic.
Picard, R. G., & Dal Zotto, C. (2006). Business models of newspaper publishing companies. In *Where news? The media future research initiative*. Ifra Report No. 1.
Picard, R. G., & van Weezel, A. (2008). Capital and control: Consequences of different forms of newspaper ownership. *The International Journal on Media Management, 10*, 22–31.
Porlezza, C., & Russ-Mohl, S. (2013). Getting the facts straight in a digital era: Journalistic accuracy and trustworthiness. In C. Peters & M. Broersma (Eds.), *Rethinking journalism, trust and participation in a transformed news landscape* (pp. 45–59). New York: Routledge.
Rayport, J. F., & Jaworski, B. J. (2001). *Cases in e-commerce*. Boston: McGraw-Hill, Irwin.
Rheingold, H. (2000). *Smart mobs: The next social revolution* (pp. 157–182). Cambridge, MA: Perseus.
SIG-eMedia, Special Interest Group (SIG) on Information Systems and Management in eMedia and Creativity Industries, Association for Information Systems (AIS). http://aisnet.org/group/SIG-eMedia
Stöber, R. (2004). What media evolution is. *European Journal of Communication, 19*(4), 483–505.
Tameling, K., & Broersma, M. (2013). De-converging the newsroom: Strategies for newsroom change and their influence on journalism practice. *International Communication Gazette, 75*(1), 19–34.
Thilmann, B., & Dowling, M. (1999). Convergence and innovation strategy for service provision in emerging web-TV markets. *The International Journal on Media Management, 1*, 4–9.
Thurman, N., & Myllylahti, M. (2009). Taking the paper out of News. *Journalism Studies, 10*(5), 691–708.

Urban, A., & Bodoky, T. (2013). The impact of Facebook on news consumption. In M. Friedrichsen & W. Mühl-Benninghaus (Eds.), *Handbook of social media management*. Berlin: Springer.

Valcke, P. (2011). Looking for the user in media pluralism regulation: Unraveling the traditional diversity chain and recent trends of user empowerment in European media regulation. *Journal of Information Policy, 1*, 287–320.

Valcke, P., Picard, R. G., Sükösd, M., & Sanders, J. (2009). *Indicators for media pluralism in the member states—Towards a risk based approach* (Vol. 1). Final Report, Lulu.com.

Wirth, M. (2006). Issues in media convergence. In A. Albarran, S. M. Chan-Olmsted, & M. Wirth (Eds.), *Handbook of media management and economics*. Mahwah, NJ: Lawrence Erlbaum.

Wirtz, B. (1999). Convergence processes, value constellations and integration strategies in the multimedia business. *The International Journal on Media management, 1*, 14–22.

Wirtz, B. (2011). *Media and internet management*. Wiesbaden: Gabler Verlag.

Wirtz, B., Schilke, O., & Ullrich, S. (2010). Strategic development of business models. Implications of the Web 2.0 for creating value on the Internet. *Long Range Planning, 43* (2–3), 272–290.

Yoffie, D. B. (1997). Introduction: Chess and competing in the age of digital convergence. In D. B. Yoffie (Ed.), *Competing in the age of digital convergence*. Boston: Harvard Business School Press.

Rethinking Convergence: A New Word to Describe an Old Idea

Germán Arango-Forero, Sergio Roncallo-Dow, and Enrique Uribe-Jongbloed

1 Introduction and Problem Discussion

The present chapter addresses the idea of convergence as both an abstract concept and as a word ascribed to certain specific phenomena. The first part of this chapter focuses on the former aspect, whereas the second will try to cover the latter. Convergence, the topic of this book, has become a common word used in many different settings, yet it is very unclear what it actually defines.

As a technological issue, convergence suggests the digital revolution that eases the share of media and peer-to-peer content throughout one single platform (Manovich, 2006). From an industrial and economical perspective, convergence also refers to the mergers, acquisitions and strategic management alliances made during the first decade of the twenty-first Century between traditional media companies (print, radio, TV, film) and new media firms focused on the Internet and online markets (Albarran, 2010). As a social phenomenon, convergence also reflects a big change in the way audiences interact, participate and respond, across multiple digital platforms, to both media and personal content flow (Jenkins, Ford, & Green, 2013). But Convergence also has been used as a term to depict industrial, economic, regulatory, global, cultural and political characteristics of the contemporary world (Dwyer, 2010).

The very current trend of turning a word into a loaded concept makes it at once useful and pointless. When too many phenomena are given a single word to unite them, the meaning loses clarity, it even fractures, leaving different ways in which to use the term according to specific settings. What are those settings in which convergence makes an appearance, then, and does that simplify, or blur, its space of action?

G. Arango-Forero (✉) • S. Roncallo-Dow
Universidad de La Sabana, Chía, Colombia
e-mail: german.arango@unisabana.edu.co

E. Uribe-Jongbloed
Universidad del Norte, Barranquilla, Colombia

A. Lugmayr, C. Dal Zotto (eds.), *Media Convergence Handbook - Vol. 1*, Media Business and Innovation, DOI 10.1007/978-3-642-54484-2_2

The next pages will try to answer that question following an analysis of communication theory inspired by the works or Marshall McLuhan, followed by a look on specific uses of the term *convergence*. The process hopes to present the space upon which convergence has been used to describe the phenomenon of multiplicity of media sources and content, and disentangle the conceptual tool for the nominal aspect it normally takes when defining specific practices.

2 State of the Art and Literature Review

This part of the text addresses convergence from the tetrad proposed by Marshall and Eric McLuhan in *Laws of Media* (1988). It tries to evidence how the idea is not at all new, as it tends to be presented nowadays, and shows that the work of media cannot be conceived otherwise; it can be understood from the four vantage points explained by the McLuhans.

With this idea in mind, it becomes necessary to stop for a second at McLuhan's concept of the *medium*. Marchessault puts it quite clearly when he states that:

> McLuhan developed a phenomenological understanding of culture [. . .] We can note that McLuhan moves from a notion of culture as landscape to one of environment, from spectatorship to immersion, and from the cinematic as an analogy for human cognition to television as the new reality and a new methodology. (2005, p. xii)

Now, this idea of *environment*, according to McLuhan, is quite curious, for it is not some sort of container; when McLuhan mentions in *Couterblast* (1969) that an environment "is a process rather than a container", he tries to show that "environments are not passive wrappings, but active processes which work us over completely, massaging the ratio of the senses and imposing their silent assumptions. But environments are invisible. Their ground-rules, pervasive structure, and overall patterns elude easy perception" (Innis, p. 386, cited in Kroker, 2001, p. 58).

McLuhan sees *culture* as an environment, that is, as a sort of ecosystem in which all spheres of human action *converge*. Thus, McLuhan's vision is that of a dense and complex culture—similar to the thick description presented by Geertz (1973)—not limited to concepts of tools and objects, but rather in the ways in which they de-localize human ways of perception.

3 Methodology and Approach: Movement as the Principle of Convergence

Convergence has been brought up in contemporary media debate, as a concept, thanks to the work of Henry Jenkins (2006). Jenkins has revisited—at least in principle—Ithiel de Sola Pool's work, trying to present a definition of convergence:

> A process called the 'convergence of modes' is blurring the lines between media, even between point-to-point communications, such as the post, telephone and telegraph, and

> mass communications, such as the press, radio, and television. A single physical means—be it wires, cables or airwaves—may carry services that in the past were provided in separate ways. Conversely, a service that was provided in the past by any one medium—be it broadcasting, the press, or telephony—can now be provided in several different physical ways. So the one-to-one relationship that used to exist between a medium and its use is eroding (de Sola Pool 1983, cited in Jenkins, 2006, p. 10).

This idea from de Sola Pool seems to agree with the notion of *arti-fact* discussed by the McLuhans (McLuhan & McLuhan, 1988):

> All of man's artifacts—whether language, or laws, or ideas and hypotheses, or tools, or clothing, or computers—are extensions of the physical human body or the mind. Man the tool-making animal has long been engaged in extending one or another of his sense organs in such a manner as to disturb all of his other senses and faculties (p. 93).

The idea of the artifact means here what Marshall McLuhan had often referred to as *medium*. However, there is an interesting issue at hand here. What does it mean to describe language, laws and tools, all at once, as artifacts? Has McLuhan rendered objectification as meaningless? In the terms that the definition is presented, a first answer would seem to be anchored in the idea of extension, which does not only suggest an increase in physical and sensory capabilities but also an ability to restructure the meanings themselves upon which humans structure their cosmological vision. Therefore, the idea of artifact is quite suggestive because of the *logos* which underscores McLuhan's statement.

Let's consider here, for a moment, what the word artifact implies. Its etymology stems from the Latin roots of *arte factus*, which mean *made with art*. This *art* present in the *arti-fact* is, precisely, what the Greek *techne* suggests. Thus, the artifact is more than an object, it is a way of thinking or, maybe, what Simondon (2008) has considered a technical-object: "the individual technical object is not one thing or another given *hic et nunc*, but rather that from which there is a genesis" (p. 42).

It is within this idea of *genesis* that we find the arti-fact. Talking about genesis implies thinking of an origin, a place from which something stems and becomes what it is whilst, all at once, the genesis remains anchored to the idea of a succession or concatenation of facts or causes that would drive to an outcome. This dualism is presented by Simondon in his technical-object which, according to de Vries (2008), would hold its ontological value in sequentiality. De Vries arrives to this conclusion when he revisits Simondon's steam engine:

> for [Simondon], the steam engine as a technical object is not what we usually call a technical artifact but a sequence of engine designs that displays a certain development. Ontologically speaking, for him, a technical object is not a material entity, but an event, or rather a sequence of events (p. 25).

Although in the terminology used by de Vries the concepts of technical object and arti-fact seem to split apart, what becomes key to his interpretation of Simondon's notion is that it evidences the sequentiality inherent to the technical object thought in the light of genesis. The idea of genesis-sequentiality evidences the mobility of technicality itself: when Simondon addresses the steam engine as an event, and not as an object, what he displays is the continuous work of the techno-logical part of technicity which, far from becoming *concrete* or *reduced* as an ultimate object, highlights the constant mutation of the technical in terms of its own complexity.

Thus, McLuhan's arti-fact would match Simondon's, because they both overcome simple objectuality. Both Simondon and McLuhan evidence a strong interest in what we could call a technical-logos[1] and which highlights the poietic features that frame the development itself of media and mediated reality. Thus, beyond de Sola Pool and Jenkins, it is striking how already in McLuhan, and even in Simondon, the dynamics of the technical implies thinking in convergent themes. The apparent novelty of Jenkins' concept seems to be nothing of the sort. In fact, Jenkins (2006) himself states that just "as Pool predicted, we are in an age of media transition, one marked by tactical decisions and unintended consequences, mixed signals and competing interests, and most of all, unclear directions and unpredictable outcomes" (p. 11). Again, what we believe to be considerably prophetical in de Sola Pool's statements and the renewed interest on the issue developed by Jenkins regarding participative-convergence, had already been addressed by McLuhan in his analysis back in the 1970s.

Let's return to McLuhan for a moment. Quite simply, the tetrad developed by the McLuhans works on a considerable basic structure and upon the four fundamental tenets that function as a single cog, creating what we would later call the *tetradic movement*.

The tetrad constitutes, from the beginning on, a dynamic system which explains the way in which arti-facts "come to life"—if we are allowed to use such idiom—right at the center of interdependent wholeness. McLuhan and McLuhan (1988) state: "Our *laws of media* are observations on the operation and effects of human artifacts on man and society, since a human artifact 'is not merely an implement for working upon something, but an extension [. . .]" (p. 94). The four tenets upon the tetrad rests are:

> [. . .] a heuristic device, a set of four questions [. . .] They can be asked (and the answers checked) by anyone, anywhere, at any time, about any human artifact. The tetrad was found by asking, "What general, verifiable (that is, testable) statements can be made about all media?" We were surprised to find only four, here posed as questions:
>
> - What does it enhance or intensify?
> - What does it render obsolete or displace?
> - What does it retrieve that was previously obsolesced?
> - What does it produce or become when pressed to an extreme?[2]

[1] For a further debate on the techno-logy (or techno-logos) and its relation with the technique, see Roncallo-Dow (2012).

[2] About the tetrad, Marchessault (2005) adds: "[. . .it] is a creative tool for thinking about objects historically and phenomenologically. [. . .]. The tetrad is not a scientific instrument that will reveal hard facts but it is an experimental technique that may open up uncharted relations. *Its dynamic structure seeks to translate the histories embedded in objects into a creative metaphor*" (p. 223).

The four verbs used in the Laws (i.e. enhance, displace, retrieve and produce) are surrounded by semantic markings which grant the idea of movement, a characteristic which becomes more evident in the way they use to render them visual:

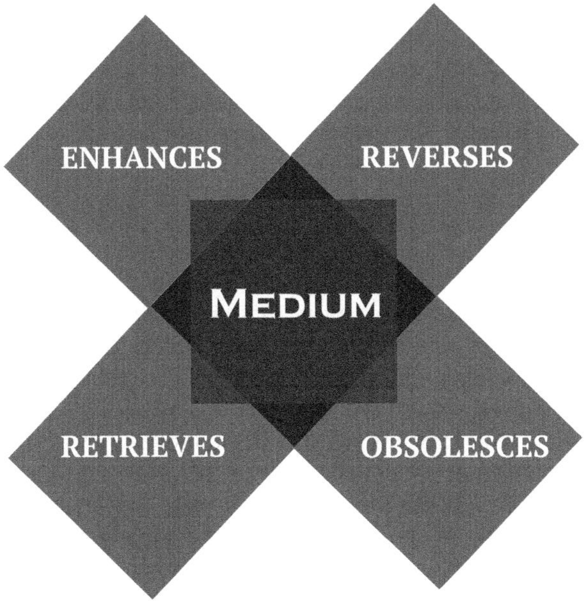

Source: Authors; based on McLuhan and McLuhan (1988, p. 129)

This type of structure suggests a multi-directional vectoring of the four tenets, whose directionality traverses the graphical space horizontally, vertically and diagonally.

The way itself in which these interdependent relations between the four tenets are presented evidences the dynamics of a non-linear perspective, which does away with the apparent independence of technical objects, hence, the media. The tetrad highlights that **convergence is not solely the way in which media contents are given new meanings by the users**—something we will address below—**but the way itself in which they hold an interdependent narrative and technicality**. Every 'new' medium assumes those which it apparently overrides, and consequently becomes part of the same whole.

This idea is paramount to the development of the argument presented by the McLuhans for their tetrad. In fact, this dialectic suggests that the complexity of technical endeavors can only be conceived within the framework given or, in other words, thinking of all the possible effects such complexity has upon whole. A good example of this point can be found in a 1978 text called *Multi-Media: The Laws of the Media*–(co-authored by E. McLuhan and K. Hutchton), in which Marshall McLuhan (using the first person singular), talking about the automobile, states that

> [. . .]what is referred to in the phrase, 'the medium is the message' is not the car, but the complicated ground of services engendered by the car; for example, macadam highways and roads, traffic signals, auto insurance, and the entire world of oil and gasoline, mechanics and parts, and motor car manufacture in general. (p. 93)

In his *Laws of Media*, McLuhan (1988) had already tried to make explicit the metaphorical and relational character of the tetrad. The background argument being what he considered the metaphorical and linguistic nature of all artifacts, which he openly compared with the *outerings* and *utterings* of humans:

> all human artifacts are extensions of man, outerings or utterings of the human body or psyche, private or corporate. That is to say, they are speech, and they are translations of us from one form into another form: metaphors. (p. 116)

For McLuhan what was truly relevant was the grammar woven in the arti-facts themselves, in the ways in which they generated meanings. What makes the tetrad show itself is the grammar and syntax of each arti-fact and the way in which they are seen as mobile, and imbued within historical dimensions.

The tetrad is, then, a device (built upon key questions) which displays the truly fundamental grammar upon which the technical fits like a cog on machinery. It is a key element inasmuch as the movement it displays is a movement that—in an Aristotelian way, perhaps—generates change. It evidences the complexity of the technical connected into a synchronic temporality which, dense and complex, leads to a repositioning of the reflection upon the technical from a holistic perspective. That is, it leads to a new ontology (see Berg Olsen, Selinger, & Riis, 2009).

4 Viewpoint on Convergence: A Word to Describe a Process

We have suggested, so far, without really saying it, that every process as an output of technological development represents a dynamic which provides meaning to the things themselves. In the field of communication these dynamics may find a new way to undertake actions, defined by the term convergence which, we have pointed out above, goes beyond the artifact itself and brings it into a dynamic-meaning relationship between those who recognize its value and develop around it a culture of meaning.

It is in the area of culture, then, where we should rethink the possibilities and reach of convergence, to overcome the technological bias. We agree with Rafael Alberto Pérez (2012), who argues that technological gadgets (arti-facts in McLuhan's discourse) may amaze us more than scientific discoveries have done so far, "which has led the majority of people, whilst enjoying the latest iPhone version, to remain anchored to a perspective typical of the 17th Century"[3] (p. 130).

[3] Our translation.

This idea presented by Perez leads us to rethink convergence as an absolute phenomenon, and gears us to consider different levels in the communication dynamics from the interpersonal to the mediatized. At the end of the first decade of the twenty-first Century, no one could deny the penetration of instantaneous digital communication in various processes, yet at the same time the is no denying the value given to the messages despite the arti-facts, although they may entail the transformation and restructuring of the languages they use.

Thus, we can ask ourselves whether we can understand convergence as a phenomenon with no equal in the dynamic and evolving process of human communication. Let us not forget the ideas of Walter Benjamin (2008), in his statements about the age of technical reproduction, where direct contact between the communicating object and the interpreting subject becomes mediated by a technological device—which we have come to know as a medium—and convergence would thus become nothing but a new arti-fact that sets the dynamics in motion of an already traditional way of making the human communication process into something tangible.

We could think, as maybe Benjamin would have done, of the issue of convergence in terms of the *sensorium*. This idea is particularly enlightening because, on the one hand it lets us grab ahold of the perspective of technical logos inherent to humanity itself as once and always convergent; and, on the other, it brings us back to the *ecological* perspective held by McLuhan about reality, which would find its key aspects in understanding *media* as environments.

Eric McLuhan (2008) remembers how, for his father, the problems regarding *media* always found their focal point in the reconfiguration of the human *sensorium*:

> Commenting on something he had written or said earlier, McLuhan (1967) offered these remarks: 'It is now perfectly plain to me that all media are environments. As environments, all media have all the effects that geographers and biologists have associated with environments in the past. Environments shape their occupants. One person, complaining about my observation that 'the medium is the message,' simply said: 'McLuhan means that the medium has no content.' This remark was extremely useful to me because it revealed the obvious, namely, the content of any medium is the user. This applies equally to electric lights, any language whatever, and, of course, housing, motor cars, and even tools of any sort. It is obvious that the user or content of any medium is completely conformed to the character of this man-made environment. His entire sensory life arranges its hierarchies and dominance in accordance with the environment in which he operates.' (p. 27).

Once more, McLuhan (McLuhan & Fiore, 2002) offers a few hints on how to learn to disentangle the possible meanings of the apothegm (i.e. the medium is the message) but he also offers new pointers to understand more clearly his perspective on the *sensorium*:

> All media work us over completely. They are so pervasive in their personal, political, economic, aesthetic, psychological, moral, ethical, and social consequences that they leave no part of us untouched, unaffected, unaltered. The medium is the massage. Any understanding of social and cultural change is impossible without knowledge of the way media work as environments.

All media are extensions of some human faculty—psychic or physical (p. 26).

The words of the McLuhans allow us to understand more clearly the modifications which technological revolution had brought to communication at the beginning of the new century by being determined by the transition from analogue communication to digital, which would then favor one of the main concepts used to describe it: convergence (Grant, 2009).

Although we have already warned that Jenkins (2006) follows the arguments presented by de Sola Pool in order to think of the idea of convergence itself, it is evident that the discussion does not stop there. The reference to de Sola Pool (whom he distances from McLuhan[4]) would enable a step forth, beyond the instrumental, and in a considerably McLuhanesque way, that convergence is not a process that can only be shown through the use of technological applications, but that it takes place, rather, in the minds of those consumers as individuals—that is, as a particular *logos*—and through their social interactions.

According to Niklas Luhmann (2000), what really matters about communication, even in the era of convergence, is the proof itself of the communication process, because the materiality of communication—the technical specifications which make it possible—will always be less relevant: "Communication only comes about when someone watches, listens, reads—and understands to the extent that further communication could follow on. The mere act of uttering something, then, does not, in and of itself, constitute communication" (p. 4).

It is precisely thus that we should address the definition of a phenomenon related to convergence, or to convergent communication. We start from Manovich (2006) by recognizing the content digitalization process which has enabled the erasure of boundaries between content distribution and exhibition, commonly experienced by the traditional media outlets, specially the press, radio and television.

Contents originally produced for one particular medium have become packaged in different formats for their consumption in convergent systems. This industrial dynamic has motivated a good deal of media corporation integrations, acquisitions and mergers with other multimedia firms, or the projection of convergent media groups traditionally regarded as specifically single-medium based in the media industries (Dwyer, 2010).

In a convergent communication environment, audiences have participated further and have taken advantage of the opportunities opened up by new media technologies. In a sense, technological changes are more than just instrumental changes for communication (Orozco Gómez, 2007); they are true transformations which slowly draw new boundaries to a communication culture which was before segmented by bound media.

[4] "If *Wired* magazine declared Marshall McLuhan the patron saint of the digital revolution, we might well describe the late MIT political scientist Ithiel de Sola Pool as the prophet of media convergence. Pool's *Technologies of Freedom* (1983) was probably the first book to lay out the concept of convergence as a force of change within the media industries" (2006, p. 10).

Thus, convergence, interactivity and user-generated content become relevant in as much as there are audiences willing and able to interact and participate in its development.

Therefore, one of the main aspects in the construction of a theoretical framework on communication for the new century is the change in definition of the concept of an audience, beyond mere technological issues, in order to avoid the pitfalls foretold by Jesús Martín-Barbero (2007) about "the fascinating achievements of technology that promise the re-enchantment of our disenchanted and unflavored lives"[5] (p. 72).

But media convergence in the twenty-first Century has led audiences to provide instant and simultaneous consumption, through multifunctional devices, for which traditional barriers between media industries, at least from the consumer end, tend to disappear. In their stead, now there is but a moment and a set of circumstances for the contact with the media message. It is the age of the audience.

We share the sentiment of Robert Picard (2003) in his perception that information revolution has kick-started an increase in the speed, flexibility and integration of current forms of communication. This is the true consequence of the process derived from media convergence, from the perspective of the media industries: "The digitalization, new media and information and communication technologies are part of an evolutionary rather than revolutionary change in communication ability. No real new communication ability is being created" (Picard, 2003, p. 154).

Convergence in the media industries, it would seem, is only a reversal of the original media segmentation of the twentieth Century. Convergence is new only because these different media outlets, much like Onassis Oil monopoly, were set apart at the beginning of the technological development, although their contents had not been similarly separated—they remained interconnected through a mixture of adaptations, modifications and, even, the participation of media figures jumping from one medium into the other.

Media convergence, then, does refer to a distinct process of mixing media outlets, languages and finances thanks to technological devices and international commercial practices that allow for both content and financial integration of ideas and capital.

5 Conclusion: Convergence as a Reloaded Term

The first pages of this chapter have shown that convergence as an idea is nothing new. The later few, however, have given certain evidence that, at least from the media industries and the audiences, convergence has implied a change in conception of what media are supposed to provide as outlets. It then begs the question of why do we believe convergence to be such a new trend. The answer, to put it shortly, is in the new ways in which convergence has been used, as a word rather

[5] Our translation.

than as an abstract concept, in current literature. Since words can develop many meanings, convergence has been used to a variety of aspects in recent years.

Thussu (2002) and Wirth (2006) use convergence to define a *specific* issue of media economics: the expansion of certain media companies into different media outlets, usually stemming from one medium and delving into others quite different in space or scope. This goes along the ideas cited above from Dwyer (2010) and Picard (2003). New international trade organizations and rules, which have promoted horizontal integration, and new market interests, which have set the base for vertical integration, are quite clearly the economic trend of the last 30 years (Sánchez-Tabernero, 2006).

Jenkins (2006), Jenkins et al. (2013) and Deuze (2007), on the other hand, refer to convergence as a process through which new players have started to develop content in media spaces that were not usually available to them. Media companies have begun using, adapting and modifying more material from different sources and have started to combine them in ways that had not been experienced before, or the people that used to be at the receiving end of media outputs have begun to develop their own personal output and set it in motion in open spaces which, on occasion, grant them a larger audience than ever before. Elsewhere, Uribe-Jongbloed (2013) has even expanded that specificity of the word *convergence* by claiming that the institutional bringing together of different source materials and separate media can be considered one extreme of the continuum, under the name of *hybridity*, whereas at the other end we have user participation in which they bring their collective intelligence to produce their own content, and this process could be truly entitled as proper *convergence*.

Convergence as economic expansion into new domains is nothing new, since artists as well as producers have constantly hopscotched from one medium to the next, even though company names have not behaved in the same manner. For instance, Orson Wells was a radio voice, turned actor, turned director, for three different media outlets, just to name one example out of the myriad that can be found worldwide. At the same time, convergence as the inclusion of different modes and styles into single products, and the use of various sources (including crowdsourcing) into the creation of cultural items, or the participation of former static media consumers—or audiences—in the media creation process, using new channels and forms of communication, although considerably larger now than ever before, thanks to the Internet, is as old a phenomenon as mail correspondence, scientific journals, and many classical forms of art–including the Celtic bard tradition.

The issue at hand, it seems, is not novelty in the genesis of the products, but their *visibility*. Yet, visibility fits into the tetrad presented by McLuhan above. The extension of the human capability of sight (and hearing) through all sorts of media technology is part of the dynamic process of the whole media environment. To use the visual metaphor cited above, it is not the length of the ribbon or band that ties the four tenets which amazes us now; its width does.

Therefore, the way in which Thussu (2002), Wirth (2006), Dwyer (2010), Picard (2003), and Sánchez-Tabernero (2006) or Jenkins (2006), Deuze (2007), and even more so Uribe-Jongbloed (2013), use the term *convergence* is only a reduced form of the more complex meaning highlighted above from the works of McLuhan and set upon more measurable instances where the process is evident. This multiplicity of uses of the word seems to create a *divergence* in the way in which convergence, as an idea now, is conceptualized. In an open discussion about convergence, which has been key to many academic plights, our first effort should be put into disentangling the weighed meanings of the word, to separate the abstract concept from the working definition of many an academic text. Convergence suffers from the same struggle another common word in our contemporary lingo experienced a couple of years ago: globalization. It becomes entangled into a definition that includes what it is, where it comes from, and what it produces, leading to a loaded term that is at once an object, a cause, and a consequence.[6]

Thus, convergence *is* nothing new. Its discussion, however, *has become* relevant more recently because of the interest in pinpointing the word to a meaning that is less abstract and more defined by current geo-economic and culturally bound ideas. But let us not think that convergence is a new issue in media theory. It was an issue settled by McLuhan long ago.

References

Albarran, A. (2010). *The transformation of the media and communication industries*. Pamplona: Ediciones Universidad de Navarra.

Benjamin, W. (2008). *Sobre la Fotografia* (J. Muñoz Millanes, Trans. Cuarta ed.). Valencia: Pre-Textos.

Berg Olsen, J., Selinger, E., & Riis, S. (Eds.). (2009). *New waves in philosophy of technology*. Basingstoke: Palgrave Macmillan.

de Vries, M. (2008). Gilbert Simondon and the dual nature of technical artifacts. *Techne*, 12. http://scholar.lib.vt.edu/ejournals/SPT/v12n1/devries.html

Deuze, M. (2007). Convergence culture in the creative industries. *International Journal of Cultural Studies, 10*(2), 243–263. doi:10.1177/1367877907076793.

Dwyer, T. (2010). *Media convergence*. Glasgow: McGraw Hill.

Geertz, C. (1973). *The interpretation of cultures: Selected essays*. New York: Basic Books.

Grant, A. (2009). Dimensions of media convergence. In A. Grant & J. Wilkinson (Eds.), *Media convergence. The state of the field*. New York: Oxford University Press.

Jenkins, H. (2006). *Convergence culture: Where old and new media collide*. New York: New York University Press.

Jenkins, H., Ford, A., & Green, J. (2013). *Spreadable media. Creating value and meaning in a networked culture*. New York: New York University Press.

Kroker, A. (2001). *Technology and the Canadian mind. Innis/McLuhan/Grant*. Toronto: Ctheory.

Luhmann, N. (2000). *The reality of the mass media*. Stanford, CA: Stanford University Press.

Manovich, L. (2006). *El lenguaje de los nuevos medios de comunicación: la imagen en la era digital*. Buenos Aires: Paidós.

Marchessault, J. (2005). *Marshall McLuhan*. London: Sage.

[6] See Rantanen (2005, p. 5) and Robertson (1992, p. 8) for a lengthier discussion on how the concept of globalization has suffered from this open usage.

Martín-Barbero, J. (2007). Tecnicidades, identidades, alteridades: desubicaciones y opacidades de la comunicación en el nuevo siglo. In D. De Moraes (Ed.), *Sociedad mediatizada* (pp. 69–98). Barcelona: Gedisa.
McLuhan, M. (1969). *Counterblast*. London: Rapp Whiting.
McLuhan, E. (2008). Marshall McLuhan's theory of communication: The yegg. *Global Media Journal-Canadian Edition, 1*(1), 25–43.
McLuhan, M., & Fiore, Q. (2002). *The medium is the massage. An inventory of effects*. Corte Madera: Gingko Press.
McLuhan, M., Hutchton, K., & McLuhan, E. (1978). Multi-media: The laws of the media. *The English Journal, 67*(8), 92–94.
McLuhan, M., & McLuhan, E. (1988). *Laws of media. The new science*. Toronto: University of Toronto Press.
Orozco Gómez, G. (2007). Comunicación social y cambio tecnológico: un escenario de múltiples desordenamientos. In D. De Moraes (Ed.), *Sociedad mediatizada* (pp. 99–117). Barcelona: Gedisa.
Pérez, R. (2012). *Pensar la estrategia*. Buenos Aires: La Crujía.
Picard, R. (2003). Business issues facing new media. In J. Servaes (Ed.), *The European information society. A reality check* (pp. 149–164). Bristol: Intellect Books.
Rantanen, T. (2005). *The media and globalization*. London: Sage.
Robertson, R. (1992). *Globalization: Social theory and global culture*. London: Sage.
Roncallo-Dow, S. (2012). Técnica, Tecno-logía: Más allá de la Sinonimia y la Objetualidad. *Universitas Philosophica, 29*(58), 39–65.
Sánchez-Tabernero, A. (2006). Issues in media globalization. In A. B. Albarran, S. M. Chan-Olmsted, & M. O. Wirth (Eds.), *Handbook of media management and economics* (pp. 463–491). Mahwah, NJ: Lawrence Erlbaum Associates.
Simondon, G. (2008). *El Modo de Existencia de los Objetos Técnicos* (M. Martínez & P. Rodríguez, Trans.). Buenos Aires: Prometeo.
Thussu, D. K. (2002). *International communication: Continuity and change*. London: Arnold.
Uribe-Jongbloed, E. (2013). Minority language media studies and communication for social change: Dialogue between Europe and Latin America. In E. H. G. Jones & E. Uribe-Jongbloed (Eds.), *Social media and minority languages: Convergence and the creative industries* (pp. 31–46). Bristol: Multilingual Matters.
Wirth, M. O. (2006). Issues in media convergence. In A. B. Albarran, S. M. Chan-Olmsted, & M. O. Wirth (Eds.), *Handbook of media management and economics* (pp. 445–462). Mahwah, NJ: Lawrence Erlbaum Associates.

The 2nd Convergence: A Technology Viewpoint

Marie-José Montpetit

1 Introduction: What Is Convergence?

Convergence? Why is it popping up in contexts from networking to cuisine? Do we really know what convergence means, when and how it originated and where it is heading? This chapter is going to attempt to define convergence from a technology point of view and will propose that there is not one but many convergences: the current trends in next generation media across the real of human activity are defining the *Second* convergence, that follows the network, device and media convergences of the last 20 years taken together as the *First Convergence* but in fact having developed interdependently.

According the dictionary *converging* means to meet or focus on a common and similar goal, which applies as well to phones as to Asian fusion recipes. In the media and technology context, however *convergence* may be defined in a narrower fashion but with still different interpretations. Henry Jenkins in his article in the (Massachusetts Institute of Technology) MIT Technology Review in 2001 (Jenkins, 2001) defines:

- *Technological Convergence*: The cross-platform digital information flow created by the Internet and how content and our relationship to it are always expanding.
- *Economic Convergence*: The horizontal integration of the [entertainment] industry and the definition of new value chains.
- *Social or Organic Convergence*: The multi-screen multitasking environment created by device ecosystems or the connectivity between different groups of users

M.-J. Montpetit, Ph.D. (✉)
MIT Media Laboratory, Cambridge, MA, USA
e-mail: marie@mjmontpetit.com

A. Lugmayr, C. Dal Zotto (eds.), *Media Convergence Handbook - Vol. 1*, Media Business and Innovation, DOI 10.1007/978-3-642-54484-2_3

- *Cultural Convergence*: Novel formats and platforms for creativity using various media technologies across different industries and consumer groups.
- *Global Convergence*: The experience of being a citizen of the *global village*

So if indeed we are witnesses to not one but many convergences, we may want to focus on one. This chapter proposes to focus on *Technological Convergence* in its numerous historical and current embodiments. Technological Convergence gave us wired and wireless network integration and the explosion of cellular communication. It created the fixed/mobile integration that lead to ubiquitous Internet connectivity. This in turn produced convergence in devices, the results of which are smartphones, connected televisions and tablets, amongst others. The convergence in our means of communications changed the way we consume content at home, at work and on the road. It redefined how we communicate with one another: not by phone only but via a web of interconnected service and social network applications.

It will be the hypothesis of this book chapter that the convergence in technology drives the economic, social, cultural and global convergences, creating in turn the media convergence of today and the years ahead: the *Second (2nd) Convergence*. This *2nd Convergence* is melding technology, business models, social networks and culture and catalyses media and story telling innovation. It is breaking through traditional design silos and is displacing traditional linear value chains and some the predominant business models and creates a new generation of innovators, users and creators across skill and age barriers.

And technology is pushing forward with the even more diversified *converged solutions*. These include of cloud-based computing and application, content-centric networking and big data, and adding social networking and crowdsourcing to traditional content production to produce novel methods of acquisition and dissemination of content.

Technological convergence was born of the dislocations between the personal and business space and the professional and entertainment realms. It has lead to business model disruptions and maybe to more dislocation in the perception and consumption of media. For example, the smartphone combines elements that were very disjointed (home and business phone, computing device, gaming console, e-reader etc.) in a single platform for which millions of applications were created: this abundance of apps fragments our attention.

In investigating many facets of convergence from the technology perspective, and present successes and challenges, this chapter uses a historical perspective. It presents the evolution from networking hardware and computing devices to current application-centric, mobile and user-centric service and applications. The next section starts this presentation with a cross-section of relevant literature. Technological convergence is a wide field that has generated a large body of academic and industrial publications over the last 10 years. Some important work will most likely be omitted, but the presented works intend to encourage the reader to look further into the field. Section 4 highlights the methodology of convergence research, based on experience and multifaceted investigation. The next three sections present the bulk of the chapter namely the *technological* convergence of the last two decades: from networks, to device and to media. Network convergence, gave us the Internet of today, by, firstly, combining the data and telephone networks and, secondly, incorporating the fixed and mobile networks. While this is still happening, it will be seen that what at first seemed like a natural evolution essentially entailed a disruption of media business models and the current market frictions between traditional operators and the over-the-top (OTT) community. The device convergence is showcased in Sect. 5. In the undistinguishing set of Internet-connected devices one can ask what will become the differentiation models to keep user loyalty. The multiscreen applications, the ubiquitous Internet and social networking that forms Social media and Social Television are the topics of Sect. 6. These three sections taken together will clearly show that the Internet is now more than a network and a diffusion medium for content: it is morphing into a service and a platform for the latest innovations to be deployed. A personal viewpoint of the future of convergence will be provided in Sect. 7, proposing the 2nd Convergence as encompassing the economic, social and global convergence of media and the creation of new communities. As a consequence, convergence forces a reinvention of the way we communicate and the need for *available and sustainable* connectivity. Finally, we conclude with a reflection on managing convergence and, in view of the previous sections, if it needs to be managed at all: convergence is happening now, continually reinventing itself. With the melding between the social and the physical networks, between locations and real and virtual reality convergence is becoming an *ideation* platform.

2 Technology Convergence: A Literature Review

Convergence has been duly documented, and both glorified or decried over the years. It is disrupting. It engendered dislocation and divergence in the way next generation media dissemination and consumption are perceived and marketed. It fostered creativity and generated economic growth especially in the content and application fields. A number of Media focused Programs including the MIT Media Laboratory were born out the perceived necessity of convergence. And convergence challenged established business model in the media industry, from newspaper to television.

In this section some seminal publications on the convergence will be reviewed. It is not an exhaustive list as the field is vast and it reflects the research and interests of the author. These papers establish a timeline from the strictly network views of the early 2000s to the more media focused recent past and the current revisiting of the business. In subsequent chapters, publications that are specific to the topics under discussion will also be referred to.

In the networking industry convergence started with wireless networks and devices in the early years of the new millennium spurred by the fast melding of the technologies of cellular communication, Internet, television (TV), computers and fixed and mobile phones. The Internet was becoming the converged network of the future to offer all communication services over a common platform. The 3rd Generation Partnership Project (3GPP), the leading standardization body of the wireless industry, standardized the *Internet Multimedia Subsystem* or *IMS* in the mid-2000s as reported in the book "The IMS: IP Multimedia Concepts and Services" (Paulson, 2010). 3GPP standardized the signalling and the related protocols that enabled wireless communication to integrate into the Internet. The European Telecommunication Standards Institute (ETSI) TISPAN[1] then expanded the reach of IMS into the fixed networks with the *Next Generation Networks* (NGN) seen as the future universal network based on Internet Protocols (IP). The final NGN architecture was published in 2008 (Bulkeley, 2010; Cisco, 2015) and its impacts will be reviewed in the next section of this chapter. In addition, the NGN family of networks and services have and still are been standardized by the International Telecommunication Union (ITU) and the Internet Engineering Task Force (IETF) and the Alliance for Telecommunication Industry Solutions (ATIS) that extended NGN in the television realm (Piokselka & Mayer, 2009).

In parallel to the work performed in the standardization groups, other aspects of technology convergence were also being investigated especially its impact on media and the rising media convergence. The rise of interactive television and its impacts on the broadcasting industry was studied by Arthur Lugmayr and his collaborators in 2004 in their book entitled: Digital Interactive TV and Metadata: Future Broadcast Multimedia (ETSI 2007). Their conclusion, that the use of metadata and technology to enhance the interactivity between the user and the content was creating new opportunities for innovation in the TV industry, rightly predicted the rise of user-centric television models and the use of the social commentary to enhance the television experience. Pablo Cesar and his team further defined "Human-centred television" in their 2008 article where they pushed the concepts of television and human experience to new levels (ATIS 2004).

A few years earlier, in his seminal 2006 book "The Convergence Culture" professor Henry Jenkins of the University of Southern California introduced how the different media outlets and available devices increasingly influenced media consumption and changed consumer behaviour. *Transmedia,* the telling of stories across media platforms was made popular by the book that also spawned a successful series of conferences and panels on the *Futures of Entertainment* held every November in MIT.

Hence, the rise of the Internet and in particular the wireless Internet was a catalyst to the technology convergence and in turn to the media convergence. The advent of Facebook created an opportunity to combine the new social networks with other media. This was particularly true in for television, which has always been at the centre of the social discourse. While the idea of combining television with some form of social networking was not new,[2] the facility of creating the social group brought by Facebook (and later Twitter) created the current trends in Social TV or STV. This will be further discussed later in this chapter but research

[1] Combination of TIPHON (Telecommunications and Internet Protocol Harmonization over Networks) and SPAN (Services and Protocols for Advanced Networks).

[2] For a good survey of the history of Social TV see the excellent 2011 presentation by David Geerts and Pablo Cesar (ATIS 2004).

performed as early as 2007 and published in 2008 (Jenkins, 2006) by Mariana Baca of the MIT Media Lab showed the value of the approach to integrate traditional TV and DVR into a social framework. STV was also the case study for a paper from Natalie Klym and the author "Communication at the Edge" also published in 2008 (Cesar, Bulterman, & Gomes Soares, 2008). This paper looked at the converged networks value chain with social television as one example of a converged service. It is also highlighted the increased competition from traditional operators, the new operators and introduced the *virtual operator*, the media consumers themselves. This concept started a pushback against the established TV grid and is now pushed further with channels and programs becoming applications that can be downloaded individually based on personal preferences.

STV has also emerged as the perfect example of convergence from a technology point of view and a next generation IP Television (IPTV). An architecture for such a converged TV system is presented in Cesar and Geerts (2011) with an emphasis on both wired and wireless devices. In this publication, the technology aspects of the move to the Internet from rights management to channel changes are addressed. STV really showed that the future of television was social, mobile and IP based as was proposed by the author and her team in 2010 when the concept was still controversial (Baca & Holtzman, 2008). In a perfectly *converged landscape*, all services should be available anywhere as content consumers, not just their devices, are moving through an ever expanding universe of content. This universe was created by the many convergences that are the focus of the rest of the chapter.

3 Convergence Research: A Methodology and Approach Based on Experience

How does one approach convergence research beyond literature searches? One methodology is actively monitoring the individual elements of a potential converged technology to discover, or predict, where and when they will have enough overlap to become one. For example, the behaviour of Internet traffic has changed greatly over the past years. It is less and less about unidirectional flows moving from a source to a destination over a wire but becoming information disseminated bi-directionally across a large number of nodes most of them wireless. In this environment it is clear that the convergence of the wired and wireless networks happened when cellular networks deployment literally exploded. They created the platform a whole net set of applications that also created a huge demand for even more Internet connectivity of all kind. The experience of the growth of the wireless industry has driven and continues to drive large investments in research and development and provides a large amount of the academic research in engineering and computer science. It is in fact the first convergence that will be described in this chapter.

Another methodology is to be a participant. Experience in network and technology design and implementation is essential to approach convergence. Network convergence emerged from the standardization bodies and resulted from the work, the common consensus, one would say the converged consensus, of a large number of participants from industry, telecommunication operators and academia. The author having been part of the mobile industry and having been a collaborator to standards got a first hand experience of the process of convergence in networks. This will be important the description of network convergence in this chapter.

Back in 2005 when convergence was a vague concept and as a result of standardization and the design of new mobile phone it became obvious for some leading engineers in a number of manufacturers that a cell phone was very much a TV screen with (then) a keyboard, a TV set was getting a new life as a computer screen and video could be consumed on a laptop. As a result of that early development in converged television were started. The approach there was more experimentation than experience but created some of the early multiscreen applications that are now ubiquitous and the further development of the converged devices because development of these applications on un-related devices was not sustainable in terms of development costs. This chapter will profit from these developments as they defined how television distribution evolved to the Internet and why smartphones, tablets and other converged devices became so popular.

Social Television (STV), the combination of both traditional and over the top content with social networking, has gone from a laboratory concept to a boardroom topic with an incredibly swift pace over the last 10 years. Social TV was the focus of an IEEE Networks feature in 2010 (Montpetit, Mirlacher, & Ketcham, 2010) and was rewarded with a MIT Technology Review TR10 in May 2010 (Montpetit, Mirlacher, & Klym, 2010), as one of the ten (10) influential technologies that will change the way video is consumed. It is now regularly reported on in newspapers and numerous blogs. But in the context of this chapter, STV as a converged service will inherit from the experience of 5 years of the MIT Media Laboratory graduate level class on Social Television. Since 2009, the class, which is multi-disciplinary, has allowed to navigate the evolving landscape of television and its relationship to social media. The students produce one final project every year and these projects have encompassed the evolution and the convergence of the social media and television beyond entertainment. Some of these projects are mentioned in this chapter as they embody some aspect or another of STV from recommendation to metadata.

And finally the methodology and approach to this chapter is just to keep abreast of the technological developments that are happening at an accelerated pace: there was about 450 years between the invention of the printing press and that of the telephone. The last 25 years have seen the personal computer, the Internet, the mobile networks and the cell phones, WIFI, social networks etc. In particular the expansion of wireless networks and of the wireless Internet is exciting: it provides new opportunities for converged services and applications. As can be seen in Fig. 1, adapted from the CISCO Virtual Network Index most recent 2013 predictions, while all type of traffic will increase, video and data services will continue to grow significantly faster than any other traffic type. The Internet is now wireless and video-centric and the combination of wireless and video just confirms that convergence in devices will continue, that new video applications will emerge and that

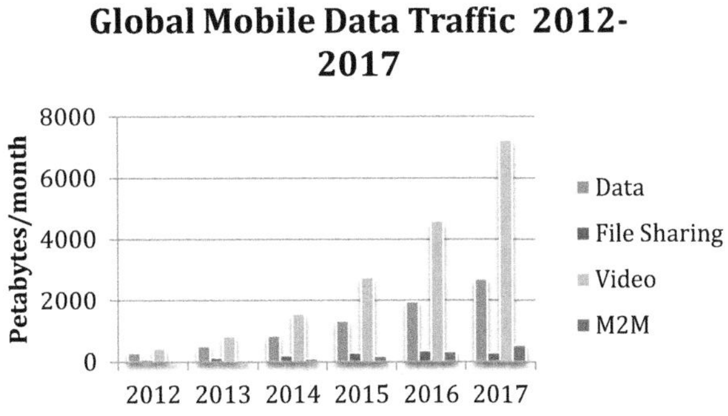

Fig. 1 Predicted growth of mobile traffic 2012–2017 from Paulson (2010)

differentiation in these application will force to rethink the way video is produced and consumed, giving life to more convergence.

4 Network Convergence: The Internet

In the technology world, convergence is often driven by a major innovation creating a business disruption that in turn engenders more innovation. This was first shown by the personal computer in the 1980s with the combination of word processing, spreadsheets and computing on a single convenient platform. Later, in the early 2000s, it was becoming clear that the Internet and its series of innovative services, from the world-wide-web (WWW) information searches to real-time communications, would become the network of the future, connecting all these personal computers together. Traditional voice telecommunication operators were noticing a fast growing application, Voice over Internet Protocols (VoIP). It was suddenly recognized as a mean to reduce operating costs. And the same time, wireless demand, driven in large part by low cost and feature-rich handsets, was growing fast. The new "triple play" offerings were born (voice, video, data) and with need to connect Internet Protocol (IP) services to the wireless handsets for data service like remote access to corporate email. The combination of VoIP and wireless services increased the need for jointly managing all networks.

But there was then in essence, four parallel networks: (1) the connected devices supporting the Internet infrastructure, the switches and central office element supporting the telephone system which in turn was divided into (2) fixed services and (3) mobile services and, finally (4) the television networks were totally independent entities with different operators with their own regulatory and business environments. Some television networks like those managed by the cable operators were offering broadband data and phone services but as different services. The only common feature of all these disjoint networks was the fact that they all could support some form of Internet protocols or interconnect at Internet points of presence (POP). Fixed-wireless convergence was greatly catalysed by the wide availability and rapid adoption of Internet technology as will be seen below.

4.1 Wireless Meets Wireline: Fixed-Mobile Convergence

Fixed-mobile convergence (FMC) came to define (1) the joint interconnection and management of traditional fixed (digital voice and data) and mobile services using the general-purpose computers running the Internet and (2) the adoption of Internet protocols to ensure seamless communications between the heterogeneous architectures. The minimization of operational costs and the enhancement of the edge network performance for both consumer and enterprise markets was used to justify the move to these converged networks. But in reality, operators were quickly realizing that the networks of the past could not support the new applications and services of the Internet without some changes to underlying protocols (see Paulson (2010) for a good discussion on the origins of FMC and subsequently IMS).

Hence FMC was born out of the necessity to jointly manage traditional fixed and wireless voice and VoIP over the Internet. FMC started with softswitches [again the reader is referred to Paulson (2010)]. A softswitch is software that allows a telephone call from one phone to another to be connected via the Internet. Softswitches represented a major disruption from the traditional systems that used hardware-based or firmware-based switchboards since they could run on more general-purpose computers. Softswitches became the *convergence point* between the IP world and the traditional telephone services. For VoIP calls the switch connected the calls directly with IP protocols. For traditional fixed and mobile calls the switches were associated to gateways that converted the calls to and from IP protocols, to and from the legacy systems protocols.

Softswitches allowed the management of voice calls across different media. But this is not where the evolution would end. Even feature (mobile) phones could be used for email and web access, and the emerging services like IP Television, moving traditional broadcast to the Internet, was creating further demands for co-management of the heterogeneous networks. Hence softswitches quickly evolved into the Internet Multimedia Subsystem and Next Generation Networks in order to extend the offered services set of the jointly managed networks.

4.2 Legacy Networks Add IP: IMS and the Next Generation Networks

As mentioned in the literature review of Sect. 2, IMS was first developed in the 3GPP to define the wireless elements to support Internet services and the NGN architecture was defined in the ETSI TISPAN to unify the wired and wireless networks. The main NGN features are available in Table 1, taken from the ETSI architecture documents.

NGN supports a set of end-to-end services using IP protocols over a network composed of heterogeneous sub-networks. The main characteristics of the NGN architecture are the uncoupling of services and underlying transport functions, in principle allowing services and networks to be offered separately, to offer different quality of communication and to evolve independently. Provisioning of existing

Table 1 ETSI TISPAN NGN capabilities (Social, 2010; Cisco, 2015)

Capability	Description
Subscriber nomadicity	Decoupling the subscriber from specific access and specific terminal equipment
Application ubiquity	Application availability from any access network. Content 'tuning' to match access and terminal capabilities
Resource control	Authorization and availability Accounting: measuring resource usage, revenue assurance Policing resource usage; fraud prevention
Subscriber identity and authentication	Common model for all devices, access and applications
Service blending	Service brokering enables applications to provide adaptive behaviours based upon subscriber events and states
Billing and settlement mechanisms	Especially beneficial for scenarios crossing multiple providers boundaries

and new services can then be independent of the network and the access technology. Hence NGN relies on a set of physically connected underlying network that use the transport of packetized information in the form of IP packets and share common signalling: the Session Initiation Protocol (SIP) defined by the IETF in the Request for Comment (RFC) 3651 (IETF 2002) with IP protocols to create and control individual sessions. The Internet is extended all the way to the end-user devices allowing photo sharing and television, video conferencing, gaming etc. to become (operator) managed services. To provide these novel services, because of network heterogeneity, network service providers need to perform additional tasks during the establishment of Internet sessions. They include application selection and routing services, session authorization services, session monitoring services, session detail recording and billing, network resource and admission control services. This involvement of the operators in the management of the Internet has generated a backlash from the IP community, as we will see in Sect. 4.3.

4.3 Networks Converge: Business Models Diverge

Fixed-mobile convergence and the development of the IMS and NGN architecture in standardization bodies were pushed by traditional operators and their equipment providers and derived from the needs to *manage* the IP based networks. But by establishing points of contact within the Internet for policy, access control and billing, NGN creates a conflict with a main tenet of the IP community: network neutrality. Network neutrality (net neutrality) is essentially ensuring that all traffic in the Internet is treated equally; hence no traffic flow, in principle, can be submitted to a different set of policies. It is recognized that net neutrality has enabled the recent innovation economy of the Internet from behemoths like Google to small applications development start-ups.

The reaction to NGN/IMS from the Internet community was very negative. For leading members like Scott Bradner of Harvard (Bradner, S. O. (2007). *Private communication*), NGN/IMS can put tolls across the sub-networks of the Internet backbone, the IP islands. Many traditional operators on the other hand welcomed the NGN/IMS because it allows offering better services to applications that pay for it or to users that have requested (and paid for) a better service; NGN created the concept of policy management in the Internet.

The divergence of thinking between the IP community (the application developers and OTT providers) and the traditional operator community is continuing today. Technological convergence contributed to economic and cultural divergence in the way of thinking about the future of connectivity in the two communities. Both camps however want to claim the consumer and the new multimedia applications users request. IMS has been used for consumer services, like television, to offer Web 2.0 services along the traditional broadcast (enhanced television) and to enable IPTV on smartphones. It is also at the heart of new services part of the Rich Communication Suite, n.d. (RCS) recently standardized by the GSM Association (GSMA) that enables real time exchange of content between users (instant messaging, video and picture sharing and some social exchanges) and has already been deployed in many markets notably in Europe. But this is still dwarfed by the growth of the over the top giants like Netflix, Hulu and Amazon, the picture sharing sites like Instagram and the new video clip sharing applications like Vines. It is undeniable that the Internet has changes the communications network landscape and it will not turn back; the friction between managed and unmanaged services will continue.

4.4 A Network Convergence Success: Television Distribution

While the NGM and IMS networks were being defined, the television industry was looking into the Internet protocols in order to distribute content. While traditional cable operators were firmly established in the broadband and voice service delivery, phone operators with the availability of Digital Subscriber Lines (DSL) were seeing television services as a growth area. Internet Protocol Television (IPTV) had (and has) the potential to provide a much richer user experience because of its potential of combining traditional broadcast with Internet services and wider distribution.

A testimony to this is exemplified in an ATIS IIF (IPTV Interconnection Forum) recommendation that defined the IPTV of the future as early as 2004 (Piokselka & Mayer, 2009):

> Going forward, IPTV is seen as a broader application than today's definition encompasses. [...] This view of IPTV extends beyond the home delivery model that is the focus of today's IPTV and also includes additional options for distribution of IPTV to wherever the consumer may be.

While IMS or NGN-based television services were hotly debated in standardization bodies, highlighting the friction between legacy operators and new entrants, the idea of combining traditional television channels with enhanced Internet content got and still has wide appeal.

With network convergence, television can be delivered over any combination of cable, DSL, Fibre to the Home (FTTH), wireless or mobile networks. These capabilities imply that media encoding and transport formats must adaptable to different device types and different access network capabilities and bandwidth. The emergence of the interlinked ecosystem of access networks and end-devices have allowed the creation of services for this *connected television* experience. They form the core of the *multiscreen* viewing experience.

As a result in recent years, video consumption has changed radically. Multiscreen television is delivered over triple-play (voice, video and data) or quadruple-play (by adding mobile) broadband access networks and managed and unmanaged WIFI. For example, a viewer with a subscription to fixed-line IPTV service may access subscribed content for display on a TV set, a Personal Computer (PC), a smartphone or tablet (3-screens), delivered over a DSL access network into the home and distributed within over WIFI. Convergence is shaping the TV experience.

Many companies such as Intel (Bourdonnec, 2010), with its television experience group, have dedicated entire teams to evaluate the user experience and dramatic changes to video delivery when it becomes multiservice. In addition, improved transmission performance is needed over any network in and out of the home to ensure consistency across viewer groups. With content moving to the cloud for easy access everywhere, the viewing experience no longer conforms to pre-defined broadcast schedules or channels. Instead, content is personalized, reflecting the viewer's individual content and display preferences, access permissions and session status, and mobile, reflecting the consumers change of location during the day. The use of social networks for video distribution and recommendation also figures prominently in this evolution. *Converged television* is nowadays associated with a television service offered on a diverse set of devices and augmented by ancillary services such as widgets or web content over a combination of wired and mobile wireless networks, managed and public. While providing a familiar and simple user interface that masks this aggregation, converged television requires a reliable and comprehensive system and network architecture for content management and device interoperability.

The challenge of TV in the next decade necessitates a comprehensive end-to-end and top to bottom strategy that continues to moves away from the current design silos, a systemic approach that defines the needs for better user experience and interaction. Figure 2 presents a simple overview of a solution to this challenge from the distribution chain point of view. What is highlighted by this architecture is the combination of operator controlled services with public Internet information, the distribution over many possible networks to a variety of end-points and the fact that while the content providers still mostly deal with operators they could themselves use the Internet for distribution (the dotted line) and become their own OTT. It is interesting to reflect on the fact that this converged TV network seems more complex than the TV of the past where content provision, operation and (over the air) distribution was done by a single entity.

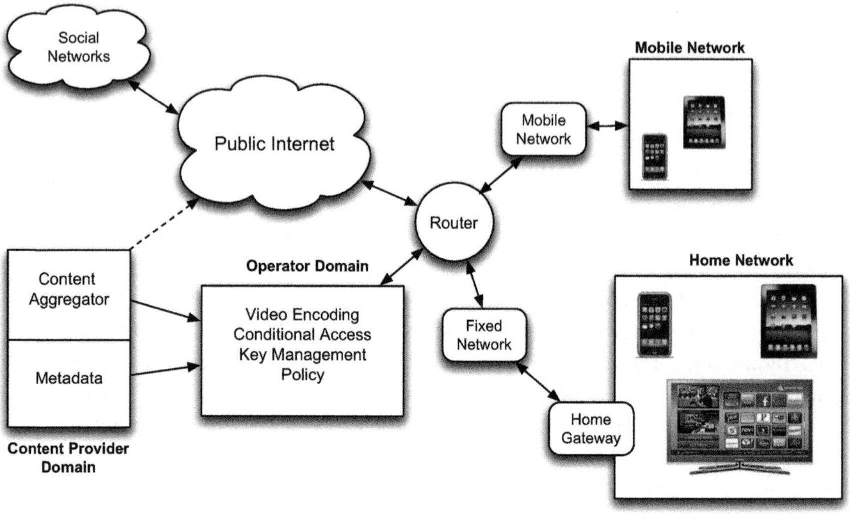

Fig. 2 Converged video delivery network

In addition, the wireless portion of Fig. 2 should not be ignored. A phenomenal growth in mobile video is predicted in the next few years reported in the CISCO Virtual Networking Index (Montpetit & Klym, 2008) already introduced in Sect. 3. In standardization circles video is considered the *killer application* for the Long Term Evolution (LTE) networks. In response there is a flurry of activity to ensure that the network infrastructure will meet the demand.

While some critics have declared that television was more or less in decline, television content provision has become an active innovation area and testament to the power of convergence to create new opportunities: the commoditization of the TV set is a consequence of device convergence but the Internet has allowed innovation in creation, distribution and consumption of TV concept. Not content to just deliver content some OTT providers have started creating it with Netflix leading the way. With more and more platforms available for content consumption, wireless (IP) distribution and the plurality of content sources, TV is becoming a personal video delivery system, available everywhere and curated by the users themselves.

The TV set of the 50s has cut its cords and antennas and moved to cyberspace.[3] This move in turn is spearheading a content convergence: a combination of real-time, on-demand and user-generated content of many forms including context and location specific information for museums, tourist sites, smart-city and art projects as will be addressed in Sect. 4. The opportunities provided by the Internet demonstrate that the *balkanization* of devices and networks is soon to be over. Television

[3] It is interesting to note that the US service Aero uses dedicated user antennas to provide real-time television services over the Internet to circumvent regulatory aspects.

is being redefined: the unidirectional broadcast from an operator to a device is now a diverse service offering and *television services* of today are truly the children of network convergence.

5 Device Convergence: Access, Services and Applications

In the days when computing hardware was very expensive it made sense to minimize functionality and develop highly optimized devices: music players, books (before and with e-readers), wireless access, television, voice etc. Even when they were physically co-located they were still discrete entities. This is not the case anymore: convergence in devices is the result of developments in microprocessors, Moore's law and user interfaces. Driven by applications and services offered by traditional operators and new entrants alike smartphones, tablets and connected televisions are starting to look the same. The Internet has created a "blank" network platform that allows concentrating much functionality on a single device, for home, enterprise and industrial uses, since in the end they all transit through a common network. From the user point of view this allows single points of contacts to the operators and simpler billing. The device convergence in turn creates a convergence of the means of acquisition, creation and dissemination of content: movies are directed using cell phones[4] and news events are captured on smartphones and distributed on Twitter feeds[5] (more in the next section).

The move of web experiences onto the converged device ecosystem has become one of the great innovation catalysts of the past few years: widgets appear everywhere and more on more on laptop and desktops; the wireless in a reversing of roles now drives the wired network. Users are now enjoying live interactions with content, other machines and ancillary devices and of course other users. User interactions are transforming the design, implementation, and use of those devices. As we move through our busy days, our devices coordinate and link to maintain the continuity of our communication events. For example as a consequence of both network and device convergence, the seamless video experience is born: it allows to start watching the Olympics on a computer at work, switch to a tablet for the commute to home, and finish watching on a web-enabled television all the while be connected to social commentary and ratings.

Converged devices are more and more video and rich-media centric. And they provide the platforms for innovative services and applications that are socially engaging with operator-based, over-the-top, and user generated media. In this section, the converged device ecosystem is reviewed from three points of view: access, services and applications.

[4] www.directr.com

[5] www.stringwire.com

5.1 Access: Communications in a Box

It is still usual to deal with multiple accounts for fixed and cellular services and with extra fees for services like texting and mobile data or international calls; his is what the triple and quadruple-play offerings are trying to reduce to a single bill. But it is even more common to use different equipment for cable or IP television, for wireless phone, for broadband access and wireless (WIFI), for Digital Video Recorders (DVRs) and for game consoles; we all know the resulting in the device clutter in the living room. To counter this trend, device convergence is happening in the home network: there is convergence toward the *box*, the home communication and media gateway, that aggregates services and provides connectivity within and outside the house and that is being adopted by many operators and equipment providers and wished for by the cluttered consumer.

The regulatory environment that allowed newcomers to enter the broadband market spurred the box paradigm. For example, in Europe it was the opening of the local loop that created competition in DSL services; in the US it was the offering of the voice and data services by the cable operators. As a result for example, the French service FREE came on the market in 2002 as new DSL offering, with triple and quadruple play bundles and very low prices to the consumers possible by the integrated Freebox.[6] Once a single operator can offer a number of services there is an incentive to provide a more integrated equipment offering.

But that was just the beginning. Gaming consoles, like the Microsoft Xbox, are now becoming the main entertainment hub for the house and are moving into the multiservice provision and seamless integration with the Microsoft tablets and phones. WIFI access points combine direct broadband access as well as embedded backup and home data storage; operators are adding home security features these existing broadband equipment and allow connection and personalization via smartphones or tablets. Apple TV offers video services but also efficient mean to transform a large television screen into a business projector. Connected televisions (large screen TVs with direct broadband access) are now at the centre of the connected home entertainment ecosystem. These integrated devices are dislodging the home computer as the main communication and computing hub and the traditional television set as the centre for all entertainment.

It does not stop there: whole new ecosystem of personal connected devices is starting to appear that use the smartphones as gateways. These like Fitbit or Nike Fuel and the new smart watches from Samsung and others extend the capability of the phones to provide, for example, better health care monitoring but also other lifestyle services from personalized workouts, diet tips and meal suggestions.

[6] The Freebox combines an Internet Protocol Television (IPTV) set-top box with recording capabilities to voice over Internet Protocol (VoIP) and WIFI access as well as gaming. This reduces operational and provisioning costs for the operator and provides a single entity in the consumer's home.

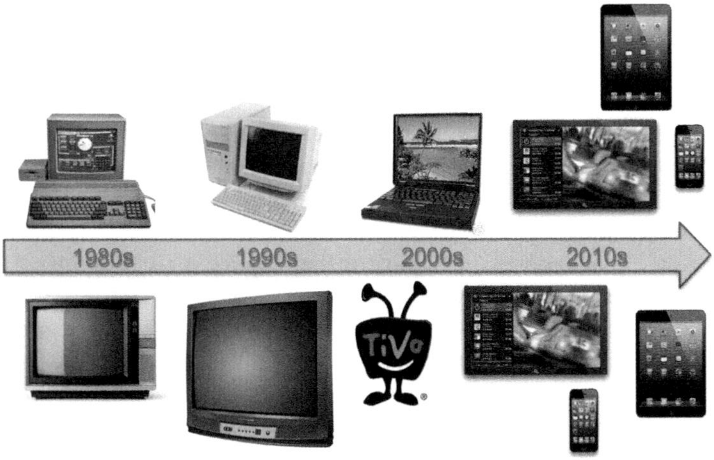

Fig. 3 Device convergence: the *top timeline* shows the computing evolution and the *bottom timeline* the entertainment evolution—they become the same devices even if their usage is still different

5.2 Services: Morphing Computing and Entertainment

How have our home devices evolved over the last 30 years? We get a graphical overview of this evolution in Fig. 3. Information and entertainment used to use vastly different platforms. The computer was in the office at work, the den or the home office (with little connectivity besides disk copies and dialup). We had a "business" relationship with it: it was work. The television, a more convivial device, was in our livingroom, bedroom or kitchen. Phones were apart and often in every room since they were fixed.

Then services like Skype appeared and the computer became the phone; laptops moved the computer into the bedroom and the kitchen and outside the house. A Digital Video Recorder (DVR) like TIVO liberated the TV content from the traditional scheduling grids. While it allowed skipping commercials and it also created a demand for TV everywhere, a TIVO in the cloud. With smartphones, the web services moved to the mobile networks and tablets moved the video experience away from the livingroom and the bedroom or even the house. We now phone people on our laptops, read books on our tablets, videoconference on our TV and buy dinner on our phones. It is now common that content of any kind is created and consumed on an ever evolving but at the same time very much look alike set of "screen": TVs, laptops, tablets and smartphone are sharing similar user interfaces and common widgets. The device convergence has evolved into the n-screen. While the 1st screen is often associated with the television, more and more the true 1st screen is the one we carry and look at first. Television is not a device anymore; it is a service and, even, a set of applications. Computers well are now phones and web access device much more than software development platforms. Smartphones and tablets are applications delivery engines.

In particular, a smartphone is becoming our main interface to the Internet. Nowadays, teenagers everywhere and many users in developing economies forgo the computer/laptop entirely, relying on their phones for all their data and media services. We all have now have a variety of Internet access devices in our lives, but the cell phone has become the primary means by which 25 % of those in the ages of 12 to 17 access the Internet. Among teenage smartphone owners, 50 % said they use the Internet mostly through their mobile and smartphones, according to the study.[7] These teenagers are now gaining a behaviour that they will most likely keep in the future because smartphones fulfil most of the requirements of the typical users: texting, email, connecting with friends on social networks, listening to music, reading books, accessing the web for information and for using cloud applications for document management, myriads of lifestyle applications and of course phoning.

The phone also provides a controller for navigating the media space and receiving or creating relevant information to be displayed elsewhere. The smartphone is used both as the remote and as the viewing screen. While continuing to watch videos on a TV-set or computer, a user can interact with friends, share opinions and ratings, and look up related information on the smartphone (or tablet). In addition, the phone can provide information and interactions that are contextually related to one's activity from mapping to sports statistics to smart city services. A large number of adults are also now connected to Internet via smartphones and the popularity of e-readers and tablets is growing.

Media offerings and delivery are adapting to the converged devices. Personalized services require interactive User Interfaces (UIs) and fine-grained information models to capture and manage viewer preferences. Instead of only managing access or *subscriptions*, which are based on business relationships, this enables personalization based on individualized preferences. Because of both user and device mobility, delivery is not localized to the consumer's home or service-provider's network. Content and applications can be accessed from any location that has Internet connectivity. As a consequence, advanced security capabilities for security, Digital Rights Management (DRM) and protection of viewer identity (privacy) have become major concerns for content providers, operators and users alike.

Because content of any kind can be consumed anywhere on converged devices, the experience is not limited anymore to entering an address and wait for the content to be displayed in a uniform manner. Instead, the device UIs and application widgets enable interactivity and allow viewers to customize displays, banners and the arrangement of information. The new devices like Google Glasses offer capabilities for advanced interaction and ubiquitous connectivity. Intelligent gateway devices (combining network facilities and home gateways) can handle message exchange between user-controlled device and the larger networks to provide

[7] http://www.networkworld.com/news/2013/031313-pew-teen-smartphone-use-soars-267647.html

in-home and outside the home services like home security. These transformative capabilities combined with Social Networks enable social interactivity at the device level; take a picture from your glasses and post it on your Facebook. Our devices are *social*: media can be delivered anywhere to other members of the same social group on any combination of devices the members prefer.

5.3 Applications: Abundance and Scarcity

Both Apple and Google are leading providers of converged devices. More and more these are becoming application delivery devices. The developer ecosystem has proven to be a very effective model for Apple and Google. Close to 90,000 applications are available on the Apple Store and Google Play claims closer to 1 million. Hence while the networks and devices are converged, the applications are following a complete different pattern of use. Seemingly there is more than one application for about any activity ranging from the serious (like remote medial diagnostic) to the completely frivolous (the large number of zombie "detectors" and cute cat pictures). As the World Wide Web (WWW) and the wireless multiscreen screen are becoming equivalent, questions about the infrastructure of the new Internet remains: can the current wireless protocols and networks support the quality of the wireless experience that are not only required for the development of new services and the ubiquitous connectivity required by the users but also to provide reliability, privacy and secure connections? The major applications like Facebook, Twitter or YouTube still generate the most traffic but the sheer number of applications is pushing the limits of the converged networks and require new approaches to ensure the Quality of Experience (QoE) of the users.[8] And the companion devices ranging from lifestyle monitoring to home security are exacerbating the trend: there are 10 billion connected devices today to become 20 billion in 2020.[9]

In particular, the abundance of always on, video and other rich media on the wireless Internet is becoming a challenge to the QoE, creating indeterminate performance especially in homes and public areas alike. There are strong incentives to investigate novel solutions to improve the wireless experience. The wireless industry has known for over 20 years realized early the impact of errors on wireless performance. This includes not only the throughput, the bits delivered to the device, but also the goodput, the bits delivered to application, and the consequences of directly applying wireline solutions in the wireless domain. Interruptions and delays are providing a poor and unwanted quality of experience. In reaction, the users

[8] For the purpose of this book chapter we will define Quality of Experience as the subjective evaluation of a service by its users. QoE encompasses measurable parameters like delay when playing a video or opening an application, service interruptions and overall application availabilities but also user interface inefficiencies, poor screen layouts and more and more application overload.

[9] http://techcrunch.com/2013/05/09/internet-of-everything/

disconnect: this is not what content providers, user communities and advertisers wish for.[10]

Novel approaches, from bandwidth sharing with device-to-device (D2D) or peer-to-peer (P2P) architectures and application layer error corrections mechanisms and application accelerator like the Google SPDY,[11] Qualcomm Raptors[12] and QFactor's Dynamic Packet Recovery[13] are being developed and deployed. P2P in particular offers the promise of creating local consumption groups that take into accounts the availability of close-by resources like storage to create the community networks described in the Innovation at the Edge paper already mentioned in the literature search of Sect. 2. P2P video on demand has been studied extensively in academic circles as a replacement for centralized remote disk farms and taking advantage of hyper-local video preferences; one such solution, a push-to-video system was described by the team from University of Massachusetts and Technicolor in 2007 (Suh et al., 2007) and the author presented architectural improvement to P2P to make it more acceptable in the wider content dissemination community by adding mechanisms to ensure content protection (Montpetit, 2008).

And finally infrastructure and bandwidth are only two aspects of the device convergence impacts. Another one is sustainability. Upgrading a converged device is becoming a frequent event (yearly for some) with the impact on the environment due to the recycling of these electronics. Compared that with the life of an old telephone or television, which could be repaired. The environmental consequences of device convergence, beyond the scope of this chapter, could spur a return to more focused devices, but this is still to come. In the mean time however their immediate impact is on media consumption, which has experienced dramatic changes in the last few years. It is the topic of the next section.

6 Media Convergence: Content meets Social Networking

The WWW were initially used as a one-way communication system. Viewers consumed static content from the content providers and couldn't interact beyond selecting which hyperlinks to follow. And this is very much what is still the model followed by a number of websites today. But starting with e-commerce sites and now social networking, the one-directional model now allows to incorporate, commenting, micro-blogging and user generated content. Interactivity enables new models for content consumption and it impacts the whole content industry

[10] http://gigaom.com/2012/11/09/online-viewers-start-leaving-if-video-doesnt-play-in-2-seconds-says-study/?utm_source=General#43;Users&utm_campaign=81ff9e61ba-c%3Amob%2Ctec%2Cvid&%2343;d%3A11-10&utm_medium=email

[11] http://www.chromium.org/spdy/spdy-whitepaper

[12] http://www.qualcomm.com/solutions/multimedia/media-delivery

[13] www.qfcomm.com

including newspapers, books, movies and television. Social Media is another phenomenon made possible by the convergence of networks and devices.

6.1 Social Connectivity: A Result of Convergence

Social networks have greatly impacted many areas from personal communications, becoming less personal when posted online, news-gathering with real-time commentary and micro-blogging from newsworthy events, to advertising and its focus on micro-blogging to provide audience measurements. Social networks are reshaping the way people find and consume content of every kind, providing major disruption in the media industry: the question is still open about its positive or negative influence.

User and business communities now create and engage in digital social innovation using platforms from Facebook to Twitter, Instagram and Pinterest. These are essentially social platforms combining social networking systems, cooperative creation software and mechanisms for the sharing of knowledge and real-time gathering of content. Applications range from more traditional gaming and micro-blogging to social video consumption, fundraising such Kickstarter,[14] distribution and rating, the health and well-being, environment and sustainability such as energy monitoring programs and smart grid applications, always-on traffic monitoring and directions, home security systems and controls and the emergence of the connected do-it-yourself (makers community).

Social media applications require that the members of the communal experience be connected via some Internet technology. But inherently, social connectivity should not imply physical connectivity, but instead should allow changing the physical connectivity to offer the best quality of experience. Our need to connect socially should drive our physical connectivity; we should be able to enable social connectivity it on demand independently of the actual platform, device or network being used. One answer is to define *content as a service* leveraging current cloud computing. This is very much aligned with content delivery networking and content-centric concepts being proposed for the next generation Internet. And of course it defines the Social Television experience.

6.2 Social Networks and Video: Social Television

As web-based social networking is becoming more and more prevalent with more than half a billion Facebook users, its impact of the TV experience is huge and barely starting to be measured. Most television programming now includes some links to the social networks either directly from the programmer or via companion application.

[14] www.kickstarter.com

Video on the web dates back to the mid to late '90s with the first versions of Apple's QuickTime and Real Networks RealPlayer. It really took off as a streaming service of decent quality with the availability of broadband and the development of better devices with cameras and powerful graphical engines. YouTube became the video archive of humanity in just a few years. And of course the mobility of both the users and their devices liberated the content from the network: crowdsourcing provides live video information from virtually anywhere, at any time. Online communities and social networks have shown that the most efficient way to create and ensure the quality of user-generated content is to leverage social capital.

The use of social networking with traditional television linear or on demand programming is creating a tremendous opportunity for a paradigm shift for TV viewing. Commentary, video sharing and multimedia interaction can be added to the TV shows to promote content, encourage viewer loyalty and measure engagement. According to Yahoo and the Nielsen Company,[15] 86 % of mobile Internet users (and 92 % of the 13–24 year olds) are using their mobile devices simultaneously with TV. Updating/reading social networking sites while watching a show are the most popular activity; the goal of programmer is to encourage this activity to be centred on the show being watched. Micro-blogging activities are now considered the best way for content producers and advertisers to promote shows, gather commentary and measure audience and their reaction to programming. Activity on Facebook about popular shows and show-specific group are both growing; promotion of upcoming programming on social media is now the norm, not the exception.

With end of the traditional TV channel brought by converged television architectures, the gates to the social channel and the re-discovery of TV and video viewing in general as a shared social experience are re-opening. The living room of the 50s is being replaced by the "global" and "cyber" living room where content is consumed together or recommended by the group. The water cooler is also moving to cyberspace: activity in social networks encourages viewing and the creation of communities around television content. One's social network allows specifying peers with whom to share video experiences: common viewing, sharing comments, posting ratings, discussing content etc. Cloud-based content, aggregation sites, OTT and broadcaster applications ensure the availability of the same content on different devices and at different times that suit the individual viewer while enabling social interaction. The TV experience moves from a single person, to a family, to a friend group and finally the wider the social network. Social TV is not just the result of aggregating social networking streams around television content but it is creating communities and social engagement around television content.

One of the original goals of STV back in 2002–2005 was to enable viewers in different locations to socialize around television content using Internet technologies. By allowing people to synchronously communicate with others while watching TV, early applications wanted to remediate to the social dislocation of modern families and provide a familiar environment to engage family and

[15] http://advertising.yahoo.com/industry-knowledge/mobile-shopping-insight.html

friends. For that reason, pioneer work on social TV focused in linking separated living rooms and create virtual viewing rooms with embedded cameras and microphones, where remote users could communicate with others, while watching television content together. While those were valid experiments they suffered from both technical issues like maintaining synchronicity across different locations but also from the fact that it was disruptive on the overall viewing experience. Hence user acceptance was fairly low but still showed a need for the togetherness created by television.

The newer implementations of STV are much more associated with direct social and community interactions via microblogging and annotation tools. They benefit from the deployment of IP connected set-top boxes and connected television. The use of smartphones and tablets for interactivity has generated a number of companion applications that provide interaction features complementary to Twitter microblogging. Popular applications use audio fingerprinting or recognition to synchronize to the watched show and applications like Zeebox[16] or Viggle[17] allow to link extra content directly to what is being watched.

To be successful STV also needs to take the viewing behaviour into account. In "Convergence Culture" already mentioned in Sect. 2, Professor Jenkins discusses the fact that:

> Different genres of entertainment provoke different degrees of social interactions. (ETSI 2008)

Hence the current reliance on real-time micro-blogging to measure social engagement is insufficient: a lot of commentary happens before of after the show especially drama that is very poignant or action-filled sequence that need attention. For these shows the successful social interaction will include interactivity before or after the show. From Professor Jenkins comments it is easy to realize that a *one size fits all* approach to STV does not satisfy cultural and artistic goals. There is still a need for a comprehensible framework for how media pieces will be combined together to create a seamless user experience, containing the right mix of social connectivity and content. The current direction toward *shows as applications* could provide customized STV that is specific to maintain the immersive television experience and provide the right amount of socialized information to enhance that experience. The use of metadata could be used for these purposes. In her paper, entitled "CommenTV" (Hwang et al., 2012) Jee Yeon Hwang suggests that metadata could be used to interact with viewers especially when they watch at different times.

There is also a need to keep some of the accepted viewer behaviour when designing STV. Many STV applications favour the *lean forward* mode of television associated with the web experience. Many viewers appreciate getting more

[16] www.zeebox.com

[17] www.viggle.com

information from their extended social network as well as ancillary information about their show. Groups of viewers can exchange comments while watching the same show or leave messages on screen for later viewing (such as provided by the viki application[18]) and suggest related content. However, some of these applications tend to create an information overload: many times microblogs are not properly filtered and as a result, viewers are exposed to all the comments, which in most of the cases are irrelevant for them or could even be interpreted as breach of privacy. The curation of the commenting via the social group is still in its infancy and will improve, as better tool are developed to take advantage of social graphs and personal preferences.

But one should not forget the *lean back* mode. One criticism that is common when discussing STV experiences is that one may just want to enjoy a show without distractions. This is a good argument for moving the interaction to the personal space of the phone and tablet or at a minimum to allow a back and forth between lean back and lean forward. In their project named NeXtream published in 2010 (Martin, Santos, Shafran, Holtzman, & Montpetit, 2010) Reed Martin and his team devised an ingenious use of the iPhone accelerometer for moving between active and passive mode: when the phone is on the table and in the horizontal position the main screen is in the non-interactive mode, when the phone is held vertically (in the user hand) it triggers the interactivity menu on the screen for commenting and recommendations.

Content recommendation is probably the most popular application linked to social television. Interactive guides like the NextGuide[19] use media related posting from a social group to promote content one could be interested in watching. Social features in applications and in self publishing sites like YouTube become virtual operators or curators that suggests the content friends should be watching, enables ratings and creates content lists influencing the group's viewing behaviour. Tablets also provide the perfect screen to create joint viewing and commenting areas in the personal space. This has the advantages of ubiquity but also of leaving the main television screen for the more public experience, the one involving other people. More recent research projects have also looked into cross-media, using what you buy or the music you listen to, as a means to recommend related video content. Cross-media is different from transmedia in the sense that it does not relate to storytelling but to media consumption in general. Freely available profiles are valuable assets for content distributors and aggregators like Netflix or Amazon, since they can use social interactions for providing more accurate content recommendations. Cross-media recommendations were utilized in the Tubely project in 2012 (Chan and Narula, 2012) (Fig. 4).

STV is also now viewed as a game changer in the way the content itself is produced and consumed to promote an immersive experience and increase user engagement. Even in the age of convergence, producers have not taken full

[18] www.viki.com

[19] www.dijit.com

Fig. 4 NeXtream: multidevice STV experience (Martin et al., 2010)

advantage of the opportunity to radically affect television content production using converged technologies. Some forays in academia have attempted to integrate television in its environment favouring the whole-room experience and the use of ambient elements; the evolution of such approach is described in Lugmayr and Zheng (2012) as a feature for the future. Some shows allow television viewers to vote off contestants (e.g., American Idol) in an aggregated manner but while this is possible for reality shows it remains limited overall. Other more interesting experiments include a television drama created in Finland, Accidental Lovers where viewers could influence in real-time by sending text messages (Ursu et al., 2008), but they are still very scarce because of their potential disruption over the show's storytelling. While there is an obvious opportunity in using STV in the creative process it remains to be seen how professionally produced content will address it. For the moment YouTube and other user generated content site are starting channels and promoting commercially produced content around social commentaries and the boundaries between the commercial and personal spaces are starting to meld and to provide a rich base for the creation of future programming.

But STV is not and should not be only using a linear process from content production by one to content consumption by many. Crowdsourcing on mobile phones is used to create a new form of content acquisition and distribution for journalism. It creates a large amount of data to be filtered for authenticity, privacy and lawfulness. The previously mentioned Stringwire new gathering application,[20] recently acquired by NBC provides curation by real journalists and a rating

[20] www.stringwire.com

mechanism for the contributor to filter the received information and ensure truthful reporting. The applicability of such approaches to TV drama is still under question but the relative success of the MySpace teenage series Freak in 2009 is promising: its viewers were asked to participate in the show's plots and tapings as well as providing personal content relating to the show and they reacted very positively.

The growth of STV is a testament to the power of convergence in changing traditional media to take advantage of the new technologies but also a reflection of social trends. Most current Social Television implementations combine social interaction and personalization features but in fact STV can be much more: it can promote the creation of community around the viewing experience. And convergence can be seen as the creator of communities: networks, devices, content and of course people.

7 A Viewpoint on Convergence: Disrupting the Content Consumption Experience

At the end of August 2013 a meeting in Montreal was held to discuss the different aspects of convergence under the theme of "community, audacity, authenticity: managing the convergence of media" (f.&co 2013). The goal of the meeting was to position convergence into the media creation process focus and emphasize the emergence of new means of connectivity to achieve community building. The meeting also wanted to re-enforce the idea that convergence and its associated technologies are creating opportunities for not just the end of the mass media experience. Too many traditional providers, from broadcasters to newspapers to book publishers continue to deliver content constrained by antiquated regulations, artificial programming schedules and single platforms, rather than tailored to the individuals and taking advantage of connectivity. In the innovation circles this state of mind is changing fast. The large variety of available content and delivery mechanisms has disrupted the industry status quo. Hence convergence in Technology, which was discussed throughout this chapter:

> Has now moved the media discourse from the simple transposition of similar content from one platform to the next, the notions of transmedia storytelling have opened a vast array of creative expressions in cable and web television, adding to the creations of app developpers, authors and video game producers to create entire new universes (f.&co 2013).

The media experience is becoming richer and more exciting as our media becomes available on an increasingly diverse set of multi-purpose devices as was described throughout this chapter. The move to the cloud allows users to choose online content they're interested in, and even share content recommendations through online social networks. But coincidentally these many options introduce complexity and frustrations for some users. Access to broadband connectivity at an affordable price is still not universal; while free municipal WIFI is appearing in many communities it is more the exception not the rule. The fragmented and confusing experience the abundance of applications widens the generational and technological gaps. Divergence emerges from convergence.

But does it have to be that way? The challenge for developers is clear: how to navigate this diverse ecosystem of content and devices while providing a satisfying user experience? For example, traditional TV viewers usually expect a passive, lean-back interaction as was described previously. Mobile device and computer users are used to more actively seek out online information. So while interactive viewing may be an advantage, it is also more complex for viewers accustomed to the traditional experience. So for truly building communities convergence must allow for diverse behaviour to not only being accommodated but to become part of any new offerings. This requires a re-thinking of existing architectures for networks, new creation and distribution models for content and novel user behaviour metrics; it is happening.

Convergence encourages a reinvention of content acquisition as we saw in the preceding sections as devices are now capable of recording and transmitting a large variety of multimedia content almost from anywhere an Internet connection is available. While its impact of movie making is still to be assessed it is easy to see how smartphone could contribute not only images and sound for the movie but via social networks a direct interaction between the characters and the viewers. This is already being used in some TV programming. The biggest impact of convergence according to his author is its changes to person-to-person communications and its promise of forming virtual communities. To communicate with one another we can now use our device of choice, at our location of choice and at our time of choice. We can use voice, especially when dealing with older people used to the phone, we can use text with our teenagers, video and pictures with our friends and often all of those together when hosting online events. Our connectivity is not defined by anyone but us. Convergence allows us to get together in an un-precedented manner.

And convergence continues to, well, converge, moving to new domains. It profits from the Information Ecology[21] and the interdependencies between users, application development, and the availability of appropriate networks as well as underlying cultural trends. These forces will shape the future of convergence. The *power of convergence* consists in:

> The integration of disciplinary approaches that were originally viewed as separate and distinct. This merging of technologies, processes and devices into a unified whole will create new pathways and opportunities for scientific and technological advancement. (MIT News, 2011)

Convergence is truly talking advantage of the *authenticity* of human interaction and the *audacity* of technical and media innovators to create the *community* of tomorrow.

---

[21] Thanks to Henry Holtzman of the MIT Media Lab for coining the word.

8 Conclusion: A Note on Managing Convergence

In "Convergence is King" (Cass, 2011), Stephen Cass proposes that companies like Apple have been the winners of the convergence economy because since the iPod in 2001 it has managed to "deliver all kinds of content to you in a way that is so seamless that you cannot pass it up" hence creating a lucrative market out of their converged platforms.

The impact of the wireless (converged) Internet in on the economy is huge. Smartphones, themselves the result of convergence are one of the reasons, along with the emergence of tablets as the communication platforms of choice, why landlines are being cancelled as was first reported in the Business Insider in 2010.[22] And the new leaders of large operators like AT&T and Verizon in the United States are now being selected from their wireless carrier subsidiaries, not from their wireline businesses. Convergence has shifted the balance in strategy to the wireless business.

A key challenge in the next decade will be for the industry to follow a comprehensive strategy to end-to-end and top-to-bottom systems to move away from the remaining silos, and to encourage innovation across networks, devices, and services; combining content and social interactions is a goal. Convergence has given us the anywhere/anytime/any device immersive world. The challenges of convergence are to move from closed devices and independent departments into a more interdependent and open ideas environment where the technology, the services, the applications and the user interface merge. In the business area this creates in the words of Andre Hagiu of Harvard a true *multisided application*, one that results for these many inputs and in turn profits from its diverse component.

As defined in the Introduction, this *2nd Convergence* about technology, business models, social networks and culture. It has moved beyond entertainment and mass media to provide comprehensive solutions from medicine to smart cities as is reflected in the wide range of applications available in app stores. The innovative use of devices and soon of the augmented reality of Google glasses for example add to these application to create immersive experiences. The human skills are evolving and are being combined, another convergence? Artists are now technology-savvy, technologists are discovering the needs of end users and we are all becoming involved in sustainable development and *living laboratory*. The new convergences is moving away from basic functionality into a richer set of interdependent elements of hardware, software, content and user interaction. Hence the managing of convergence means a return to the engineering principles of the past: the realization of the need of the different fields involved to collaborate and interact at all points in the realization of a common project. Can the second convergence lead to convergence of skill and a new Renaissance enabled by technology? The future may tell.

[22] http://www.businessinsider.com/chart-of-the-day-almost-a-third-of-us-households-have-cut-the-landline-cord-2010-8

Acknowledgements None of the work presented in this chapter would have been possible without the collaboration over the years of Henry Holtzman, Pablo Cesar, Oscar Martinez-Bonastre and the MAS 571 Social TV Class at MIT Media Laboratory. Recent interactions with Francis Gosselin, Sylvain Lafrance, François Bédard and Gabrielle Madé on the use of convergence to create communities were invaluable. Thanks also go to my daughter Mathilde Montpetit and my friend Michael Mercurio for their support through the years.

Appendix

Acronyms

3GPP	3rd Generation Partnership Project
AF	Access Function
ATIS	Alliance for Telecommunication Industry Solutions
D2D	Device to Device
DRM	Digital Rights Management
DSL	Digital Subscriber Line
DVR	Digital Video Recorder
ETSI	European Telecommunication Standards Institute
GSM	Global System for Mobile (Communications)
GSMA	GSM Association
IETF	Internet Engineering Task Force
FMC	Fixed-Mobile Convergence
IEEE	Institute of Electrical and Electronics Engineers
IIF	IPTV Interoperability Forum
IMS	Internet Multimedia Subsystem
IP	Internet Protocol
IPv4	Internet Protocol version 4
IPv6	Internet Protocol version 6
IPTV	Internet Protocol Television
ISP	Internet Service Provider
MAC	Medium Access Control
MIT	Massachusetts Institute of Technology
NGN	Next Generation Network or New Generation Network
OTT	Over the top
P2P	Peer to Peer
POP	Point of Presence
QoE	Quality of Experience
QoS	Quality of Service
RACS	Resource and Admission Control Subsystem
RFC	Request for Comments
SIP	Session Initiation Protocol
STB	Set-top Box
STV	Social Television

TISPAN	TIPHON (Telecommunications and Internet Protocol Harmonization over Networks) and SPAN (Services and Protocols for Advanced Networks)
TV	Television
VoIP	Voice over IP
XoIP	"Anything" over IP

Definitions

Architecture:	abstract representation of a communications system
Control plane:	plane that has a layered structure and performs the call control and connection control functions; it deals with the signalling necessary to set up, supervise and release calls and connections
Flow (of IP packets):	traffic associated with a given connection-oriented, or connectionless, packet sequence having the same 5-tuple of source address, destination address, Source Port, Destination Port, and Protocol type
Goodput:	the number of bits delivered to an application
Forwarding:	process of relaying a packet from source to destination through intermediate network segments and nodes
Management plane:	the management plane provides two types of functions, namely Layer Management and plane management functions
Throughput:	the number of bits delivered from a network to an attached device
User plane:	plane that has a layered structure and provides user information transfer, along with associated controls (e.g. flow control, recovery from errors, etc.)

References

ATIS. (2004, November). ATIS TOPS council NGN Focus Group has defined the NGN in *ATIS Next Generation Network (NGN) Framework Part I: NGN Definitions, requirements, and architecture*, Issue 1.0.

Baca, M., & Holtzman, H. (2008). Facebook Meets TV. *EuroITV 2008*.

Bourdonnec, F. (2010). *The changing TV experience*, Intel Consumer Experience Group. Private communication.

Bulkeley, W. M. (2010, May-June). Social TV. *MIT Technology Review*. http://www.technologyreview.com/communications/25084/

Cass, S. (2011, December 29). Convergence is king. *MIT Technology Review*.

Cesar, P., Bulterman, D. C. A., & Gomes Soares, L. F. (2008). Human-centred television: Directions in interactive digital television research. *ACM Transactions on Multimedia Computing, Communications, and Applications, 4*(4), Article 24. doi: 10.1145/1412196.1412197
Cesar, P., & Geerts, D. (2011). Understanding social TV: A survey. In *Proceedings of the Networked and Electronic Media Summit* (NEM Summit 2011).
Chan, J., & Narula, N. (2012). *Tubely: A personalized group watching experience.* MIT MAS 571 Final Projects. Available from nehanaru.la/Tubely.pdf.
Cisco. (2015). Cisco visual networking index: Forecast and methodology, 2009–2014. http://www.cisco.com/en/US/solutions/collateral/ns341/ns525/ns537/ns705/ns827/white_paper_c11-481360.pdf
ETSI. (2007, November). ETSI ES 282 001: Telecommunications and Internet converged Services and Protocols for Advanced Networking (TISPAN); NGN Functional Architecture.
ETSI. (2008, May). ES 282 007 Telecommunications and Internet converged Services and Protocols for Advanced Networking (TISPAN); IP Multimedia Subsystem (IMS); Functional architecture.
f.&co. (2013, August 23). Community, audacity & authenticity: Managing the convergence of media, Discussion between Marie-José Montpetit and Sylvain Lafrance, at Technicolor Montreal. http://fandco.ca/en/news/article/2013/08/community-audacity-authenticity-managing-the-convergence-of-media-1/
Hwang, J. -Y., Pla i Conesa, P., Holtzman, H., & Montpetit, M. J. (2012, January). CommenTV: A time-sensitive social commenting system for audiovisual content. In *Proceedings of the 9th IEEE Conference on Consumer Communication and Networking (CCNC2012).*
IETF. (2002). IETF RFC 3261 "Session Initiation Protocol". www.ieft.org
Jenkins, H. (2001, June 1), Convergence? I diverge. *MIT Technology Review.*
Jenkins, H. (2006). *Convergence culture: Where old and new media collide.* New York, NY: New York University Press.
Lugmayr, A., & Zheng, H. (2012). Forecasting the usage of ambient media in TV broadcasting in the year 2016. In A. Lugmayr, T. Risse, B. Stockleben, J. Kaario, B. Pogorelc, & E. S. Asensio (Eds.), *Proceedings of the 5th international workshop on Semantic Ambient Media Experience (SAME).* Tampere University of Technology (TUT).
Martin, R., Santos, A. L., Shafran, M., Holtzman, H., & Montpetit, M. J. (2010, January). NeXtream: A multi-device, social approach to video content consumption. In *IEEE CCNC10.*
MIT News. (2011, January 4). The power of convergence. *MIT News.*
Montpetit, M. J. (2008, January 18). *Community networking: Getting peer-to-peer out of prison.* Presentation at CISCO, MIT Communications Futures Program (CFP).
Montpetit, M. J., & Klym, N. (2008, September). *Innovation at the edge: Social TV and beyond.* MIT Communications Futures Program (CFP). Available on demand at cfp.mit.edu.
Montpetit, M. J., Mirlacher, T., & Ketcham, M. (2010). IPTV: And end-to-end perspective. *Journal of Communications*, Special Issue on Multimedia Computing and Communications. JCM, ISSN 1796-2021.
Montpetit, M. J., Mirlacher, T., & Klym, N. (2011, Spring). The future of TV: Mobile, IP-based and social. *Springer Journal of Multimedia Tools and Applications, 53*(3), 519–532.
Paulson, L. D. (2010, July 1). New approach combines TV and social networking. *IEEE Networks Magazine, 43*(7), 16–19.
Piokselka, M., & Mayer, G. (2009). *The IMS: IP multimedia concepts and services.* New York: Wiley.
Rich Communication Suite. (n.d.). http://www.richcommunicationsuite.com/
Suh, K., Diot, C., Kurose, J., Massoulié, L., Neumann, C., Towsley, D., et al. (2007, December). Push-to-peer video-on-demand system: Design and evaluation. *IEEE Journal on Selected Areas in Communication, 25*(9), 1706–1716.
Ursu, M. F., et al. (2008). Interactive TV narratives: Opportunities, progress, and challenges. *ACM Transactions on Multimedia Computing, Communications, and Applications, 4*(4), Article 25.

Convergence Policy and Regulation: A Free Speech Perspective

Shantanu Dey

1 Introduction

The technological and social phenomena of 'convergence' has been accompanied by attempts at introducing 'Converged Legislative Frameworks' by various countries, which have been successful in implementing policy changes in varying degrees. The premise for most of these legislations has been the perception that the technological possibility of 'convergence' of erstwhile differentiated media streams now necessitate 'convergent' laws and regulations that should replace the plethora of distinct and discrete legislations that were targeted at differentiated media.

Henry Jenkins has rejected any perception of Convergence as a mere technological process; he prefers a view of convergence as a 'cultural shift' where consumers are encouraged to seek out new information and make connections among dispersed media content (Jenkins, 2006).

In registering convergence as a social, cultural and political process that is not a mere outcome of a technological phenomena, we appreciate the possibilities of 'divergence' and dissipation of corporate power, especially as it manifests itself in a scatter and flow of 'communication power' across multiple 'platforms' through a multitude of 'networks'.

According to Vinton G. Cerf, the properties of cyberspace that have led to its tremendous success are: freedom of expression, transparency and openness, participatory policy and technology development (Cerf, 2013). We believe the struggle to assume 'communication power' is being played out in a dynamic, fluid triad that is broadly organized around three nodes of players—governments, corporations and citizens/citizen-groups. The key challenge from a Constitutional and Human Rights perspective is to use regulatory means to thwart any attempts by any individual or

S. Dey (✉)
Indian Institute of Management, Calcutta, India
e-mail: dey.shantanu6@gmail.com; shantanu@iimcal.ac.in

A. Lugmayr, C. Dal Zotto (eds.), *Media Convergence Handbook - Vol. 1*, Media Business and Innovation, DOI 10.1007/978-3-642-54484-2_4

collective, whether governmental, corporate or citizen-centric to assume 'communication power' in a way and to an extent that it chills the Free Speech of others.

1.1 Problem Discussion: 'Free Speech' in the Convergence Era

The UNESCO Constitution states that it is dedicated to the 'free flow of ideas by word and image'. The UNESCO Report asserts that 'Freedom of expression is not an 'inevitable outcome of technological innovation.' Such freedom can be reduced or reinforced by the 'design of technologies, policies and practices—sometimes far removed from freedom of expression.' There is a 'wider ecology of choices' that is shaping freedom of speech and expression in the digital age; we believe the implications of convergence need to be understood in this context (Dutton, 2010).

In the Convergence era, the issue of whether the constitutional value of freedom of speech is sufficiently guarded by legal and regulatory means needs to be interrogated afresh at two levels: (a) are the paradigms guiding such instruments of regulation, especially those in an economic mould relevant, appropriate and effective in the convergence era; (b) is there a need to review and rework such instruments of regulation that have historically been used to achieve constitutional values in the pre-convergence era; (c) is there a need for a new public policy paradigm that is appropriate and effective in achieving freedom of speech in the convergence era?

In this context, we will revisit competition laws as well as structural regulation in the media-communication domain and analyze them from a Free Speech perspective, which we define and develop in Sect. 3.

We present a literature review in Sect. 2 that discusses various ideas and concepts which we have draw upon to offer our own legal and policy analysis. The comprehensive literature review covers the experience of convergence; economic aspects of convergence that includes business models of communication industries and concentration of ownership; economic regulation; and contemporary regulatory and policy challenges. In Sect. 3, we present our preferred methodology and approach, which is a 'Free Speech' perspective applied to interrogate policy-regulatory issues in the media-communication domain in the convergence era. This section also provides a brief review of some of the relevant theoretical foundations of economic regulation to relate our Free Speech perspective to the concept of competition. We discuss the concepts of 'networks' and 'platforms' as well as the 'network layers regulation model' in Sect. 4.

In Sect. 5, we offer a doctrinal analysis of the major 'Convergent Legislative Frameworks' in various jurisdictions; also, analyze the issues of structural regulation, market failures and policy challenges in the convergence era.

In Sect. 6, we suggest an approach to Convergence Policy and Regulation; this includes our proposal for 'affirmative public ownership' as a means of achieving effective freedom of speech in the convergence era. We conclude with some final remarks in Sect. 7, which includes our views of convergence as a multi-hued, complex process that also demonstrates divergence. We claim that regulation and policy in the Convergence era needs to be alert to this complexity and duality that arises in the enactment of convergence as a process that is, perhaps, driven by technology but not necessarily limited or ruled by it.

2 Literature Review

In this section we provide a literature review of the key ideas concerning guiding our analysis: a review of ideas concerning economic regulation in the Convergence era; and a review of specific convergent laws and regulations in the media-communication domain.

2.1 Economic Aspects of Convergence

We now briefly discuss a few studies that demonstrate the nature of 'convergence' from an economic perspective. We will discuss those aspects that pose regulatory and legislative challenges in this realm; in particular, we are interested in horizontal and vertical integration as well as concentration of ownership, which impact freedom of speech and expression.

2.1.1 Convergence and the Evolving Business Model

Simon McPhillips and Omar Merlo, who have studied the current business model in the American and UK markets, reject the notion of a revolution due to convergence; rather, they believe an evolution is occurring where old models co-exist with new models for a while till they converge (McPhillips, 2008).

2.1.2 Horizontal Integration

The technology, media and telecoms sectors (TMT) have all merged to some extent. Media convergence is redefining industry boundaries by facilitating horizontal integration.[1] Media owners are leveraging potentially profitable business opportunities to a single platform given the possibilities unleashed by convergence (McPhillips, 2008). So, BSkyB in the UK is now a local loop unbundler, offering telecommunication services along with its cable package. Again, Google is moving into television. Verizon is competing with cable operators for movie content in the US. Firms are combining existing and new services in innovative ways—e.g. triple or quad-play that incorporates cable TV, internet, fixed line telephony and mobile. In the UK, Virgin Media and SKY TV are competing in the quad-play space.

[1] In this connection see Bennett and Strange (2008), Caldwell (2006), Doyle (2010) and Vukanovic (2009).

2.1.3 Vertical Disintegration

We are likely to see vertical disintegration within each sector. Earlier, media owners dominated the entire value chain from creative inception to production, marketing and distribution. Now, producers do not need to depend on media owners for either production or distribution; production costs have dropped with digitalization and the Internet is offering opportunities for an open-source route to market.

Media owners are worried by vertical disintegration, but they can still find some opportunities by decoupling the value chain and focusing on the more profitable parts of it (McPhillips, 2008). This is a positive trend from the point of view of freedom of speech; it means that there may be a lesser need for oversight of ownership across vertical levels within a layer. However, this needs to be reconciled with our suggestion[2] that there may still be a need for 'structural regulation' to govern the vertical 'layers' of networks—ownership between the layers—which, we believe is a different matter.

2.1.4 Concentration of Ownership

According to Graham Murdoch, digital convergence has highlighted 'ownership convergence' because digitalization has been one of the key projects for media companies seeking their ways to rationalize the production and distribution of media content; so to speak, to minimize their costs (Murdock, 2000).

However, the conglomerate media structure has not been working well too in the convergent era.[3] Time Warner has been under pressure to divest itself of some of its divisions, Viacom has been split and Disney has sold holdings like its radio division. However, the Internet itself seems to be becoming more concentrated than its image suggests. For instance, Eli Noam believes that the concentration in the Internet industry is much higher now than it was 10 years ago; but the concentration level is lower now (Noam, 2008). Many major components of the Internet—broadband access, backbones, portals, search engines, etc.—are highly concentrated.

2.2 Economic Regulation in the Convergence Era: Literature Review

Since we believe that instruments of economic regulation are important to attain free speech objectives, especially in an age marked by privatization and convergence, we flag them in our present analysis.[4] In this section we review the 'competition' paradigm that rules much of economic regulation; in particular, we consider the problem of 'market failures' and the relevance of 'structural regulation' in a Convergence era.

---

[2] See Sect. 6.1 on suggestions for structural regulation across vertical layers.

[3] In this connection see NewsCorp (2012)

[4] Please discussion of 'Free Speech' perspective in Sect. 3.

2.2.1 Competition

Gentzkow believes that the traditional definition of 'competition' focuses on 'competition in the product market' while the relevant notion of competition in the context of news markets is that of 'competition in the information market' (Gentzkow, 2008). This implies that even in media markets where there is high concentration of ownership or huge market-share is enjoyed by a few major players it is entirely possible that the market for information is competitive.

Market Failures The rationale for the economic regulation of markets is most often cast in terms of 'market failures'—i.e. regulation needs to step in when the markets fail to be competitive and realize all desirable objectives. The Office of Communications (Ofcom) of the United Kingdom has done a fair amount of work with the notion of 'market failures' as a means of achieving its regulatory goals. There may be many causes for 'market failure' such as market power and externalities.[5] Ofcom defines it as follows:

> Economic theory suggests that perfectly competitive markets deliver efficient outcomes that maximize social value. However, economists recognize that certain factors, such as the presence of market power, can prevent this. To describe the outcome where factors are present that stop the market working efficiently, economists use the term market failures (Ofcom, 2004).

To discuss the issues surrounding the new digital era, Ofcom conducted a *Digital Dividend Review* to 'set out a framework for considering market failure with particular reference to the broader social value of some media and communication services (Ofcom, 2007).' It argued that when a market fails to maximize 'total value', this can indicate the presence of market failure. The notion of total value has been explained as:

> Within the concept of total value there are two key elements of value. The first is private value, which is the value consumers and firms derive from their interaction in a market. The second is external value, which is value that results from services provided by a market but which is not fully reflected in the choices of consumers or firm (Ofcom, 2007).

Ofcom has intervened in the broadcasting market on grounds of market failures.[6] Public service broadcasting has traditionally seen policy intervention to solve market failures. Further, competition law in the UK has been extended to the realm of broadcasting regulation to achieve 'structural regulation' of the sector. The Government has termed this as a 'competition plus' approach that allows it to retain industry-specific rules where it considers it necessary.

[5] According to Ofcom, an externality is said to occur when a transaction between two parties has an impact on a third party, which is not involved and typically does not take it into account.

[6] See the *Public Service Broadcasting Review (Phase I)* by Ofcom for a detailed discussion of its approach to market failures in public service broadcasting (Ofcom, 2004).

2.2.2 Structural Regulation

The regulatory and anti-trust policy of the USA has traditionally attempted to secure a diversity of voices, structurally, through rules about ownership; such rules—viz. cross-ownership rules and ceilings on ownership within a market to defuse 'concentration of ownership'—have been sometimes upheld by the constitutional courts, and also, compelled by them (Horwitz, 2004).

The idea that diversity in media output can be achieved through diversity in media ownership is presented as an objective of regulation. According to Jerome Barron, we encounter three diversity rationales for multiple ownership rules in the revised duopoly rule report (Barron, 2000). One of the rationales is that multiple ownership of media outlets will lead to diversity of viewpoint in programming; and another one is that media de-concentration rules will prevent 'undue concentration of economic power contrary to the public interest (FCC, 1999).'[7] Such 'diversity rationales' behind ownership rules and structural regulations of mass media have been diluted by legal trends as well as political developments in the last 20 years (Horwitz, 2004). Gillian Doyle has argued that 'the issue of who owns the media, and how much of it they own matters (Doyle, 2002).'[8]

Therefore, it is perhaps appropriate to consider 'structural regulation' in terms of the extent to which it serves successfully as a surrogate for imperative content-based regulation which might be expected to pursue more directly the objective of diversity in output (Feintuck, 2006).'

2.3 Contemporary Regulatory and Policy Challenges: Literature Review

In a 'Broadband Age', the regulatory and policy challenges are vastly different:

> Future media policy and regulation... will have to address the entire "media ecosystem," viewed as a "regulatory space" in which self-regulation and the market are all part of the basket of regulatory tools. Its goal should be to maintain and strengthen the public sphere (Hitchens, 2011).

Hitchens points out that convergence involves changes in the 'public's engagement with the media and transformation of audience into content creator'.[9] The convergence era poses special challenges to regulatory regimes that used to make the mode of transmission the primary determinant of the scope of regulatory obligations. Convergence even when nascent enables 'previously distinct systems

---

[7] 'The greater the diversity of ownership in a particular area, the less chance there is that a single person or group can have an 'inordinate effect, in a political, editorial, or similar programming sense, on public opinion at the regional level' (FCC, 1964).

[8] In this connection see Bagdikian (2004) for a detailed description of how a cartel of five giant media corporations that owned most of the newspapers, magazines, books, radio and TV stations, and movie studios of the United States controlled their content.

[9] See detailed discussion of 'Convergence Legislative Frameworks' in Sect. 5.2.

of production to engage in a form of intermodal competition, such as the emerging rivalry between wireline and wireless telephony (Yoo, 2009).'

Since convergence opens up technologies that were once the preserve of natural monopolies to 'intermodal competition', some commentators have suggested that it also implies a narrower role for sector-specific regulation and a greater role for conventional antitrust law (Geradin & Kerf, 2003). This has to be considered in the context of 'the U.S. Supreme Court's recognition of antitrust court's limited liability to engage in longstanding and ongoing oversight of the access mandates and other traditional attempts to curb anticompetitive practices through behavioral rather than structural relief (Verizon, Inc v. Law Offices of Curtis V. Trinko, & LLP, 540 U.S. 398, 2004).'[10]

Cuilenburg and Verhoest believe that the emerging policy paradigm for media and communications is mainly driven by an economic and technological logic, although it retains certain values. They write:

> In general, the philosophy of European communications policy is attuned to the idea that a large and dynamic market, with open frontiers, should within a clear and agreed framework, be able to provide for the current and expanding communication needs of society...However, the EU is also committed to maintaining human rights (as embodied in the ECHR), which has significant implications for the performance of media in some respects (Cuilenburg, 2003).

According to Cuilenburg and Verhoest any new communication policy model will need to be built around freedom of communication, access and control/accountability. Horwitz believes that although legislative and judicial trends have been quite hostile to any possibility of a mixed media system that is governed by structural rules and curbs on media concentration, there are legitimate grounds for such a policy (Horwitz, 2004).

Jon Garon believes that 'treating the Internet as analogous to print rather than broadcast leaves the most dynamic market ever created to be regulated under the common law tradition[11] (Garon, 2013).' The model of increasing regulation and differentiating among the various media has led to what has been called 'technological convergence' and 'legal Balkanization.' The elimination of cross-ownership rules was based on a premise that 'the rapid change of technology might result in convergence of industries, but this convergence need not eliminate the number of participants (Garon, 2013).' The optimism surrounding the Internet for providing a multitude of information sources is not universally shared.[12] The efforts to deregulate cross-ownership restrictions have been opposed by an 'emerging information-activist movement that includes activists both on the Left and Right, who are concerned about access and diversity within the media (Noam, 2008).'

[10] Cited in Yoo (2009).

[11] Common law is based on case law and precedent rather than codified law. It is followed in England and most of the Commonwealth of nations (The Robbin Collection, 2010).

[12] See discussion on the difference between competition in the product market and information market in Sect. 2.2.1 on 'competition'.

Blackwell has argued 'certain regulatory controls like that over newspaper ownership and media cross ownership...have existed historically in various countries. They do not necessarily cease to have a rationale in the digital age...ensuring pluralism and cultural objectives may not simply be a matter of market share (Blackwell, 1998).'

3 Methodology and Approach

The core element of our present analysis is an inter-disciplinary (doctrinal and policy) analysis of laws and regulations in the media-communication domain in the convergence era. We approach this analytical task from the vantage point of free speech—that is, a 'Free Speech' perspective—that is brought to bear on various issues such that this human right is preserved and realized even in the era of convergence. Our literature review persuades us that a Free Speech perspective may be implicitly present in certain policy and regulatory approaches; however, a 'Free Speech' perspective has not been coherently and systematically framed, derived or applied to critically examine policy and legislation in this domain or develop it in a manner that sustains human rights.

Our preferred methodology comprises of a rigorous review of convergent legislative frameworks from a critical perspective; followed by a detailed scrutiny of particular aspects of policy and regulation from a Free Speech perspective; and finally, an attempt to develop and present a suggested approach to Convergence Policy and Regulation based on our views on convergence.

3.1 Theoretical Approach: A Free Speech Perspective

We believe that the underlying leitmotif and intent of laws, policies and regulations relating to Communications need to be derived from human rights principles like 'Freedom of Speech' variously enshrined in Constitutional values and the international human rights frameworks (Beitz, 2009).

A 'Free Speech' perspective may be understood as one that interrogates every legal and regulatory issue from the point of view of 'freedom of speech', i.e. whether or not any particular policy, law or regulation helps achieve the public value of 'freedom of speech' in any meaningful way, or does it abridge 'freedom of speech.'

In the next section, we consider the relationship between the 'competition' paradigm and the public value of 'freedom of speech' at a conceptual level. We register that the competition paradigm is a driver for many regulatory paradigms of economic regulation; but believe that while regulation designed to foster competition carries some overlap in purpose and effect with free speech driven legislation, both are not the same. Therefore, it is important to understand the difference between the competition paradigm and 'free speech' in the Convergence era, especially because it is also marked by increasing privatization.

3.2 Free Speech and Competition: Conceptual Aspects

While 'perfect competition' is an economic objective it does not necessarily reflect the constitutional intent of 'freedom of speech' as a human rights value; the notion of 'competition' offers an 'economic' lens on communication law and regulation while our preferred Free Speech perspective stems from a concern for human rights and constitutionalism in a democratic context. We are privileging Free Speech as a frame that helps us focus on 'communication rights' or 'communication entitlements' as recognized by the UDHR in the Convergence era.

We do register that the 'marketplace of ideas' metaphor has governed free speech jurisprudence since Justice Holmes had written that 'the ultimate good desired is better reached by free trade in ideas—that the best test of truth is the power of the thought to get itself accepted in the competition of the market (Abrams v. US, 1919).' The belief that perfect competition in the market will lead to free speech is implicit in such a conception. The metaphor can be traced back to John Milton in *Areopagitica*:

> Let [Truth] and Falsehood grapple; who ever knew Truth put to the worse, in a free and open encounter (Milton, 2006)

But, the iconic 'marketplace of ideas' metaphor has faced serious critique:

> 'It is a particularly loaded metaphor because not only does it tie ideas to the market, but essentially conveys the sense that the marketplace is natural and unproblematic...In its rhetorical deployment, the marketplace of ideas metaphor unites and reduces the manifold justifications of freedom of speech to a simple talisman that ignores the reality of the large, concentrated ownership of the electronic means of communication—a point that has been made by many commentators in recent years (Horwitz, 2004).

Alleman and Rappoport critique the belief that 'utopia has been achieved with perfect competition. Is it any wonder that the policy makers point to competition as a panacea with which to solve all economic ills (Alleman and Rappoport 2005).'[13] We endorse such critiques, and believe that a Free Speech perspective that is rooted in democracy, constitutionalism and human rights and cannot be held co-terminus with an economic perspective rooted in the ideologically invested ideal of perfect competition.

Our preferred 'Free Speech' perspective compared to the competition paradigm can also be considered in the wider framework of Robert Entman and Steven Wildman who compare a 'social value' approach[14] with a 'market economics' approach (Entman, 1992).[15] Our preferred Free Speech perspective is analogous to the 'social values' approach, and the 'competition paradigm' is an articulation of

---

[13] See Kumar (2012) for a discussion of cultural policy issues that need to be considered in addition to market issues while responding to convergence.

[14] In this connection also see Sect. 2.2.1 for discussion by Ofcom on 'private value' and 'external value'.

[15] See (Horwitz, 2004) for a detailed discussion of these two approaches.

the 'market economics' approach. According to Horwitz (2004), the market economics approach seems to have trumped the social values school at the Federal Communications Commission of the USA as well as its courts. While acknowledging the power of the market—'the market shapes programming to a tremendous extent'[16]—Horwitz clarifies that this is 'not to say that such forces are so commanding that ownership is irrelevant and that the concern with who owns the media is fundamentally misguided.'

4 Networks and Platforms: The New Context for Regulation

Before considering specific convergent legislative frameworks in Sect. 5, this section presents the key concepts of 'networks' and 'platforms' that provide the technological as well as economic, social and cultural context for regulation in the era of digital convergence. The notions of 'networks' and 'platforms' had existed even before the advent of the digital era; the digital era has provided the technological facility to extend and express these notions in real ways to an extent that is perhaps, greater than ever before. At the end of this section we discuss the 'Network Layers Regulatory Model' which is based on the key technical ideas governing the design and structure of the Net. Our 'Suggested Approach' to Convergence Policy and Regulation is based on the concepts of networks and platforms.

4.1 Networks

The layered organization of the computer networks has allowed for the open nature of the Internet. The least number of layers is three: the layer of network infrastructure, the layer of transport and operations and the layer of application services. According to van Dijk, the workings of these three layers—which are divided into seven layers as per the standard open systems interconnections (OSI) model—are very important in understanding the media sector, both from the technical perspective as well as the perspectives of business and power (van Dijk, 2012).[17]

The seven layers comprise of two bottom layers that contain infrastructure specifications for networks and their equipment—viz. the 'physical layer' and the 'data layer', the latter being on top of the former—and four layers on top of the infrastructure layers that define the software required to operate networks; here, a standardized code is provided for the network layer, the transport layer, the session layer and the presentation layer. The top network layer is the 'application layer'

[16] Justice O'Connor's noting in her Metro Broadcasting (1990, p. 626) dissent.

[17] We will draw upon the layered organization of networks, and the network layers regulatory model to present our suggestions about ownership regulations from a Free Speech point of view in Sect. 6.

which enables the content of network applications and the synchronization of network communication.

The need for regulation from a Free Speech perspective can be best appreciated by understanding the workings of the 'vertical' columns between the various layers and within the 'horizontal' layers. Specifically, the 'verticals' across the layers apply to the various networks of telephony, computer or Internet and broadcasting—while the 'horizontal' layers carry the functions performed by them. In van Dijk's formulation the layers of supply in public networks have been presented as a grid that maps the various 'verticals' (telephone networks, Internet and Broadcasting networks) vis-à-vis the 'horizontal layers', which have been identified as:

1. Producers of infrastructure
2. Services: construction and maintenance
3. Services: transport and management
4. Services: content (information and communication)

We are particularly interested in the first and fourth layers in the context of our suggestion in Part 6 for a differential regulatory treatment of these two crucial layers. Here, we note van Dijk's conceptualization of the horizontal layers that are mapped vis-à-vis the three verticals—viz. Telephone networks, Internet and Broadcasting networks. The role of 'network operators and carriers' is held to be supremely important by van Dijk since they are the 'gatekeepers' for the networks. 'So the telephone operators, Internet service providers and broadcasting operators largely decide who and what has access to networks and how expensive particular applications on networks are' (van Dijk, 2012).[18]

We now discuss the notion of 'platforms' to understand the workings of the top layers; this is important from a Free Speech perspective since 'platforms' may act as locations for domination and control in a convergence age.

4.2 Platforms

A platform is the set of components used in common across a product family whose functionality can be extended by third parties and which is characterized by network effects (Parker & Van Alstyne, 2012). Some examples of platforms are:

1. Desktop OS: Unix, Mac, Windows
2. PDAs: Palm, Psion, Newton
3. Game Consoles: Wii, Xbox, Playstation
4. Network Switches: Cisco, IBM, HP
5. Multimedia: Adobe/Flash, MS/Silverlight, Google-Apple/HTML5

[18] See Sect. 6.2.1 for our discussion of Blevin's notion of 'infrastructural scarcity' in the context of our suggested regulatory approach.

6. Payment Systems: Paypal, Google Checkout, Visa, Apple, Mobile Felica
7. Mobile Devices: iPhone, Android, Symbian, Blackberry
8. Enterprise Systems: Salesforce, Oracle, i2, IBM, SAP
9. Social Networks: Facebook, MySpace, LinkedIn, Monster, Twitter
10. Voice over Internet Protocol (VOIP); Skype, Nextiva, Yahoo!
11. Web Search: Google, Bing + Yahoo!, Baidu
12. Ebooks: Kindle, iPad, Nook, Sony

The components of a 'platform' are not usually developed by any single firm. The success of 'platforms' from a business point of view depends on the extent to which any platform is able to spawn an ecosystem in and around it. The specific issue that is of concern from a Free Speech perspective is whether regulatory oversight of ecosystems of platforms is required since 'closed' platforms may inhibit other players in particular markets.

4.3 Network Layers Regulatory Model

The layered regulatory approach treats the Internet and IP-enabled services in a 'technology-neutral' manner by structuring the regulations around the service function. It is based on identifying various network layers and the services that are carried over those networks; The layered regulatory model does, in fact, signify a paradigm shift since it is based on the network layers of next-generation networks or IP-based technologies[19] compared to the earlier regulatory models based on a paradigm of vertical sectors of telecommunications, broadcasting and IT each of them treated differentially.

The real difference between the two paradigms is that the earlier legacy paradigms of regulation were based on a 'vertical' conception of services and industries that considered communication services and the underlying technology that delivers them to be one and the same. In contrast the layered network regulatory paradigm is based on the architecture of the Internet itself.

The new paradigm tries to adapt the new IP-centric services and applications to the 'reality of how the Internet is fundamentally changing the very nature of the business and social world' (ACMA, 2011). This is different from attempting to fit the new services and applications into pre-existing legal and public policy constructs. The layered regulatory model has found play in various countries like Korea, Malaysia and the EU.

[19] The network layers framework has been earlier discussed in Sect. 4.3.

5 Legislation in the Convergence Era

We consider the 'convergence' legislation in various countries in this section, presenting and discussing their essentials; then we specifically analyze those policy and regulatory developments in the convergence era that are most significant from a Free Speech perspective—viz. competition law, market failures and cross-ownership rules.

In recent years various countries have enacted 'Converged' Legislative Frameworks. We begin this section by briefly reviewing and discussing some of the policy and regulatory approaches with a view to understanding the paradigms underlying them from a Free Speech perspective (ACMA, 2011). Some of the key jurisdictions that have integrated their media and communication laws into a 'Converged Legislative Framework' include the European Union (EU) and its member states—the United Kingdom, Finland, Sweden, Italy, South Africa and Malaysia. In addition, Korea, Japan and Taiwan have also initiated steps towards achieving convergence in their laws. India introduced a Convergence Bill a decade back but it has not yet been approved by Parliament. Such legislative reforms in different countries have primarily been a policy response to convergence, although other contextual factors have played a part—viz. the need to reform licensing regimes, regional harmonization goals and the desire to promote the competitiveness of national communication markets (ACMA, 2011).

The 'Converged Legislative Frameworks' that represent the confluence of three different regulatory traditions—telecommunications, media and the internet—demonstrate a partial integration of media and communications regulation in most jurisdictions, which is due to the difference in approach marking content regulation and carriage regulation. The content/carriage distinction is apparent in most jurisdictions that have converged legislative frameworks. A layered regulatory approach is evident in the matter of electronic communications networks and services while a sector-specific approach is still evident in the domain of content regulation.

In Sects. 5.1 and 5.2 we present and discuss the convergent legislative frameworks of Malaysia, which was the first jurisdiction to enact a converged legislation, and the European Union, which is regarded to have a 'best practice' framework.

5.1 Malaysia

The Communications and Multimedia Act (CMA) was passed by Malaysia in 1998, the first jurisdiction to enact a converged legislation. This set up the regulatory framework for telecommunications, radio communications, broadcasting and online activities in Malaysia. It also passed the Communications and Multimedia Commission Act in 1998 to enable the creation of a new converged regulator to administer the CMA—the Malaysia Multimedia and Communications Commission (MCMC).

The CMA provisions economic regulation—in particular, it includes provisions for the promotion of competition and prohibition of anti-competitive conduct. Also, it provisions the development and enforcement of access codes and standards. This is an instance of the paradigms of 'access' and 'competition' continuing to find play in convergent legislative frameworks. However, the Malaysian legislation continues with content regulation as well as the licensing of network and application providers, demonstrating the transient nature of paradigms still characterizing convergent legislative frameworks.

5.2 European Union

The EU's *electronic communications regulatory framework* (ECRF) of 2003 is often considered to be a best practice model. The ITU regards it as a 'paradigm legislation aimed at addressing convergence and its challenges'. It provides a framework for the regulation of electronic communication services and networks that member states can use to enact their domestic laws. Significantly, the ECRF does not apply to content regulation; there is only broad guidance in the form of the Audio-Visual Media Services Directive (AVMD Directive).

The following aspects of the ECRF are particularly relevant from a Free Speech perspective:

- *The Framework Directive* requires national regulatory authorities (NRAs) to define relevant national competition markets and analyze whether there are any operators with significant market power (SMP) in that market.
- *The Access Directive*[20] establishes the rights and obligations of operators and undertakings seeking interconnection and/or access to their networks. The EU Access and Universal Service Directives are designed to enable some principles of public service broadcasting like 'universality of access' and pluralism in an era of digital convergence.

5.2.1 Competition Law

The EC termed the ECRF 2003 as a 'light regulatory touch' that helps to make the markets become more competitive and protects the basic rights of consumers. The convergent legislative frameworks have been harmonized with competition law in the EU to respond to the challenge of convergence between the telecommunication and information sectors. The ECRF requires NRAs to oversee any emergence of market power by carrying out regular market analysis. It uses a 'horizontal' approach to market definition—i.e. define markets based on a functional assessment of the services provided—and checking the extent to which any market player enjoys market power.[21] Again, a telecommunications reforms package was passed

[20] The issue of 'access' is discussed in detail in Sect. 6 on 'A Suggested Framework'.

[21] In this connection see de Steel (n.d.)

in 2009 that has made certain changes in the ECRF which are relevant from a Free Speech perspective. It has introduced new remedies for dealing with dominant companies, improved spectrum management and mobile broadband services. Also, the 2009 reforms package incorporates new rights for citizens and consumers—the most significant of which is the concept of a 'right to internet access.'

5.3 Market Failures in the Convergence Era

In some post-convergence markets like broadcasting, the nature and extent of traditional market failures is changing due to technological change. The growth of multichannel and pay television has mitigated many problems related to market failures; and the growth of online content markets may eliminate some of them altogether (Ofcom, 2004).[22] A wide range of revenues and suppliers creates a competitive marketplace, so market power becomes less likely. However, while the growing penetration of multichannel platforms may make it seem that structural problems have disappeared, Ofcom believes that the nature of broadcasting continues to provide enough reasons for intervention.

5.4 Cross-Ownership Rules in the Convergence Era

Julius Genachowski, the former Chairman of the Federal Communications Commission of the USA revived a proposal to "streamline and modernize media ownership rules" after the courts had denied the FCC's last effort to loosen the prohibition in 2007 (Dyer, 2013).[23] The new FCC Chairman Tom Wheeler took Genachowski's controversial proposal off circulation at the end of 2013 (Reuters, 2013). While initiating the 2014 Quadrennial Regulatory Review the FCC sought comment on "whether to retain the prohibition on the cross ownership of newspapers and television stations"; and further asked "if so, should we reform the restriction to consider waivers for newspaper/television combinations (FCC, 2014)?"

The Communications Act 2003 of the United Kingdom liberalized cross-ownership rules affecting commercial media organizations. Sector-specific media ownership rules were relaxed with rules preventing non-EU media companies from holding UK broadcasting licenses and most cross-ownership rules being abolished. The government wanted to 're-base broadcasting regulation upon modern competition Act principles' with the communication regulator, Ofcom given concurrent powers with the Office of Fair Trading, one of the UK competition authorities.

[22] In this connection see Swanson (2013)

[23] The proposal was to revoke a 30-year-old cross-ownership rule that does not allow a company to own one newspaper, two television stations and upto eight radio stations in the same media market.

The current ownership structure in Canada reflects the convergence of telecommunications and broadcasting, but the converged industry is still regulated by three separate Acts. Amidst new concerns about media ownership since 2010 regarding the concentration of ownership in Canada of wireless and Internet networks, the chairman of the Canadian Radio-television and Telecommunication Commission (CRTC) has argued for a unified and comprehensive legislation (Theckedath, 2012).

Argentina has recently passed a media diversity law that limits the number of broadcast channels a single company can own. Bolivia passed a new media law in 2011 that divides the broadcast spectrum between the state, private sector, indigenous/peasant organizations and community groups. Porfirio Lobo, the Honduran President has announced his intention to propose a plebiscite on media democratization (Dyer, 2013).

6 Convergence Policy and Regulation: A Suggested Approach

We recognize the effectiveness of instruments of economic regulation like 'structural regulation' like 'ownership regulations' that can defuse market power, and in turn, ensure 'public access', an essential requisite of freedom of speech in privatized communication and media environments; such 'structural regulation' is different from competition law. Further, we register that since 'market failures' impede competition, regulatory measures to counter market failures may facilitate free speech in media markets. However, we are wary of any suggestion that all matters of democratic deficit involving Free Speech deficits can be attributed to 'market failures.'

We appreciate the continuing role that 'structural regulation' can play in the convergence era where divergence and the concomitant dispersal of power is as possible as convergence involving coagulation of power embedded in media conglomerates or coagulating around emerging nodes of power that settle around issues on a dynamic basis; however, we need to recognize that such nodes of power do not always form on the basis of ownership in a post-convergence world.

6.1 A Suggested Approach: Ownership Regulation Across Vertical Layers

The network layers model of regulation does not quite register the reality of industry players often crossing layers. We draw upon this critique to suggest that there may still be a need for regulatory oversight over 'vertical' cross-ownership across network layers. The possibility of vertical cross-ownership between the layers—especially, between the network infrastructure (carriage) layer and application layer—needs to be scrutinized from a Free Speech perspective as control over both makes the system vulnerable to domination over both content creation and its distribution.[24]

Joseph Farrell has discussed what economists call 'vertical relations' using the notion of 'internalizing complimentary efficiencies' (ICE); such important efficiency benefits can stem from a 'vertical relationship closer than an arm's length one.' For Farrell, the possibilities of ICE imply that the FCC must develop 'a framework for regulating what economists call "vertical relations": how a firm relates to other firms in adjacent markets, and whether it integrates into those markets (Farrell, 2003).' He points out that even a platform monopoly often has incentives to make efficient choices about when to maintain modularity and when to get involved in an adjacent market.

6.2 A Suggested Approach: Structural Regulation within Horizontal Layers

There is no need for ownership regulation within any layer—that is horizontal 'convergence' now possible with convergent digital technology that needs to be encouraged and not resisted since that would mean battling the tide of technology. It is better for the various players, especially, the larger ones like Google, Facebook, Apple and Microsoft to fight their competitive battles. The regulatory aim as far as horizontal layers are concerned should be to maintain the structural openness of the regime and the possibility of fair competition.

6.2.1 Network Infrastructure and Network Services

The horizontal layers of 'network infrastructure' and 'network services,' the former providing the vital backbone and the latter being in a position to erect 'gateways' should be dealt with in differential ways.

We endorse the proposal to apply a paradigm of 'infrastructural scarcity' to treat the 'network infrastructure' layer that is vulnerable to domination by a few players (Blevins, 2012).

Blevin has used his conceptualization of 'infrastructural scarcity' to argue that courts should defer to access regulations of uncompetitive *network-layer* platforms such as broadband access infrastructure.[25] He has recommended that regulations targeting *application-layer* platforms such as search engines, by contrast, should be subjected to higher First Amendment scrutiny.[26]

---

[24] See discussion of 'networks' in Sect. 4.1.

[25] Here, 'network layer' refers to physical transmission infrastructure—the "roads" upon which traffic is transmitted. The 'application layer' refers to platforms like search engines and social networking sites that are themselves (like postal packages) transmitted upon these physical roads.

[26] In this connection, also see Baxter's law—'where the platform (the core monopoly) is subject to regulation but the applications market is not (Farrell, 2003).'

Blevin is critical of the conflation of 'network-layers' with 'application-layers' in considering regulatory measures from a First Amendment perspective since they display distinct economic and technological characteristics; the abundance of 'application-layers' is mistakenly held to be true for 'network-layers' as well although the latter are vulnerable to domination by a few major players. For instance, broadband cable players are now seen to dominate the market in comparison to wireless broadband players; they are expected to continue to do so in the near future. Therefore, there is the real possibility of domination in this layer.

6.2.2 Gateways

The 'gateway' layer of 'network services- transport and management' controlled by 'network operators and carriers'[27] needs to be approached differently. Here, the various network operators may be allowed to work without any ownership restrictions as long as they maintain 'openness'; specifically, regulatory oversight may be imposed through the tested paradigm of 'network neutrality' that will ensure 'access'. Further, competition laws can ensure the continuing possibility of competition between the various players in this layer. It may not be necessary or useful to require 'public ownership' in this layer as market competition and the stipulation of 'net neutrality' should be enough to secure 'access' and provide a safeguard against a confiscation of 'gateways' by any player.

6.2.3 Application Layer

We believe the 'application layer'—or rather, the layer that houses operating systems, browsers and 'applications'—need not be subject to ownership restrictions at the horizontal 'layer' level since the problem of 'infrastructural scarcity' does not affect this layer; in fact, there is an abundance of 'application-layer' platforms.

There are various markets in the 'application' layer that need regulatory oversight to ensure 'access', of course subject to Free Speech scrutiny. For instance, we support Blevin's proposal for assuring 'search neutrality' in the 'search' market. We believe that within this 'application' layer the major players must be allowed to compete subject only to competition laws, with the qualification that the 'open' character of any platform needs to be maintained, at least in as far as key 'applications' are concerned.[28]

[27] This layer is identified as such in Van Dijk's formulation of network layers which we have discussed in Sect. 4.1. The network layers have been detailed here as those of 'network infrastructure' and 'network services'; in contrast, Blevin uses a broader categorization of 'network-layers' and 'application-layers' to construct his argument.

[28] This is important in the context of increasing concentration of the Internet. See discussion on this in Sect. 5.4 on post-convergence policy challenges.

6.3 Proposal for Affirmative Public Ownership

Although it may be useful to subject the 'network-layer' platforms to some structural regulatory oversight from the perspectives of 'concentration of ownership' and competition laws, we believe that instead of applying ownership limits to this vital layer where infrastructure and development are crucial it will be more effective to explore affirmative policy choices that structurally help develop 'voice' and 'agency' on behalf of public values.

In this context, we propose a policy of 'affirmative public ownership' of network infrastructure companies as one of the ways of developing such 'voice'. This may be done in the form of requiring various public[29] entities (e.g. trusts or non-profit organizations) to demonstrate investment up to a specified limit in network infrastructure companies; the latter may be compensated through tax incentives or subsidies from a dedicated public fund.

Further, it may be useful to engage in 'affirmative public ownership' even in the 'applications' layer to construct 'public media' that would be meaningful and substantial in the convergent era—this may produce 'public interest' applications.

Such a policy of 'affirmative public ownership' will structurally create a space for the articulation of public 'voice'; also, it will strategically, enable better regulatory oversight of the design and play of the crucial infrastructural 'backbone'. This would be one way of using structural regulation beyond restrictions on ownership to develop 'public media' in a digital world. Various commentators have discussed modes of constructing Public Media in the digital era. Our proposal for 'affirmative public ownership' is aligned to these initiatives in providing substance to public media in the digital age.[30]

7 Conclusions

The Free Speech problematic of coagulated 'communication power' impeding 'public access'—depleting the public values of diversity, plurality and deliberation—does not necessarily disappear in a digital era because the production and distribution of content as well as its cultural form and manner of consumption is not totally transformed with the advent of the new possibilities of convergence.

[29] We distinguish between 'public' and 'government' ownership, although the latter is not necessarily ruled out.

[30] Malcolm Matson has celebrated the emergence of 'hundreds if not thousands of examples of what are becoming known as *open public access networks* (Matson, 2009).' Yochai Benchler has discussed the possibilities—and advantages—of 'open wireless' systems compared to systems of 'licensed spectrum' (Benkler, 2012).

7.1 Perspective on Convergence

We believe that in a convergent world, 'communication power' flows through a checkerboard of networks and digital platforms that are fluid, and often, ephemeral, coagulating in moments and dispersing in seconds to reconfigure again at alternative locations. This is happening because the paradigm and practice of mass communication is being replaced by a new found paradigm of 'mass self-communication' (Castells, 2011). The digital revolution has amongst other things created a 'digital culture' that works against any eruption of unitary power (Balkin, 2004).

Significantly, convergence is not necessarily leading to an emergence of 'convergent' power—that is 'economic power' that is rooted in ownership is not necessarily being superimposed on 'cultural power'. The structural shift in favor of a convergent world has led to the emergence of diversified pockets of content. This has occurred in spite of content aggregation. In such circumstances, even powerful, media conglomerates or content aggregators may find it difficult to dominate the inherently disaggregated structure of the Net.

In the convergence age, power is as likely to form along an axis of domination privileging major corporate players and governments as it is likely to emerge at nodes based on issues and concerns on behalf of individuals and groups with no corporate or state affiliation. The existing 'open' architecture of the Internet permits such articulation of 'divergent' power which is in sharp contrast to the pre-Convergence context of 'convergent' power.

Further, the earlier 'cultural forms' of print media, radio, television and cinema have continued to exist even in the convergence era; we believe that 'convergent consumption' is still not a uniform habit or preference. The simultaneous presence of offline and online content will necessitate reviews and regulatory oversight on Free Speech grounds.

Therefore, the regulatory objective needs to register the propensity for divergence and challenges to coagulations of corporate and state power, and help maintain the structural 'openness' in the system to enable Free Speech.

Our suggestion for constructing an effective system of public media in the Convergence era that provides effective and real 'voice' and 'agency' by pursuing a policy of 'affirmative public ownership' in the two crucial layers—the 'network infrastructure' layer for strategic control and in the 'applications' layer for public interface—will lead to, we believe, an effective articulation of 'voice' in the digital era.

References

Abrams v. US, 250 US 616 (Supreme Court 1919).

ACMA. (2011). *Converged legislative frameworks—International approaches. Occasional paper.* Australian Communications and Media Authority.

Alleman, J., & Rappoport, P. (2005). Regulatory failure: Time for a new policy paradigm. *Communication & Strategies, 60*(4), 107–108.

Bagdikian, B.H. (2004). The new media monopoly. Boston, MA: Beacon Press

Balkin, J. M. (2004). Digital speech and democratic culture. *New York University Law Review, 79*, 1–58.
Barendt, E. (2005). *Freedom of speech* (2nd ed.). New York, NY: Oxford University Press.
Barron, J. A. (2000). Structural regulation of the media and the diversity rationale. *Federal Communications Law Journal, 52*(3), Article 8.
Beitz, C. (2009). *The idea of human rights*. Oxford: Oxford University Press.
Benkler, Y. (2012). *Open wireless* vs. *Licensed spectrum: Evidence from market adoption*. Working paper. Harvard.
Bennett, J., & Strange, N. (2008). The BBC's second-shift aesthetics: Interactive television, Multi-platform projects and public service content for a digital era. *Media International Australia, 126*, 106–119.
Blackwell, C. R. (1998). Convergence between telecommunication & other media: How should regulation adapt? *Telecommunication Policy, 22*(3), 163–170.
Blevins, J. (2012). The new scarcity: A first amendment framework for regulating access to digital media platforms. *Tennessee Law Review, 79*, 353–416.
Caldwell, J. (2006). Critical industrial practice: Branding, repurposing, and the migratory patterns of industrial texts. *Television & New Media, 7*(2), 99–134.
Castells, M. (2011). *Communication power*. New York, NY: Oxford University Press.
Cerf, V. G. (2013, January 28). Why the internet works and how to break it. *The Economic Times*.
Cuilenburg, J. V. (2003). Media policy paradigm shifts: Towards a new communication policy paradigm. *European Journal of Communication, 18*, 181.
de Steel, A. (n.d.). *The current & future European regulation of electronic communications: A critical assessment*. Retrieved September 9, 2013, from www.ucl.ac.uk/laws/conferences/regulated-industries/R02_destreel_paper.pdf
Doyle, G. (2002). *Media ownership: The economics and politics of convergence and concentration*. London: Sage.
Doyle, G. (2010). From television to multi-platform. Less from more or more for less? *Convergence: The International Journal of Research into New Media Technologies, 16*(4), 1–19.
Dutton, W. H. (2010). *Freedom of connection—Freedom of expression: The changing legal and regulatory ecology shaping the internet*. Paris: UNESCO's Division for Freedom of Expression, Democracy and Peace.
Dyer, Z. (2013, January 14). *Journalism in the Americas blog*. Retrieved January 20, 2013, from Knight Centre for Journalism in the Americas. The University of Texas at Austin: http://knightcentre.utexas.edu/blog/00-12551-relaxation-media-law-could-threaten-media-diversity-us-critics-say
Entman, R. A. (1992). Reconciling economic and non-economic perspectives on media policy: Transcending the 'marketplace of ideas'. *Journal of Communication, 42*, 5–19.
Farrell, J. (2003, September). *Modularity, vertical integration and open access policies: Towards a convergence of antitrust and regulation in the internet age*. Retrieved September 9, 2013, from Working Paper No. E 02-325_Rev UC Berkeley: http://escholarship.org/uc/item/4dh7q2dd.pdf
FCC. (1964). *Multiple ownership order, 45 F.C.C. 1476, para 3*. FCC.
FCC. (1999). *Revised duopoly report, 14 F.C.C.R. 12,903, para.15, 17 Comm.Reg. (P&F) 1*. Federal Communications Commission.
FCC. (2014). *Further notice of proposed rulemaking and report and order* [FNRM]. Retrieved September 30, 2015, from https://apps.fcc.gov/edocs_public/attachmatch/FCC-14-28A1.pdf
Feintuck, M. A. (2006). *Media regulation, public interest and the law*. Edinburgh: Edinburgh University Press.
Garon, J. (2013). *Media and monopoly in the information age: Slowing convergence at the marketplace of ideas*. Retrieved September 9, 2013, fromwww.trademarkfilingindia.com/case_studies/media.pdf
Gentzkow, M. A. (2008). Competition and truth in the market for news. *Journal of Economic Perspectives, 22*(2), 133–154.
Geradin, D., & Kerf, M. (2003). *Controlling market power in telecommunications: Antitrust v. sector-specific regulation*. Oxford: Oxford University Press.

Hitchens, L. (2011). Media regulatory frameworks in the age of broadband: Securing diversity. *Journal of Information Policy, 1*, 217–240.
Horwitz, R. B. (2004). *On media concentration and the diversity question*. San Diego, CA: University of California.
Jenkins, H. (2006). *Convergence culture. Where old and new media collide*. New York, London: New York University Press.
Kumar, V. (2012, December). *Internet NZ—discussion document on 'Responding to convergence in communication markets'*. Retrieved September 9, 2013, from http://internetnz.net.nz/system/files/workstreams/2012-12-convergence-paper-final-pdf
Matson, M. (2009). Disruptive technologies in communications: Observations from an entrepreneur. In R. W.-F. Kunneke (Ed.), *The governance of network industries. Institutions, technology and policy in reregulated infrastructures* (pp. 193–206). Cheltenham: Edward Elgar.
McPhillips, S. A. (2008). Media convergence and the evolving media business model: An overview and strategic opportunities. *The Marketing Review, 8*(3), 237–253.
Milton, J. (2006). *The Project Gutenberg EBook of Areopagitica*. Produced by Judith Boss and David Widger. Retrieved October 2, 2015, from http://www.gutenberg.org/files/608/608-h/608-h.htm
Murdock, G. (2000). Digital futures: European television in the age of convergence. In J. Wieten, G. Murdock, & P. Dahlgren (Eds.), *Television across Europe: A comparative introduction*. London: Sage.
NewsCorp. (2012, December 3). *Press releases. News corporation announces details regarding proposed separation of businesses*. Retrieved January 21, 2013, from News Corporation: www.newscorp.com/news/news_549.html
Noam, E. (2008, January 17). *Columbia Business School ideas@work*. Retrieved January 15, 2013, from Columbia University: www4.gsb.columbia.edu/ideasatwork/feature/131444/Media+Concentration+in+an+Era+of+Digital+Convergence
Ofcom. (2004). *Annex 11: Market failure in broadcasting*. Retrieved September 9, 2013, from Review of Public Service Broadcasting Phase 2: http://stakeholders.ofcom.org.uk/binaries/consultations/psb2_1/annexes/annex11.pdf
Ofcom. (2007). *Digital dividend review*. Retrieved September 9, 2013, from Annex 11: Market failure in broadcasting—Stakeholders—Ofcom: http://www.ofcom.org.uk/consult/condocs/ddr/
Ofcom. (n.d.). *Digital dividend review*. Retrieved September 9, 2013, from Annex 11: Market failure in broadcasting—Stakeholders—Ofcom: http://www.ofcom.org.uk/consult/condocs/ddr/
Parker, G., & Van Alstyne, M. (2012). *A Digital postal platform: Definitions and a roadmap*. Cambridge, MA: The MIT Center for Digital Business.
Reuters. (2013). *FCC withdraws proposal to relax media ownership rules*. Retrieved October 2, 2015, from http://www.reuters.com/article/2013/12/17/us-usafcc-mediaownership-idUSBRE9BGODE20131217
Swanson, B. (2013, September 8). *Society and culture, media & technology*. Retrieved September 9, 2013, from American Enterprise Institute—Public Policy Blog: www.aei.ideas.org/2013/08/cbs-time-warner-standoff-is-a-sign-of-market-dynamism-not-market failure/#mbl
The Robbin Collection. (2010). *Common law and civil law traditions*. Retrieved October 4, 2015, from https://www.law.berkeley.edu/library/robbins/pdf/CommonLawCivilLawTraditions.pdf
Theckedath, D. (2012, April 10). *Media ownership and convergence in Canada*. Retrieved January 21, 2013, from Library of Parliament Research Publications: www.parl.gc.ca/Content/LOP/ResearchPublications/2012-17-e.htm
van Dijk, J. (2012). *The network society* (3rd ed.). London: Sage.
Verizon Communications, Inc v. Law Offices of Curtis V. Trinko, LLP, 540 U.S. 398 (US Supreme Court 2004).
Vukanovic, Z. (2009). Global paradigm shift. Strategic management of new and digital media in new and digital economics. *The International Journal of Media Management, 11*(2), 81–90.
Yoo, C. S. (2009). The convergence of broadcasting and telephony: Legal and regulatory implications. *Communications & Convergence Review, 1*(1), 44–55.

Part II

Convergence Challenges in Journalism and Publishing

The Internet's Impact on Journalistic Quality

Christian M. Wellbrock

1 Introduction and Problem Discussion

The journalistic quality of media products is constantly in the focus of discussions about the media business. Especially from a democracy point of view, the decline of newspapers as a source of high quality journalism and the inclusion of conflicting opinions is bemoaned. Newspapers lose readers and—even more intensely—advertising revenues to other media outlets. In the United States, for instance, revenues from newspaper advertising have dropped from $47.4 billion in 2005 to $22.8 billion in 2010, a decrease of 51.9 % (NAA, 2011). Similarly, advertising revenues in Germany decreased by 18 %, from 4.5 billion euros to 3.69 billion euros, in the same period (BDZV, 2010; BDZV, 2005).

As a consequence, many newspapers and magazines feel the pressure to follow readers and advertisers onto the Internet. Unfortunately, generating revenues besides advertising remains difficult online. And even online advertising revenues per pair of eyeballs do not yet compare to offline rates.

Against the background of relatively high journalistic standards in newspapers, the shifting of readers and advertisers onto the Internet raises a rather important question concerning the integrity of the democratic system. What is going to happen to journalistic quality and the variety of opinions in mass media when publishers shift their business activities online?

To answer this question, this paper focuses on economic incentives for producers of journalistic content and on the question of how likely it is—from an economic point of view—that providers of journalistic content will offer (sufficiently) high quality online and provide room for competition of opinions.

The remainder of this article is structured as follows. Chapter 2 addresses and defines the concepts of journalistic quality and public goods. After that, both the

C.M. Wellbrock (✉)
Universität Hamburg, Von-Melle-Park 5, 20146, Hamburg, Germany
e-mail: christian.wellbrock@uni-hamburg.de

A. Lugmayr, C. Dal Zotto (eds.), *Media Convergence Handbook - Vol. 1*, Media Business and Innovation, DOI 10.1007/978-3-642-54484-2_5

print and online markets are investigated regarding three major weaknesses that are commonly associated with media markets and its consequences for the supply of quality and the diversity of opinions: non-excludability, non-rivalry, and quality uncertainty (non-transparency).[1] Finally, taking reasoning in favor of existing regulation into account, possible solutions on how to deal with problems in the market structures and its consequences are discussed.

2 Literature Review and Theoretical Foundations

2.1 The Concept of Journalistic Quality

In this paper, journalistic quality is defined according to McQuail's concept of media performance (McQuail, 1992). This definition is based on three core values in Western democracies that mass media is supposed to serve: freedom, equality, and order. The most important functions of mass media that can be derived from this definition are the procurement of information, participation in public opinion making, control and criticism, education and entertainment (Blumler & Gurevitch, 1990; USAID, 1999).

An important distinction from the definition of quality that is usually employed in the Industrial Organization and management literature is that journalistic quality plays a crucial role in society that goes beyond the pure satisfaction of consumer needs. Usually, product quality in economics is defined as the capability to satisfy consumer needs, i.e. quality is attached to what consumers want, and therefore an increase in quality must lead to greater demand (Garvin, 1984). This standard approach has been applied to media products quite frequently, e.g. to newspapers (Crampes, Haritchabalet, & Jullien, 2009; Mantrala, Naik, Sridhar, & Thorson, 2007, among others), and television (e.g. Anderson & Gabszewicz, 2006).

Journalistic quality, however, is driven by "objective" criteria, derived from mass media's democratic functions. Some of the criteria that determine a media product's journalistic quality are, for example, truth, relevance, neutrality, impartiality, and immediacy (Westerstahl, 1983). Since these criteria do not depend on consumer preferences, an increase in journalistic quality does not *necessarily* imply higher utility for all consumers. Although high journalistic quality may be valued by some consumers, it does not have to affect consumers' utility positively in general.

In the following, this type of quality will be referred to as journalistic quality.

[1] Collins (2008) has discussed the same causes of market failure, regarding how they might change due to technological innovation.

2.2 Public Goods

The understanding of public goods differs in economic literature. Some approaches consider non-excludability from consumption as the sole characteristic that constitutes public good. Strictly speaking, however, a public good is made up of two properties: non-excludability *and* non-rivalry in consumption. Non-excludability means that consumers who are not willing to pay for the product cannot be technologically excluded from consumption at reasonable cost. Non-rivalry, on the other hand, means that the consumption of a product by one consumer does not preclude its use by other consumers (Mas-Colell, Whinston, & Green, 1995: 359). In the following, we consider public goods to be both non-excludable and non-rivalrous.

The existence of non-excludability leads to free riding behavior and to positive externalities, because individuals can benefit from the good without having to compensate the producer of the good (Mas-Colell et al., 1995: 352). Since this incentivizes the producer to produce less of the product than is economically efficient, this leads to market failure. For example, it would be very difficult—practically impossible—to exclude potential "consumers" of a lighthouse signal from consuming it. As a result, no private institution would build lighthouses, because barely any individual would be willing to pay for the usage. This is why lighthouses are provided collectively, mostly by collecting taxes and using these means for their construction and operation.

Lighthouse signals are also non-rivalrous in consumption, because anybody can use one without precluding others from using it. Satellite television is another example. In both cases, it does not matter how many people are using the product, the utility stays the same for all consumers. They do differ in excludability, though. As already mentioned, lighthouse signals are non-excludable. The supplier of satellite television, however, can choose who to grant access to its services to, can easily exclude consumers who are unwilling to pay and can charge a price.

3 Methodology and Approach

The main purpose of this contribution is to investigate the effects of publishing houses increasingly distributing their content online rather than in traditional ways such as newspapers and magazines on the provision of journalistic quality and diversity of opinions. The approach chosen here to answer this question is to examine economic incentives companies face in the print and the online market that are caused by market structures. The following chapters therefore evaluate market failure arguments that are commonly mentioned in association with media markets—namely non-excludability, non-rivalry, and quality uncertainty—from a theoretical perspective. Comparisons are drawn between the print, online, and the TV market to highlight differences, derive probable market outcomes and discuss regulatory approaches.

4 Non-excludability and the Priority of the Advertising Market

4.1 Non-excludability in Media Markets

In general, recipients who are not willing to pay can be excluded from consumption in all media markets.[2] This is certainly the case in most print markets, where this excludability is obviously practiced via copy prices. While, admittedly, excludability is not perfect, for example because it is possible to pass along a newspaper to others (who would then most of the time benefit without compensating the supplier), newspapers possess a rather high degree of excludability.

When it comes to broadcasting and online markets, this is less obvious, because there is often no observable price. Except for pay-TV, in most countries no direct prices are charged by most television channels. This also holds true for the online market, in which the vast majority of content is available for free (Perez-Pena, 2007).

Due to the absence of a positive price as the clear signal for practiced exclusion, one might come to the conclusion that these markets are characterized by non-excludability and lead to the market failure due to positive externalities. This would mean that the production or utility function of one firm or individual is directly affected by another agent's action, i.e. without compensation via the price mechanism (Mas-Colell et al., 1995: 352). In this case of positive externalities on the consumption side, this would lead to lower output and lower prices than are favorable from a social welfare perspective (Mas-Colell et al., 1995).

From a strict point of view, however, substantial market failure cannot be diagnosed in this respect. The reason for this is that consumers who are unwilling to pay *could* be excluded at reasonable cost, but exclusion is simply not *practiced* in some markets. Reasons for this phenomenon are diverse.

One frequently mentioned argument is that in the case of the internet, people have been 'raised' to get it for free. Since the emergence of the Internet, most of the available content has been accessible at no cost (except the cost for Internet access), and consumers might have grown accustomed to this status. Changing such behavior or expectation, which has been established over a long period of time, appears difficult and very slow to accomplish.

Moreover, prices of zero are predicted by the model of perfect competition when the cost of reaching an additional consumer (marginal cost) is zero and the market is highly competitive. In this case, suppliers are price-takers and are forced to supplying their product at a price close to marginal cost. This economic explanation considers such an outcome a natural and even economically efficient and favorable outcome, which is not the consequence of lacking excludability.

---

[2] An exception might be radio. Although technologies to exclude do exist (mainly cable radio), almost all consumption occurs in cars, making it difficult to exclude (Berry & Waldfogel, 1999: 189).

Additionally, price-setting mechanisms in two-sided markets suggest that companies cross-finance the copy price with advertising revenues, due to large positive network effects from the recipient market towards the advertising market (Anderson & Gabszewicz, 2006; Blair & Romano, 1993; Ludwig, 2000). This means that advertising space increases in value as the number of readers increases. Media companies therefore face an incentive to implement a low per-copy/per-view price that might even be below marginal cost (Blair and Romano).

Altogether, considering low marginal cost in TV and online distribution of media content and the existence of two-sided markets, charging a price might simply not be worth it.

To sum up, media markets do not suffer from substantial non-excludability from consumption.[3] In contrast, media companies rather choose not to charge readers/viewers.

4.2 Non-exclusion and Its Consequences for Media Companies

Even though market failure due to non-excludability cannot be diagnosed, in some markets (e.g. free-to-air television and online) media companies obviously do not practice exclusion and therefore only generate revenues from advertisers. The fact that indirect network effects can cause inefficient outcomes in two-sided markets and therefore market failure, has been discussed in economic literature (Anderson & Gabszewicz, 2006; Armstrong, 2006; Liebowitz & Margolis, 1995). The effects depend on several aspects—such as single or multi homing consumers and the type of fee (lump-sum versus transaction-based)—and are ambiguous.

Some other very clear media specific consequences of advertising based revenue generation have also been discussed in many publications (e.g. Doyle, 2002; Owen and Wildman 1992; Spence & Owen, 1977; Sjurts, 2004). Generally speaking, the more profit-oriented media companies depend on the advertising market (the less they can generate revenues on the recipient market), the more they will focus on the advertising market and strive to meet advertising clients' needs in the form of the best possible advertising spaces. The demand for advertising space depends on

[3] It might be argued that excludability is more complex than this and that consumers could be excluded from specific presentations of news and information, but not from the news and information itself. This obviously holds true in some markets, especially online. However, this is mainly due to the fact that a different entity supplies the item of information or news for free, i.e. that a competitor chooses to do so. This also might be caused by low marginal cost in combination with intense competition, cross-financing in two-sided markets or other reasons, but it is not due to technological impossibility or extreme high costs of excluding consumers. Newspapers, for instance, still charge readers, although the information itself can be obtained elsewhere. Additionally, media products certainly cannot be considered totally homogenous. The piece of information or news does not account for the entire amount of utility a consumer derives from a journalistic piece, but qualities such as the style of presentation and how it is commented on also affect consumers' appreciation and give every article distinctive characteristics.

various aspects, but the most important one is certainly the number of recipients reached by a program or publication.

At first sight this might not be problematic, since this implies a focus on satisfying consumer preferences. However, a closer look reveals that trying to reach as many people as possible leads to two kinds of biases that threaten the fulfillment of the democratic function of mass media, namely the bias against expensive (quality) programs and the bias against minority programs.

The latter is easily explained: when looking for the greatest number of readers or viewers, it is irrational to offer programs targeted at smaller audiences (although willingness to pay and welfare might be as great or even greater for the minority program and would therefore be favorable from a welfare perspective). A similar principle can be applied to quality programs: even if as many people could be reached by a high-quality program as a low-quality program, the producer will choose the low quality program, simply because its production costs are normally smaller. Even if these higher costs could be compensated for by profiting from greater willingness to pay, this will not change the decision in markets where prices are fixed to zero on the recipient market and revenues can only be generated on the advertising market (Spence & Owen, 1977: 112).

Recalling what has been mentioned above, these mechanisms result in mainstream programming or catering to the lowest common denominator (Beebe, 1977) which is not in line with what is desirable from a democratic point of view. Mainstream content takes up too much of the space available, alternative thinking does not get a platform and different opinions cannot compete.

What does this imply for print products that go onto the Internet? In print markets, players are more likely to offer higher journalistic quality and less mass appeal compared to markets with prices of zero, since higher prices can potentially be achieved for high quality content. Niche strategies are more likely to be successful. This can also be observed in the corresponding market results of the print and (free) broadcasting markets. The online market of today does not offer the opportunity to generate revenues in the recipient markets, and the above mentioned mechanism applies, leading to biases against high quality and diversity of opinion.

4.3 Other Forms of Externalities

Besides externalities from non-excludability from consumption, externalities are also discussed as a consequence of two other phenomena. One is that media content usually has effects not only on the individual consumer, but also on society as a whole. On the one hand, this media effect can be negative, as specific kinds of content may lead to increasing levels of violence in society, for example, or at least increased fear of violence (Doyle, 2002: 65). On the other hand, specific kinds of content (e.g. high journalistic quality) may also lead to high levels of political competence in society (and better societal decisions as a consequence) or revelation of scandals. Certain individuals can benefit from that without having to consume the content themselves, which would represent positive externalities (Logan & Sutter, 2004).

The consequences of these two forms of externalities are twofold. Negative externalities caused by media effect (which the producers do not have to account for) lead to an inefficient overproduction of the good. Positive externalities from the production of high journalistic quality (for which the producer is not compensated for by everyone who benefits) lead to an inefficient under-provision of those type of goods.

These two forms of externalities indeed do cause market failure. But since they occur in all media markets to a similar extent, they should not have a significant effect on the production of journalistic quality when publishers shift the distribution of their content from print to online.

5 Non-rivalry and the Natural Monopoly

5.1 Non-rivalry in Media Markets

Media products in general are also subject to non-rivalry. This means that the consumption of a product by one consumer does not preclude its use by other consumers (Mas-Colell et al., 1995: 359). If a consumer watches a certain TV program, for instance, others are able to watch it at the same broadcasting quality, too. There is no rivalry in consumption. Differently put, a product characterized by non-rivalry can be made accessible to an additional consumer practically without any extra cost. In this case, the total cost is mainly driven by fixed costs, as marginal costs are basically zero. This is also known as the problem of "indivisibility": input factors cannot be sufficiently divided, but a certain amount of input is needed to produce the prototype in the first place. What is called "first copy cost" in the media business is referred to as "indivisibility" in microeconomics.

The result of this is quite clear: the higher the share of fixed costs in a production function, the more likely it is that a natural monopoly will emerge, since the entire amount of output can be supplied at lowest average cost by a single supplier (decreasing average costs over increasing output). This company can therefore offer the product at the lowest price. This results in a single company that can produce the product cheapest. According to this theory, if two or more companies exist in the market at one point of time, only one of them will end up taking over the entire market.

At first glance, this situation is also economically favorable, if the company chooses a price equal to marginal cost or perhaps slightly higher (e.g. to cover average costs in the long run). In the end, however, there is the danger of the monopolist taking advantage of his monopolistic power and setting prices not according to marginal cost, but in order to maximize profits. This will then typically lead towards smaller quantities and higher prices than are economically efficient. This also brings about dynamic inefficiencies, because this company has no incentive to be innovative (Mas-Colell et al., 1995).[4]

[4] All of this, of course, only holds true if the monopoly is not contestable. For contestable markets, see Baumol, Panzar, and Willig (1982).

Media products are always subject to non-rivalry and indivisibility, because the core utility generating part of a media product is the content, which is usually produced at high fixed cost and can then be consumed by many without "wearing it out". However, the extent to which a media good is subject to non-rivalry also depends to a great degree on the carrier medium. While newspapers are certainly rivalrous and create marginal costs to a certain degree (e.g., paper and printing), broadcasting or Internet distribution is much less rivalrous, or perhaps almost not at all. Consequently, concentration tendencies in broadcasting and online distribution are stronger than in the printing industries, implying that media companies in the former markets have a greater incentive to serve the mass market.

5.2 Consequences for Media Companies

The question needs to be asked whether the conclusions above have an impact on markets with fixed prices, as in free broadcasting markets and the online market, where prices appear to be fixed at zero on the recipient market. The reason that only one supplier emerges in markets with declining average cost curves is that lower costs lead to lower prices for the consumer. When prices are fixed at zero, however, cost advantages cannot be forwarded directly to consumers, who might at first glance be readers or viewers.

The concept still applies to media markets when combining the recipient and advertising markets. First, recalling that a media company's economic success depends partly (sometimes entirely) on its performance on the advertisement market, which again is significantly driven by the number of recipients reached.

From this perspective, the media product itself (newspaper, magazine, TV or radio program) can be considered an input factor in order to "produce audience" which is then sold to advertisers. Since the content part of the newspaper or program (which is here considered only an intermediate product to "produce" audience) is characterized by fixed costs only, the company which "produces" the largest audience has less cost per output unit. As a result, products that reach larger quantities of recipients have a cost advantage in the advertising market, which can here be passed on to customers, who in this case are advertisers. Therefore, this principle also holds true in online and broadcasting markets.[5]

[5] An argument that is quite frequently put forward against the emergence of natural monopolies in media markets is the following. Media companies do not offer homogenous goods on the recipient market, since the practice of journalism involves choice, presentation and commentary which makes every product unique. Since natural monopoly theory assumes homogenous goods, this is why natural monopoly tendencies are less likely when considering the recipient market for all newspapers or all TV programs, for example.

Yet on the advertising market, newspapers and TV stations offer advertising space, which is a much more homogenous entity, because it depends mainly on the media product's reach. Thus, natural monopoly tendencies can be presumed for media markets in which economically rational profit maximizers can reach an additional reader or viewer with low marginal costs. In the case of constant marginal costs, decreasing average costs per audience contact and a natural monopoly

What does this mean for print products going on the Internet? In the print world, marginal costs do exist, because only the content is non-rival, whereas the carrier (paper) is not. An additional copy of the newspaper or magazine needs to be printed to reach an additional consumer which incurs paper, printing and distribution costs. Equally importantly, marginal costs are likely to be increasing due to the law of decreasing marginal returns. For example, the further customers are geographically away from the newspaper's printing site, the more expensive it will be to provide them with the product.

Distributing the same content over the Internet, however, does not cause significant marginal costs. This is why there is a greater tendency towards a natural monopoly in online markets, meaning that companies try to reach as many people as possible in order to minimize average costs.[6] This leads not only to a bias against minority programs, but also against high-quality content, if certain people cannot draw utility from high quality products due to a lack of consumption capital, meaning particularly a good education.[7]

The market structure in online distribution and television is very similar in this respect, implying that publishers going online is bad news for journalistic quality.

6 Quality Uncertainty and the Collapse of Markets for High Quality

Characterized as experience or credence goods, media products in general are subject to severe quality uncertainty (or non-transparency). This means that the quality of a product can only be observed after consumption (experience good) or even never (credence good) (Mas-Colell et al., 1995: 436). Regarding both entertainment and informative content, it is hard for consumers to value the good before they have consumed it in both cases. Yet after consumption, there is no point to paying for the product anymore (Arrow, 1962: 615; Doyle, 2002: 65).

Additionally, the journalistic quality of informative content (news) is characterized by pre-defined criteria, whose fulfillment cannot usually be examined by the reader or viewer. The correctness of a piece of information, for instance, is impossible to verify most of the time, especially when it comes to supra-regional news. This is why information and news are generally classified as credence goods. One major consequence of this type of information asymmetry is the danger of an adverse selection process leading towards a breakdown of the market for high-

tendency are the consequence. In addition to that, the more the market's participants depend on advertising revenue, the stronger this effect will be.

[6] The effects of competition on journalistic quality and diversity from a holistic point of view are not discussed in this work. See Hollifield (2006) for an overview.

[7] Consumption capital theory goes back to Becker and Stigler (1977). It states that an individual's present consumption of a cultural product depends on its past consumption. See Leroch and Wellbrock (2011) for an application.

quality goods, although a demand and supply for these goods does exist in principle (Akerlof, 1970).

What does this mean for print products going on the Internet? It is quite difficult to clearly state whether there are great differences in the degree of quality uncertainty between the print and online markets. Potential clients can screen and inspect print products in shops before buying them, and thereby reduce uncertainty. Yet the same can be done in online markets by reading teasers, which are usually offered prior to consumption.

Reputation, as a method of signaling high quality and reducing uncertainty as well, plays an enormous role in media markets. This aspect seems to work quite well in print markets, where strong brands are established, helping to overcome information asymmetries. Some newspapers are known for offering high quality, others are known to focus on yellow press content. Even the political orientation of media products is usually an important aspect of their brand image.

Some online media firms also appear to have strong brands that provide information as to their products' levels of quality. Findings of substantial structural differences between media markets concerning reputation building are therefore hard to deduce. Although the Internet might need some time to catch up to build strong brands on its own, it is rather likely that media companies possess similar opportunities to build up a quality reputation, no matter what channel of distribution they choose.

For this reason, the extent to which the two markets examined here are subject to quality uncertainty are probably similar. Consequently, there is no clear structural reason to worry about a decline in journalistic quality when publishers move their activities online in this respect.

7 Consequences of the Market Structure

7.1 Preliminary Results

It has been shown that the market structure of online markets contains strong incentives for its commercial members to serve mass markets and neglect high standards of quality. Exclusion is practiced much more often in print markets than in online markets, which makes companies that distribute content over the Internet more dependent on the advertising market. In addition to that, online distribution is also subject to strong non-rivalry which leads to concentration tendencies. This shows that online markets are poorly equipped to supply high journalistic quality and diversity of opinions in comparison to print markets.

7.2 Regulatory Approaches

Regarding the importance of high quality content and variety of opinions in a democratic society, one has to think about ways to reduce or even prevent this problem as media companies increasingly turn away from print markets and toward the Internet.

One very extreme approach is to simply ban advertising in journalistic products. This suggestion would, however, endanger many publications' economic ability to survive, since a great amount of revenue is generated by advertising. In the case of newspapers, the share of revenues from advertising is around 50 % in Germany (BDZV, 2010) and almost 80 % in the United States (NAA, 2010). Denying media companies this portion of their income would most likely force firms into bankruptcy, thereby reducing the degree of diversity in the market, which certainly is not socially desirable.

Another less radical approach is to call on both media companies and advertisers to comply with their democratic responsibilities. In the case of media companies, one could argue that they receive the privileges of press freedom and other special treatment for journalists in exchange for the fulfillment of democratic functions. Unfortunately, this call is unlikely to succeed. A look at market outcomes in commercial broadcasting and online markets suggests that economic forces are stronger. Media ownership has changed dramatically over the last decades from rather "entrepreneurial" owners and family-owned businesses, who put great emphasis on the social responsibilities of publishers, towards investment firms, which consider media companies almost exclusively as ways to earn money (Habermas, 2007). As pointed out, economic success does not always comply with journalistic quality. A second group that could be addressed is advertisers. They could choose to give extra margins when booking in high quality products and therewith support this kind of content. The chances of success here are probably also close to zero. The major problems would be to coordinate advertisers to solve the problems of free-riding, as well as the identification of "high-quality products", since journalistic quality is quite difficult to measure objectively.

Another method of tackling these problems has been state interference by regulation, e.g. by granting subsidies or by providing the good collectively. In some cases, public service broadcasters (PSB) have been implemented, for example in the UK, the US and Germany. The next chapter deals with this approach in more detail.

7.3 Public Service Media

The parallels in market structure between the online and broadcasting markets are quite obvious. Most broadcasting is free-to-air, where consumers are not being charged by media companies. The same holds true for online distribution, where the vast majority of content is free. Yet in both cases, the absence of a positive price is not due to non-excludability. However, this condition makes media companies focus on the advertising market for revenues.

Non-rivalry is also similarly present in these two markets, as the marginal costs of broadcasting and Internet distribution of media content are basically zero, and

Fig. 1 Legal reasons for the need for PSBs, own creation

Market structure	Online	TV
"Non-exclusion"	✓	✓
Non-rivalry	✓	✓

TV's special position	Online	TV
Immediacy	✓	
Suggestive power	✓	
Broad effect		✓

this leads to indivisibility and strong sub-additive cost structures, such as constantly declining average costs.

These forms of market failure in broadcasting have been important reasons for the implementation of public service broadcasters, for example the BBC in the UK, NPR and PBS in the US, and ARD and ZDF in Germany.

In Germany, the existence of severe market failure has explicitly been acknowledged by the Federal Constitutional Court (*Bundesverfassungsgericht*) as a reason for the implementation of PSBs. Additionally, the Court assigns a special role to television and radio due to three characteristics that are especially pronounced in broadcasting: immediacy, suggestive power, and widespread effect (BVerfG, 2007, Art. 114, 116).

All of these characteristics can easily be assigned to the Internet, too. The dissemination of news is even faster online than on TV (immediacy), and the Internet offers all types of content presentation, from pure text to interactive video, which enables this medium to be even more influential than TV (suggestive power). And even if the widespread effect is not considered to be equally strong at this point in time (more people have access to TV than to the Internet and people spend more time watching TV than online), the increasing importance of the Internet in everyone's daily life suggests that the gap in this aspect is narrowing.[8] Figure 1 summarizes these findings.

As a consequence, we find that the most important reasons that are put forward in favor of regulating the television market also apply to the online market.

[8] It might be mentioned here that the German Federal Constitutional Court stresses the importance of reacting sufficiently early to changes in the media business, since failure to do so is likely to lead to almost irreversible consequences on diversity.

8 Conclusions

Media markets in general do not work very well when it comes to supplying products of high journalistic quality. In addition, online markets are worse equipped in this matter compared to print markets. To a large extent, this is due to the facts that the advertising market serves as the only source of financing, and that production is characterized by a great amount of fixed costs and almost no marginal cost.

This results in economic incentives for media companies to act mainly as "audience production plants". This implies focusing only on catching the attention of as many people as possible and serious biases against high quality and minority programs. Consequently, news need to be sensational, spectacular and easy to understand, and not necessarily well-researched and balanced.

To counter this structural deficit, some countries have decided to implement PSBs that are (for the most part) exempt from commercial pressures coming from advertisers. Main reasons for primarily regulating television in this fashion is television's unique capability to influence opinions due to high degrees of immediacy, suggestive power and widespread effect.

The online market is found to work very much like the TV market with all its strengths and weaknesses. As the same reasons that are put forward to justify public service media in broadcasting also apply to the online market in a very similar way, this speaks in favor of regulating the online market in the same way, e.g. by simply allowing PSBs to extend their activities into the online sector.[9]

A political decision concerning this proposal has already been made in both the United Kingdom and Germany. German regulation has once more followed the British example closely, and has also implemented a so-called public value test (PVT). This instrument basically measures the public value of the PSBs' supply against the damage it does to the market and its competitors (public value vs. market impact). Since the publishers of newspapers and magazines still have an inherited tradition of social responsibility and comparably high-quality products, the PVT might be a good instrument to protect them in the early stages of Internet supremacy and at least give them the chance to prove the quality of their performance on the Internet.

The future of high quality journalism certainly depends on more than only economic considerations, which have been the almost exclusive concern in this contribution. What this paper intends to lay out, however, is that the Internet provides far worse circumstances for economically successful production of high quality journalistic content than traditional print markets. As far as economic incentives are concerned, the more traditional print markets are substituted by online markets, the less likely mass media will be able to fulfill its functions in a democratic society.

[9] This stands in contrast to Collins (2008), who suggests that the implementation of the BBC is a disproportionate answer to market failure (partly due to technological innovations).

References

Akerlof, G. A. (1970). The market for 'Lemons': Quality uncertainty and the market mechanism. *The Quarterly Journal of Economics, 84*, 488–500.

Anderson, S. P., & Gabszewicz, J. J. (2006). The media and advertising: A tale of two-sided markets. In V. A. Ginsburgh & D. Throsby (Eds.), *Handbook of the economics of art and culture* (Vol. 1, pp. 567–614). North Holland, Amsterdam: Elsevier.

Armstrong, M. (2006). Competition in two-sided markets. *RAND Journal of Economics, 37*, 668–691.

Arrow, K. J. (1962). Economic welfare and the allocation of resources for invention. In R. R. Nelson (Ed.), *The rate and direction of inventive activity: Economic and social factors* (pp. 609–625). Princeton, NJ: Princeton University Press.

Baumol, W. J., Panzar, J. C., & Willig, R. D. (1982). *Contestable markets and the theory of industry structure*. Boston: Thomson Learning.

BDZV. (2005). *Schaubilder 2005: Zur wirtschaftlichen Lage der deutschen Zeitungen*. www.bdzv.de/maerkte-und-daten/schaubilder/seite/10

BDZV. (2010). *Schaubilder 2010: Zur wirtschaftlichen Lage der deutschen Zeitungen*. www.bdzv.de/maerkte-und-daten/schaubilder/seite/4/

Becker, G. S., & Stigler, G. J. (1977). De Gustibus Non Est Disputandum. *American Economic Review, 67*, 76–90.

Beebe, J. (1977). Institutional structure and program choice in television markets. *Quarterly Journal of Economics, 91*, 15–37.

Berry, S. T., & Waldfogel, J. (1999). Public radio in the United States: Does it correct market failure or cannibalize commercial stations? *Journal of Public Economics, 71*, 189–211.

Blair, R. D., & Romano, R. E. (1993). Pricing decisions of the newspaper monopolist. *Southern Economic Journal, 59*, 721–732.

Blumler, J. G., & Gurevitch, M. (1990). Political communication systems and democratic values. In J. Lichtenberg (Ed.), *Democracy and the mass media* (pp. 269–287). Cambridge, MA: Cambridge University Press.

BVerfG. (2007). 1 BvR 2270/05, 2007, 09/11/2007, passages 1-213. http://www.bverfg.de/entscheidungen/rs20070911_1bvr227005.html

Collins, R. (2008). Hierarchy to homeostasis? Hierarchy, markets and networks in UK media and communications governance. *Media, Culture & Society, 30*, 295–317.

Crampes, C., Haritchabalet, C., & Jullien, B. (2009). Advertising, competition and entry in media industries. *The Journal of Industrial Economics, 57*, 7–31.

Doyle, G. (2002). *Understanding media economics*. Los Angeles: Sage.

Garvin, D. A. (1984, Fall). What does 'product quality' really mean? *MIT Sloan Management Review, 26*(1), 25–45.

Habermas, J. (2007). Keine Demokratie kann sich das leisten. *Süddeutsche Zeitung Online*. http://www.sueddeutsche.de/kultur/21/404799/text/

Hollifield, C. A. (2006). News media performance in hypercompetitive markets: An extended model of effects. *International Journal on Media Management, 8*, 60–69.

Leroch, M., & Wellbrock, C. (2011). Saving newspaper with public grants - the effects of press subsidies on the provision of journalistic quality. *Information Economics and Policy, 23*(3–4), 281–286.

Liebowitz, S. J., & Margolis, S. E. (1995). Are network externalities a new source of market failure? *Research in Law and Economics, 17*, 1–22.

Logan, B., & Sutter, D. (2004). Newspaper quality, Pulitzer prizes, and newspaper circulation. *American Economic Journal, 32*, 100–112.

Ludwig, J. (2000). The essential economic problem of the media: Working between market failure and cross-financing. *Journal of Media Economics, 13*, 187–200.

Mantrala, M. K., Naik, P. A., Sridhar, S., & Thorson, E. (2007). Uphill or downhill? Locating the firm on a profit function. *Journal of Marketing, 71*, 26–44.

Mas-Colell, A., Whinston, M. D., & Green, J. R. (1995). *Microeconomic theory*. Oxford: Oxford University Press.
McQuail, D. (1992). *Media performance*. London: Mass communication and the public interest.
NAA. (2010). *Trends & numbers*. http://www.naa.org/TrendsandNumbers.aspx
NAA. (2011). *Annual newspaper ad expenditure*. Accessed January 15, 2012, from http://www.naa.org/Trends-and-Numbers/Advertising-Expenditures/Annual-All-Categories.aspx
Owen, B., & Wildman, S. (1992). *Video economics*. Cambridge, MA: Harvard University Press.
Perez-Pena, R. (2007). Times to stop charging for parts of its web site. *The New York Times online*. http://www.nytimes.com/2007/09/18/business/media/18times.html?_r=1&scp=1&sq=Times%20to%20Stop%20Charging%20for%20Parts%20of%20Its%20Web%20Site&st=cse
Sjurts, I. (2004). Der Markt wird's schon richten!? Medienprodukte, Medienunternehmen und die Effizienz des Marktprozesses. In K.-D. Altmeppen & M. Karmasin (Eds.), *Medien und Ökonomie, vol. 2: Problemfelder der Medienökonomie* (pp. 159–181). Opladen: VS Verlag.
Spence, M., & Owen, B. (1977). Television programming, monopolistic competition, and welfare. *Quarterly Journal of Economics, 91*, 103–126.
United States Agency for International Development (USAID). (1999). *The role of media in democracy: A strategic approach*. Washington, DC: Centre for Democracy and Governance.
Westerstahl, J. (1983). Objective news reporting. *Communication Research, 10*, 403–424.

Convergence and the Changing Labor of Journalism: Towards the 'Super Journalist' Paradigm

Lia-Paschalia Spyridou and Andreas Veglis

1 Introduction and Problem Discussion

Journalism and technology are inextricably intertwined with each other (Preston 2009b). Web 2.0 and its related technologies have profoundly disrupted modern journalism which is in the midst of tumultuous change, driven primarily by technological developments and economic uncertainty on a global scale (Picard, 2011). Scholars conceptualize journalism's transformation and explain the changes occurring at different levels under the rubric of convergence (Mitchelstein & Boczkowski, 2009). Convergence is commonly perceived as a multidimensional construct allowing for different conceptions and operationalizations (Domingo et al., 2007; Dupagne & Garrison, 2006), while inherent in the concept of convergence is the notion of integration and collaboration (Erdal, 2011).

Within the field of journalism, Dupagne and Garrison (2006) proposed a model of media convergence which rests upon three dimensions: technology, economics and regulation. These three dimensions generate effects which impact on media use, content diversity and newsroom practices. According to Domingo et al. (2007) convergence connotes integrated news production, multiplatform delivery of news and information, multimedia storytelling and participatory models of journalism. It is argued that any of these dimensions of convergence can be developed in the media on its own, but in many cases they are part of the same 'convergence project'. Again the notion of effects is noted. "Every single project has its own contextual factors that shape the development of convergence dimensions and produce positive and/or negative outcomes in the process" (p. 3). Gordon (2003) provided a

L.-P. Spyridou (✉)
Department of Social and Political Sciences, University of Cyprus, Nicosia, Cyprus
e-mail: lia@ucy.ac.cy

A. Veglis
Department of Journalism and Mass Media, Aristotle University of Thessaloniki, Thessaloniki, Greece

A. Lugmayr, C. Dal Zotto (eds.), *Media Convergence Handbook - Vol. 1*, Media Business and Innovation, DOI 10.1007/978-3-642-54484-2_6

model of convergence in media organizations, which includes five distinct forms of convergence. Ownership convergence referring to corporate strategies encouraging the ownership of multiple content or distribution channels, thus often involving media mergers and acquisitions. Tactical convergence emphasizing content partnerships and cross-promotion of content. Structural convergence which involves changes in job descriptions and organisational structures within a newsroom in order to achieve maximum performance. Information-gathering convergence requiring reporters to be multi-skilled and able to write stories, shoot video and edit it themselves. Finally, storytelling convergence connoting that journalists must rethink the way they tell stories and take advantage of each medium's unique capabilities (p. 6371). A recent definition put forward by the Infotendencias Group (2012) succinctly summarizes the notion of convergence as a complex process affecting journalism in many ways:

Convergence in journalism is a multidimensional process which, facilitated by the widespread implementation of digital communication technologies, affects the technological, business, professional and editorial aspects of the media, fostering the integration of tools, spaces, working methods and languages that were previously separate, in such a way that journalists can write contents to be distributed via multiple platforms, using the language that is appropriate in each medium (p. 2930).

Against this background, the chapter is structured as follows: first, a comprehensive account of convergence is provided, focusing on the logic, dimensions and effects of convergence on journalism. Second, the chapter presents the multitude of skills required in a converged newsroom, and argues for the development of the 'super journalist' paradigm. Finally, it contests optimistic views of convergence claiming for its potential to satisfy both good journalism and good business practices, as predicated on a misapprehension of the fundamental structures of the news industry which is corporately owned, profit driven and faced with extensive uncertainty.

2 Literature Review: From Digital Utopias to Convergence

Deuze's (1999) influential work conceptualizes online journalism as a fourth type of journalism defined by its unique technological characteristics allowing for new manners in producing, presenting, consuming and distributing news. Digital utopias concerning the technological potential and subsequent regeneration of journalism fuelled academic publications with multiple normative approaches regarding the promise of online journalism as a cutting-edge news vehicle capable to transform journalism (Scott, 2005). Following this line of thought, online journalism would: (a) increase the volume of news and information people can have access to, (b) provide sheer flexibility regarding the time and the mode of news consumption (nonlinear), (c) provide the possibility of reciprocal communication and interaction among the involved parties, (d) define 'community' in terms of common goals and interests rather than geographical, (e) question the role of traditional gatekeepers and the impact of established elites, (f) blur the boundaries among different media since they all converge into one (multimedia), (g) alter the relationship and the role of the producers and receivers of news and information allowing the latter to participate actively in the creation of content and agenda shaping, (h) enhance news diversity, immediacy and analysis (Fulton, 1996; Kawamoto, 2003; Pavlik, 2001). This new model of journalism would renew the promise of the democratic role of the press by fostering transparency and public accountability of the media (Singer 2005) as well as promoting political interest and social mobilization and participation (Hacker and Van Dijk 2000). However, empirical data (Boczkowski, 2004; Domingo, 2008; Engebretsen, 2006; Kenney, Gorelik, & Mwangi, 2000; O'Sullivan, 2005; Oblak, 2005; Salaverría, 2005; Spyridou & Veglis, 2008) showed that the web has not revolutionised journalism in such a radical manner as many theorists have predicted. Rather it reproduced the problematic logic and practices of traditional journalism.

Although innovation is declared as the main success factor for journalism (Garcia, Fernández, Bran, & Fariña, 2012; Pavlik, 2013), recent research treating online journalism as an open process rooted in and influenced by economic, social, professional and organizational factors (Domingo, 2008; Preston, 2009a), has identified several factors impacting on innovation and normalizing technology's potential: lack of financial resources (Spyridou & Veglis, 2008), the perpetuation of traditional working practices and norms (O'Sullivan & Heinonen, 2008), time pressures imposed by the 24/7 news cycle (Robinson, 2010), multimedia journalism's increased demands in skills and time resources (Brannon, 2011), work overload (Paulussen, 2012), dominant perceptions of professional identity (Spyridou, Matsiola, Veglis, Kalliris, & Dimoulas, 2013) and reluctance to surrender communication control and engage in participatory forms of journalism (Örnebring, 2013). However, despite the various factors hindering innovation, news organizations are opting for new manners in doing online journalism. The end of the advertising revenue model (Franklin, 2012), the industry operating in a state of hyper-competition (Schultz & Sheffer, 2009) and the urgency to adapt to the new portable, personalized and participatory news culture (Paulussen & Ugille, 2007; Pew Internet, 2010) have accelerated the process of convergence pointing to new modes of production (integrated production), novel delivery of news and information (multiplatform delivery) and participatory models of journalism (active consumption of news) (Domingo et al., 2007; Klinenberg, 2005).

3 Convergence Continuum

In practice news organizations tend to embrace convergence at different speeds and modes (Quinn, 2005). Dailey, Demo, and Spillman (2005) propose a dynamic 'convergence continuum' to provide a conceptual framework for understanding convergence based on varying degrees of cooperation and interaction among cross-media partners. They identify five overlapping levels of activity among news organizations: cross-promotion, cloning, coopetition, content sharing and full convergence. These five stages are defined in terms of growing content synergy and organizational autonomy and point to increased collaboration on content production and diminished organizational autonomy. The key idea behind their typology is the notion of project-driven production in order to optimize the use of resources (human, material, technological) and develop suitable content for multiple platforms. Erdal (2011) put forward a model of cross-media journalism operating on two axes, the work and the content axis. The intersection of those two axes produces four types of cross-media practice: The single-reporter multiplatform journalism which involves one reporter producing the same story for two or more platforms. The hard-drive journalism, which describes a form of journalism where a single reporter creates a new version of an already existing news report for a different platform. The intra-platform coordination representing a form of journalistic practice where reporters or editors from different platforms share information and coordinate their efforts in covering a particular news story, typically during editorial meetings or more informally. Finally, the intra-platform production which is a more complex form of cross-media journalism, and requires reporters from different platforms to cooperate extensively in covering a particular news story, sharing content and raw material.

Avilés and Carvajal (2008) argue that newsroom convergence is not a technology-driven process, but a process that uses technological innovation to achieve specific goals in particular settings and that is why each convergence project might have a different outcome. Even in digitally mature markets like the US, "convergence is usually an internal process of integration between online and print or broadcast newsrooms within the same organization" (Thurman & Lupton, 2008: 441). Quinn and Quinn Allan (2005) see the notion of culture as a decisive factor influencing the level and type of convergence. "Introducing and managing convergence involves appreciating the specific cultures unique to any organisation, and the form of convergence that evolves will be a product of those unique cultures" (p. 56).

As online journalism is trying to find its role and mission, there is evidence showing that the development of online journalism follows heterogeneous tendencies among news organizations (Deuze, 2004), among journalism professionals (O'Sullivan & Heinonen, 2008; Robinson, 2010) and among different countries (Benson, BlachØrsten, Powers, Willig, & Zambrano, 2012; Quandt, 2008; Sarrica et al., 2010). Micó, Masip, and Domingo (2013) emphasise the processual character of convergence. Following this line of thought convergence is perceived as the process of socially constructing a new technological system for news production, which calls for the adoption of novel production systems and devices, the redefinition of work practices and newsroom layouts, the rethinking of journalistic roles, values and skills, and the multiplication of publication platforms. It seems that news organizations are increasingly turning online in a simultaneous effort to bid for survival and growth. But convergence comes at a price. Despite its micro and macro benefits, convergence requires significant organizational change, complex and demanding editorial processes and costly investments (Quinn & Quinn Allan, 2005; Steinle & Brown, 2012).

4 Convergence Effects, Professionalism and Journalism Work

Regardless of the type and speed of convergence implemented in news organizations, convergence operating on multiple levels is reshaping the landscape of journalism in a variety of ways. "Newsroom structures, journalistic practices and news content are all evolving" (Pavlik, 2004: 28). Convergence processes and goals generate 'convergence effects' impacting on media labor, newsroom structure, professional values and norms, form and diversity of content produced (Domingo et al., 2007; Dupagne & Garrison, 2006; Klinenberg, 2005; Quandt & Singer, 2009). Undoubtedly the converged model raises serious implications on how journalism is done, hence a major question to be raised is: Has convergence enriched journalism or has it impoverished journalism? This is a multidimensional question.

In economic terms, convergence fosters cost-effective production by sharing information and resources and by coordinating the distribution of content (Micó et al., 2013). It facilitates synergistic revenue models and new advertising opportunities (Klinenberg, 2005), caters for effective targeting (Domingo et al., 2007) and facilitates branding strategies designed to build loyalty and increase website traffic (Vujnovic et al., 2010). Content wise, convergence emphasizes immediacy (Domingo, 2008), content-sharing (Paulussen, 2012) and leads to the 'continuously updated news story' as opposed to the idea of news as a finished product (Saltzis, 2012). At the same time, convergence dynamics reinforce mimicry among news outlets (Boczkowski, 2009), and induce 'copy-paste journalism' as journalists are increasingly forced to rely uncritically on wire material and engage in minimal fact-gathering (Davies, 2008), while stressing information provision as opposed to analytical journalism which in turn focuses on sense-making, contextual information, analysis, and interpretation (Singer, 2009a). Redden and Witschge (2009) concluded that although there is an abundance of news online, the content of mainstream news outlets is largely the same.

In professional terms, convergence is redefining well-rooted and long established notions of who is a journalist, what exactly does he do and what is his social role. Discussing the changing context of journalism work, Örnebring (2009) argues that two major trends are developing. First, the deregulation of labour markets and the rise of new forms of employment favouring flexible working conditions. Deuze and Marjoribanks (2009) describe eloquently the emerging working landscape for journalists. "Mass layoffs, off-shoring and outsourcing, and elimination of open positions are now standard managerial practice" (p. 555). The rise of casual employment and atypical media work signals labor relations that are often temporary and always contingent, non-committal (on both sides), generally without contractual or otherwise stipulated responsibilities (or accountability) for either employee or employer beyond the (informally agreed upon) deadline of the project, story, or item at hand (Deuze 2007 in Deuze & Marjoribanks, 2009).

Second, the technologisation of the workplace creates a need for new journalistic skills and competencies, while rendering large parts of the workforce deskilled. Lee-Wright (2010) posits that the current emphasis on new media is the apotheosis of the techie ascendancy. Looking at the BBC as an example, it is argued that "during times of confidence and expansion, the BBC invests in programmes and people; at other times it favours the hardware that cannot argue back" (p. 76). In a historical view of journalistic labor, technology has been used by owners and managers to increase control of the news production process and make journalistic labor cheaper (Örnebring, 2009).

Beyond labor conditions and multiskilling, a third major implication of convergence is associated with issues of professional role and authority. Anderson (2008) defines journalistic authority as "the power possessed by journalists and journalistic organizations that allows them to present their interpretations of reality as accurate, truthful and of political importance" (p. 250). Work routines and news values have traditionally been used to legitimize journalism's output and distinguish it from other sources of media work (Karlsson, 2011), while normative claims to gatekeeping and objectivity have functioned as the cornerstones of authority and autonomy (Singer, 2009b). But web 2.0 technologies, citizen journalism and participatory models of journalism connote serious changes and control shifts in all stages of news production. Thus journalism's professionalism and identity—commonly used as a form of occupational control—are seriously disrupted.

Finally, a fourth result of convergence is associated with multitasking and work overload. Amid shrinking staffs and fewer resources, "journalists find their jobs expanding and their routines vastly altered" (Gade & Lowrey, 2011: 31). The 24/7 news cycle, the reduction of personnel (Franklin, 2012) and extra duties stemming from multimedia and multiplatform delivery call for multitasking and increasingly more work to be done. As mentioned earlier, convergence connotes a social and conversational approach to journalism (Bradshaw, 2012a; Currah, 2009) as well as a new paradigm articulated in the never ending story (Saltzis, 2012). In that respect, journalists are expected to optimize the social diffusion and curation of news in an attempt to maximize traffic, brand loyalty and community-building (Vujnovic et al., 2010), while following and updating news on a constant basis. Furthermore, participatory journalism initiatives are spreading. The benefits deriving from user participation—namely alternative views, novel content and exchange of views following a habermasiam approach, are not to be contested. On the other hand, participatory journalistic models are time-consuming, requiring constant engagement of the professionals, so that the content submitted is both moderated (eg. in the case of comments) and verified (eg. in the case of photos or information received).

5 Convergence and Journalism Skills

Emily Bell (2007), formerly the director of digital content for Guardian News and Media and currently the director of the Tow Center for Digital Journalism at Columbia University's Graduate School of Journalism, is blatantly clear. "The truth, of course, is that an excellent journalist, using the best technology, is still a fantastic and powerful combination for providing the world with the best information. But even in the most expensive, challenging and complex areas of journalism, such as war reporting and investigative journalism, the creative technologist will have a far more inspiring role than previously" (p. 103). As multitasking and multiskilling are becoming the new buzzwords of the trade under convergence, a major issue to be raised relates to the set of skills and type of training required by journalism professionals. The following section attempts to highlight the emerging set of skills needed in conditions of convergence.

5.1 Journalism and ICT Skills

Information gathering in order to build news items can be defined as a structured process that extends for a certain time period (Reich, 2006). The news process consists of two fundamental stages: news discovery and news reporting (McManus, 1994). Reich (2006) suggests that the news discovery stage is preceded by another one, namely information gathering. During the gathering stage journalists obtain further data from various sources and crosscheck information. Today in all stages of news-production journalists employ ICT tools. These tools may include software applications as well as internet tools and services. In simple terms, the news production process includes five stages: Information Acquisition, Information Validation, Information Processing, Information Presentation and Dissemination and Information Curation (see Fig. 1). Next, the activities that take place in each stage are briefly discussed.

The Information Acquisition stage involves the discovery of information in raw form (Reich, 2006). As far as ICT tools and services are concerned the journalist employs various internet services and tools, such as WWW, email, search engines, RSS, and social networks. The journalist interacts with various internet sources as well as people that are involved or can express opinions about the news subject. During the Information Validation stage the journalist crosschecks the original data and obtains further data from sources in order to enrich the available information. All the tools and services mentioned in the previous stage are again employed. During the Information Processing stage, news begins to take its final form. If required data visualization techniques are employed helping the journalist to tell a complex story through engaging infographics (Bradshaw, 2012b). In most cases, media organizations relay on experienced web developers to produce data visualizations. But in some cases journalists may be asked to work on their own and use special software applications to prepare effective visualization content. This task can also be accomplished more easily by using free cloud computing

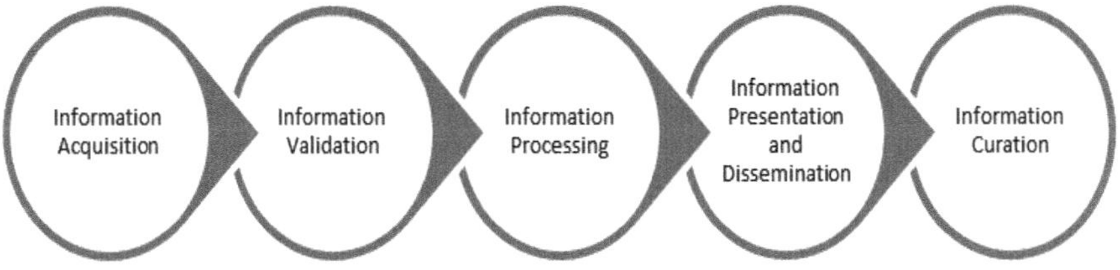

Fig. 1 Stages in the journalistic work process

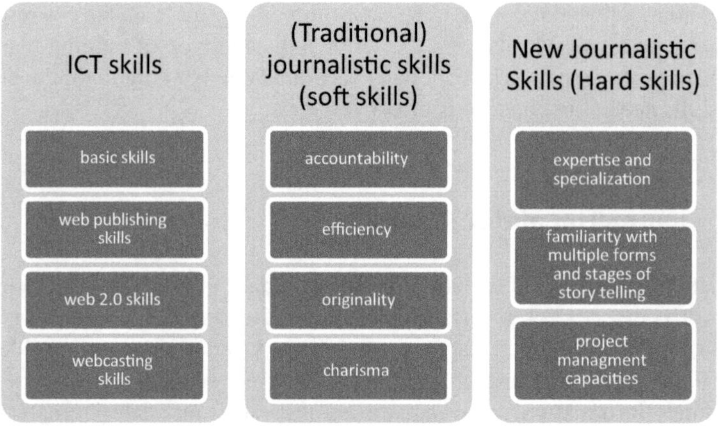

Fig. 2 Skillsets needed in a convergent newsroom

applications (like Google Docs) (Veglis, 2010). Naturally, word-processing applications are employed in order to write and store news items. During the Information Presentation and Dissemination stages, three major tasks take place: (a) the final form of the story is decided upon, (b) a decision regarding whether other publishing channels will share the story is made, and (c) content is adapted to the requirements specified by other publishing platforms. Some publishing channels, for example WWW can incorporate various forms of information, like text, images, sound, video, flash animation etc. Whereas others, like RSS or radio are limited to a single format of content (Veglis, 2012). Finally, during the Information Curation stage, journalism professionals are expected to further diffuse stories via social networks and engage in conversation with users.

Following the aforementioned workflow process, this section presents in detail the necessary ICT skills a journalist must possess in order to cope successfully with the challenges in his everyday work. Attempting to categorise the journalistic IT skills needed in today's convergent newsroom four categories can be distinguished (see Fig. 2). In the first category, entitled basic skills, journalists should have the ability to work efficiently with office automation suites (that include word processing, use of spreadsheets, presentation capacities, and database use), as well as effectively employ basic internet services (WWW, email) to accomplish tasks. This type of skillset is considered to be necessary and basic for every office worker. Journalists are expected to be able to perform basic functions in a spreadsheet and have at least a general understanding of how journalists use data to get stories. They must also be able to use relational database programs to crosscheck those data files and find various information (Peebles, 2011). As far as word processing is concerned, basic typing and formatting skills are assumed as a standard capacity.

In the case of the second category, called web publishing skills, basic knowledge of HTML is considered to be necessary as well as the ability to use Content Management Systems (eg. WordPress, Drupal, Joomla!). More precisely, journalists must have an understanding of the basic concepts of HTML and CSS

(Cascading Styling Sheet) (Peebles, 2011). Such knowledge can be proven extremely helpful when working with modern Content Management Systems. Familiarity with the most popular Content Management Systems is considered crucial since the majority of media companies have incorporated one of them in order to facilitate their web presence.

The next category, entitled web 2.0 skills, presupposes the use of blogs (both accessing and owning), RSS, Wikis, social bookmarking and social networking (Facebook, Google+, Twitter) (Franklin & van Harmelen, 2007). Journalists must possess the skills to update the media's profile on social networks, and also to interact with the audience through various Web 2.0 tools and services, like blogs and Wikis. Journalists must be able to use Web 2.0 tools like RSS, and social bookmarking, in order to be timely informed about news and information.

The last category, namely webcasting skills, includes the ability to create and publish podcasts and videocasts. Journalists must be able to record audio material, perform a simple edit on the audio recording and upload it to the WWW, in order for users to have access to the podcast (Fletcher, 2008; Peebles, 2011). Podcasts are considered to be an effective way to tell news stories using audio. They are easy to create and their size is small in comparison to other multimedia files (for example, video). Videocasts on the other hand, are more powerful tools to convey news. Of course they are more difficult to create and their size is quite large. But there is always the option of reducing their quality in order to decrease the videocast file size. Thus journalists should be capable of making at least a short video story even if it is shot with a FlipCam or a smartphone camera. Familiarity with entry-level editing software is also required (Fletcher, 2008; Peebles, 2011).

Finally, it ought to be mentioned that ICT related skills are constantly evolving and expanding. News professionals are required to follow these developments and adopt new tools, services and applications as they become available.

5.2 Soft and Hard Skills

Apart from a series of ICTs related skills, journalists much also possess traditional journalistic capacities associated with the production of good stories. A recently released report on Post-Industrial Journalism (Anderson, Bell, & Shirky, 2012) posits that journalists (should) possess four distinct characteristics which add value to journalism; the latter perceived as an efficient, reliable and knowledgeable source of information and analysis in the new digital era (see Fig. 2):

(1) Accountability. Professional accountability stems from the attempts of media practitioners to establish and follow ethical and quality standards that would render their work useful and viable for the society. Professional accountability is intimately connected to principles and practices of self-regulation (ethical guidelines, in-house ethical rules of conduct etc.). These, in turn, are expected to inform individual journalists in their daily work and foster public trust in what journalists do. This is actually in their self-interest in order to maintain their autonomy and credibility (Heikkilä et al., 2012: 7).

(2) Efficiency. Journalists can be much more efficient than machines and citizen journalists at obtaining and disseminating certain types of information (Compton & Benedetti, 2010). Access and 'exclusivity' or 'ownership' of a story is created through personal contacts and efforts. Making phone calls to the White House or the hospital, showing up at events and being receptive to feedback, sharing views and expressing doubt, collaborating with sources to get a story, all that makes news. These very personal and human activities mark journalism as a form of information performance rather than simply a process of disseminating facts (Anderson et al., 2012: 28).
(3) Originality. The origination of ideas, the formation of movements, and turning innovations into practices all require originality of thought. Journalists should be provoking change, initiating experimentation and instigating activity. Cultural literacy skills distinguish reporters, editors, designers and other journalists from other systems of data gathering and dissemination (Anderson et al., 2012: 2829).
(4) Charisma. In a networked world, the traditional notion of news judgment associated with journalism professionals' ability to recognize, find and tell a story is further expanded to include conversational and marketing skills. Adapting to this environment is a stretch for journalists who have developed their skills in newsrooms where precision and security were the key demands of the product, and where there was unity and clarity around a small set of processes—talking, writing, editing. Under convergence conditions, the ability to recognize, find and tell a story, in the most appropriate format and for a specific audience, remains a constant requirement, but the number of formats and variability of the audiences have grown. Spotting good stories and telling them in the right way to the right people via the right platform is ultimately crucial in a severely competitive and multiplatform news environment (Anderson et al., 2012). Furthermore, the ability to comprehend and incorporate the logic of web metrics into news production without compromising the essence of journalism, is important and calls for an 'expanded charisma' directed to balance journalism as a public good and economic sustainability of the outlet. Telling stories in a useful and attractive way in multiple platforms requires not only familiarity with multimedia reporting but also with storytelling formats which support transparency, plurality and openness (for instance, to embed hyperlinks and cite relevant material). On top of that, mastering content monetization strategies -for example, search engine (SEO) and social media optimization techniques (SMO)- has become a 'must' to ensure visibility and traffic within the web's attention economy paradigm (see Fig. 2).

Anderson et al. (2012) emphasise the need for journalism professionals to possess sets of soft and hard skills. The former set refers to having a mindset looking for innovation and change, being networked in terms of sources, contacts and communities, and also to develop a persona embodying integrity and judgment as these are attributes that journalists carry with them, as part of their public persona. The hard skills set refers to issue expertise and specialization, the ability to handle data and statistics, be able to code, be familiar with various forms and stages of storytelling (writing, filming, editing, recording, interviewing, designing and producing) and have project management capacities (p. 3140) (see Fig. 2).

6 The 'Super Journalist' Paradigm

Amid severe job insecurity caused by systematic staff cuts and revenue shortage (Franklin, 2012), journalism professionals are expected not only to enhance and adapt traditional journalistic skills, such as writing, news judgment and networking to the demands posed by convergence, but also to develop new skills, some of which are really hard to possess (for instance, coding). At the same time, while journalists are asked to work more, redefine news practices and even share control of the news production process with amateur journalists, a development which threatens occupational authority and identity, economic incentives are rare for the majority of journalism professionals across the globe.

From an organizational point of view, the reskilling of journalists is part of the convergence process guided by multitasking, which is driven by economic motives of cost-efficiency, greater productivity, competitiveness and profit consolidation. Seen, however, through the lenses of occupational professionalism articulated in terms of values, journalistic practices, job performance and satisfaction, the idea of multitasking calling for the reskilling of new professionals and the deskilling of midcareer ones having a really hard time following the trends and changes, is causing serious tensions within the journalism community. Huang et al. (2006) argued that multimedia training is really important and feasible for new and midcareer professionals. However, since 2006 and as shown in the present chapter, skill demands are steadily increasing, salaries are shrinking, while mergers and synergies are augmenting. Unsurprisingly, recent research shows that difficulties and tensions emerge as the industry moves towards the construction of the 'super journalist' paradigm.

Investigating the Swedish landscape, Witschge and Nygren (2009) conclude that as technical and economic changes are disrupting the established professional status, roles, and practices of journalists and remove professional control that previously existed, journalists are adopting a defensive attitude indicating that they are not ready to relinquish traditional professional standards. "They try to guard the boundaries of the profession and ensure its perceived distinctiveness. Of course, this can be seen as a direct response to the pressures put on the profession by external forces—the economic focus and the production of news by non-journalists" (p. 57). Robinson's (2010) ethnographic study on readers' comments policy revealed significant internal conflict among both journalists and readers. The 'traditionalists'—those who want to maintain a hierarchal relationship between journalists and audiences—clashed with the 'convergers'—those who felt users should be given more freedoms within the news site. Interviewing 20 journalists in UK, Saltzis and Dickinson (2008) found that while multimedia news is becoming well established, the multimedia journalist has been slow to arrive. Workload pressures on journalists daily routine and a concern over the impact on the quality of output, resulted in BBC's policy to encourage multiskilling but not to make it mandatory designating two types of journalist in the workforce: (a) the 'single skilled' specialists valued for their high journalistic standards and (b) the 'multi-skilled' professionals valued for their versatility and adaptability (p. 222). Robinson (2011) argues that changed power hierarchies in convergence conditions privilege laborers with technological skills engaged in digital domains, while reporters with print-cultural mindsets find themselves increasingly isolated in the newsroom, often excluded from new workspaces. The convergent landscape calls for expanded competencies rendering higher education a necessary qualification at entry level (Siapera & Spyridou, 2012). However, despite journalism university degrees being on the rise, Spyridou et al. (2013) found serious knowledge gaps in relation to software use, limited capabilities in using social networks as reporting tools and extensive unfamiliarity with audience metrics. According to Compton and Benedetti (2010) labour rationalization and intensification of production is fundamentally reorganizing the division of labour in newsrooms negatively affecting professional performance and content output. Similar evidence from Wallace (2013) demonstrates that the spread of multiskilling in combination with a reduction in workforce increases stress levels on journalists and affects quality of output.

7 Conclusion: Towards Better Journalism?

On January, 21st 2013, the Financial Times announced a net headcount reduction of 25 journalists—after hiring ten journalists for digital roles in an effort to save £1.6 million at that year. According to Lionel Barber, editor of the FT, his trip to Silicon Valley in September "confirmed the speed of change", and added that "the internet offers new avenues and platforms for the richer delivery and sharing of information. We are moving from a news business to a networked business" (source: Halliday, 2013). This kind of approach is not new. Hundreds of layoffs have taken place in the European and US market, especially since 2009. Employing a neoliberal rationale, convergence is the perfect recipe to save the media. The key question of whether editorial managers are embracing convergence for business reasons or to do better journalism was raised early in 2004 by Quinn, who concluded that "successful convergence satisfies the twin aims of good journalism and good business practices" (p. 109). However, his position is premised on a misapprehension of journalism's structure, which is corporately owned, profit-driven and faced with extreme uncertainty.

On an ideal basis, full convergence operating in an environment of economic stability that would secure adequate resources (both financial and human), training opportunities, and above all newsrooms committed to satisfy journalism's civic role of channeling accurate, adequate and relevant information, acting as a watchdog, and providing a forum for public discussion and deliberation, would probably work. Indeed it would be most possible to elevate journalistic output to meet both normative (democratic-oriented) and emerging technological, organizational and consumption demands.

But the institution of journalism, as both a practice and a discourse, is structured by industrial logic and its own political economy (Broesma & Peters, 2013; Compton & Benedetti, 2010; Scott, 2005). Journalism scholars have systematically critiqued news in many ways, but a central thread involves questions around truth and accuracy mainly as a result of pressures imposed by market forces (Fuchs, 2009; Herman & Chomsky, 1998; McManus, 1995) as well as constraints stemming from professional routines and organizational standards (Deuze, Bruns, & Neuberger, 2007; Preston, 2009a; Reese, 2001). Understanding thus the impact of convergence on news labour and production requires a broader consideration of the news industry structure and logic. In the context of neoliberal capitalism the news industry functions as much as any other profitmaking sector. At the same time it is faced with a severe sustainability crisis, as the traditional advertising model is not working online and the industry seems unable to monetize content (Picard, 2011). In the logic of late capitalism combined with the digitization utopia, the ultimate discourse is a creation of a sense of inevitability around changes taking place in the economy and the labour market (Paulussen, 2012). In that context, the main trait of convergence in the news industry follows the 'do more with less' dogma. On an organizational level, acquisitions and mergers and content-sharing strategies, favour, and further reinforce trends of ownership concentration monitored since the 1980s. Huang et al. (2006) has shown that mergers, a key component of convergence tends to benefit the media companies rather the general public or news professionals. On a professional level, shrinking staffs work harder without economic incentives. Additionally, professionals are drawn away from their traditional role as mediators and investigators of news and information and transformed into copywriters for multiple formats. Unsurprisingly journalists seem to be perpetually concerned with the decline of their occupation. Journalism used to be more credible and they enjoyed a stable occupational identity. A narrative of decline is at the heart of many journalists' accounts of journalism, and these narratives cannot be easily discounted (Gitlin, 2009; Örnebring, 2009). Declining job satisfactions levels stem from discontent with salaries, workload, perceived lack of promotion prospects (Paulussen, 2012), job insecurity (Spyridou et al., 2013) and higher rates of burnout (Reinardy, 2011).

Convergence projects are primarily driven by the market logic that aims to reduce costs, while increasing productivity and maximizing profit (Klinenberg, 2005). As news production intensifies, job demands and skill requirements increase. Reviewing the aforementioned skillsets on various levels, journalism professionals are expected to become some kind of 'super professionals' possessing multiple kinds of expertise and knowledge, work hard and perform well. Multiskilling goes hand in hand with wider changes in the news labour market including atypical, nonstandard employment, reduced economic rewards, functional flexibility and desk-journalism. Considering the deep compromises news workers are faced with, and the level of frustration experienced, such expectations are rather extreme and unrealistic, especially in small markets. The 'super-journalist' paradigm, at least at this point in time, probably refers to a frustrated news worker struggling to keep up with constant professional and technological changes.

Benkler (2006) argues that we are shifting from a monopolistic 'industrial model of journalism' to a pluralistic 'networked model' based on profit and nonprofit, individual and organised journalistic practices. Convergence culture celebrates the emancipatory character of citizen and participatory media culture articulating a fundamental shift in the way global media industries operate and how people interact with them (Jenkins, 2006). Yet, in practice leading legacy brands tend to have monopolized attention on the web (Benson, 2010; Curran, 2010), while still holding a dominant role in framing main political and economic news content (Chakravartty & Schiller, 2010). Citizen and alternative journalism initiatives have a hard time sustaining themselves (Pew Project for Excellence in Journalism, 2009; Spyridou, 2010), while a tendency to mythologise the role of the Web in 'mainstreaming' minority journalism is also documented (Curran, 2010). Reese, Routigliano, Hyun, and Jeong (2007) found that the blogosphere relies heavily on professional news reports, while Redden and Witschge (2010) noted that no alternative outlets make it high on search results on either Google or Yahoo.

The paradigm of the Guardian and a handful of other news organizations, is not and cannot be universal. In most cases, skilled journalists are not asked to produce better journalism but rather exploit the technological and social opportunities offered by convergence in order to enhance monetization opportunities. Commercialization and concentration of ownership have always been declared as threats to the democratic role of the press (Scott, 2005). At a time when hypercommercialization, overconcentration of ownership, intensification and flexibilazation of labour is at the centre of convergence procedures, the real transformative power of convergence should be questioned.

To sum up, convergence allows people to get news on a 24/7 basis on their tablet, receive customized news on their smartphone, get news from their Facebook account. However, the essence of journalism as described by traditional democratic theory is hardly enhanced given the material and economic constraints of the networked era. Convergence is certainly proving an efficient strategy for media consolidation and sustainability, but in this process journalism as a profession and as practice is in decline. Considering the erosion of the profession in terms of labour conditions and over-skilling pressures, it is highly questionable whether journalists can perform their job better. Job dissatisfaction and job insecurity diminish work autonomy and devotion. The emerging 'super-journalist' paradigm may reach a point when news workers exhibit high levels of familiarity with new tools and applications. But what about accountability, charisma and integrity? Historically, numerous cases have proven the interplay between news organizations, journalists, political and economic elites. Convergence, implemented primarily as a cost-effective strategy does not promise or promote better journalism. Rather it tends to consolidate the industry logic by promoting the 'super-journalist' paradigm.

References

Anderson, C. (2008). Journalism: Expertise, authority, and power. In D. Hesmondgalgh & J. Toynbee (Eds.), *The media and social theory* (pp. 248–264). London: Routledge.

Anderson, C., Bell, E., & Shirky, C. (2012). *Post-industrial journalism: Adapting to the present*. Report, Tow Center for Digital Journalism. http://towcenter.org/wpcontent/uploads/2012/11/TOWCenterPost_Industrial_Journalism.pdf

Avilés, A. G., & Carvajal, M. (2008). Integrated and cross-media newsroom convergence two models of multimedia news production—the cases of Novotécnica and La Verdad multimedia in Spain. *Convergence, 14*(2), 221–239.

Bell, E. (2007). User + tools = journalism. *Index on Censorship, 36*(4), 100–104.

Benkler, Y. (2006). *The wealth of networks how social production transforms markets and freedom.* New Haven, CT: Yale University Press.
Benson, R. (2010). Futures of the news: International considerations and further reflections. In N. Fenton (Ed.), *New media, old news: Journalism and democracy in the digital age* (pp. 187–200). Los Angeles: Sage.
Benson, R., BlachØrsten, M., Powers, M., Willig, I., & Zambrano, V. S. (2012). Media systems online and off: Comparing the form of news in the United States, Denmark, and France. *Journal of Communication, 62*, 21–38.
Boczkowski, P. (2004). *Digitizing the news: Innovation in online newspapers.* Cambridge, MA: MIT Press.
Boczkowski, P. (2009). Materiality and mimicry in the journalism field. In B. Zelizer (Ed.), *The changing faces of journalism* (pp. 56–97). London: Routledge.
Bradshaw, P. (2012a). *Model for the 21st century newsroom Redux.* http://leanpub.com/21stcenturynewsroom
Bradshaw, P. (2012b). What is data journalism? In J. Gray, L. Bounegru, & L. Chambers (Eds.), *The data journalism handbook.* http://datajournalismhandbook.org/1.0/en/introduction_0.html
Brannon, J. (2011). Maximise the medium: Assessing obstacles to performing multimedia journalism in three US Newsrooms. In C. Paterson & D. Domingo (Eds.), *Making online news* (Vol. 1, pp. 99–112). New York: Peter Lang.
Broesma, M., & Peters, C. (2013). Introduction: Rethinking Journalism: The structural transformation of a public good. In C. Peters & M. Broesma (Eds.), *Rethinking journalism: Trust and participation in a transformed news landscape* (pp. 1–12). London: Routledge.
Chakravartty, P., & Schiller, D. (2010). Neoliberal newspeak and digital capitalism in crisis. *International Journal of Communication, 4*, 670–692.
Compton, J., & Benedetti, P. (2010). Labour, new media and the institutional restructuring of journalism. *Journalism Studies, 11*(4), 487–499.
Currah, A. (2009). *What's happening to our news.* Report, Reuters Institute for the Study of Journalism. https://reutersinstitute.politics.ox.ac.uk/fileadmin/documents/Publications/What_s_Happening_to_Our_News.pdf
Curran, J. (2010). The future of journalism. *Journalism Studies, 11*(4), 464–476.
Dailey, L., Demo, L., & Spillman, M. (2005). The convergence continuum: A model for studying collaboration between media newsroom. *Atlantic Journal of Communication, 13*(3), 150–168.
Davies, N. (2008). *Flat earth news.* London: Chatto and Windus.
Deuze, M. (1999). Journalism and the web: An analysis of skills and standards in an online environment. *Gazette, 61*(5), 373–390.
Deuze, M. (2004). What is multimedia journalism? *Journalism Studies, 5*(2), 139–152.
Deuze, M. (2007). *Media work.* Cambridge: Polity Press.
Deuze, M., Bruns, A., & Neuberger, C. (2007). Preparing for an age of participatory news. *Journalism Practice, 1*(3), 322–338.
Deuze, M., & Marjoribanks, T. (2009). Newswork, journalism theory. *Practice & Criticism, 10*(5), 555–561.
Domingo, D. (2008). Interactivity in the daily routines of online newsrooms: Dealing with an uncomfortable myth. *Journal of Computer Mediated Communication, 13*, 680–704.
Domingo, D., Salaverría, R., Aguado, J. M., Cabrera, A., Edo, C., Masip, P., et al. (2007). *Four dimensions of journalistic convergence: A preliminary approach to current media trends at Spain.* Paper presented at the International Symposium of Online Journalism, Texas. http://online.journalism.utexas.edu/2007/papers/Domingo.pdf
Dupagne, M., & Garrison, B. (2006). The meaning and influence of convergence. *Journalism Studies, 7*(2), 237–255.
Engebretsen, M. (2006). Shallow and static or deep and dynamic? Studying the state of online journalism in Scandinavia. *Nordicom Review, 27*(1), 316.
Erdal, I. J. (2011). Coming to terms with convergence journalism: Cross-media as a theoretical and analytical tool. *Convergence, 17*(2), 213–223.
Fletcher, K. (2008). *Journalism skills survey.* National Council for the Training of Journalists. http://www.skillset.org/uploads/pdf/asset_13022.pdf?2

Franklin, B. (2012). The future of journalism. *Journalism Studies, 13*(56), 663–681.
Franklin, T., & van Harmelen, M. (2007). *Web 2.0 for content for learning and teaching in higher education.* Report, Franklin Consulting and Mark van Harmelen. Franklin Consulting.
Fuchs, C. (2009). Information and communication technologies and society. A contribution to the critique of the political economy of the internet. *European Journal of Communication, 24*(1), 69–87.
Fulton, K. (1996, March/April). A tour of our uncertain future. *Columbia Journalism Review.* http://www.cjr.org/html/9603 04tour.htm
Gade, P., & Lowrey, W. (2011). Reshaping the journalistic culture. In W. Lowrey & P. Gade (Eds.), *Changing the news* (pp. 22–42). London: Routledge.
Garcia, X. L., Fernández, M., Bran, C., & Fariña, X. (2012). Convergence as a renovation/conservation tool for journalism in Galicia: Strategies and uncertainties. *Brazilian Journalism Research, 8*(1), 48–62.
Gitlin, T. (2009). *Journalism's many crises. Open-democracy.* http://www.opendemocracy.net/article/asurfeitofcrisescirculationrevenueattentionauthorityanddeference
Gordon, R. (2003). The meanings and implication of convergence. In K. Kawamoto (Ed.), *Digital journalism—emerging media and the changing horizons of journalism* (pp. 31–55). Oxford: Rowman and Littlefield Publishers.
Hacker, K. L., & van Dijk, J. (2000). What is digital democracy? In K. L. Hacker & J. van Dijk (Eds.), *Digital democracy: Issues of theory and practice* (pp. 1–9). Thousand Oaks, CA: Sage.
Halliday, J. (2013, January 21). Financial times editor announces digital-first strategy. *theguardian.com.* http://www.theguardian.com/media/2013/jan/21/financialtimesdigitalfirst
Heikkilä, H., Domingo, D., Pies, J., Glowacki, M., Kus, M., & Baisnée, O. (2012). *Media accountability goes online a transnational study on emerging practices and innovations,* MediaAcT Working paper series on 'Media Accountability Practices on the Internet', Journalism Research and Development Centre, University of Tampere, Finland.
Herman, E., & Chomsky, N. (1998). *Manufacturing consent: The political economy of mass media.* New York: Pantheon.
Huang, E., Davison, K., Shreve, S., Davis, T., Bettendorf, E., & Nair, A. (2006). Facing the challenges of convergence media professionals' concerns of working across media platforms. *Convergence: The International Journal of Research into New Media Technologies, 12*(1), 83–98.
Infotendencias Group. (2012). Media convergence. In E. Siapera & A. Veglis (Eds.), *The handbook of global online journalism* (pp. 21–38). London: Wiley-Blackwell.
Jenkins, H. (2006). *Convergence culture: Where old and new media collide.* New York: NYU.
Kawamoto, K. (2003). Digital journalism – emerging media and the changing horizons of journalism. In K. Kawamoto (Ed.), *Digital journalism – emerging media and the changing horizons of journalism* (pp. 1–29). Lanham: Rowman and Littlefield.
Karlsson, M. (2011). Rituals of transparency. In B. Frankin (Ed.), *The future of journalism* (pp. 100–110). London: Routledge.
Kenney, K., Gorelik, A., & Mwangi, S. (2000). Interactive features of online newspapers. *First Monday, 5*(1). www.firstmonday.dk/issues/issue5_1/kenney/index.html
Klinenberg, E. (2005). Convergence: News production in a digital age. *Annals of the American Academy of Political and Social Science, 597*, 48–64.
Lee-Wright, P. (2010). Culture shock: New media and organizational change in the BBC. In N. Fenton (Ed.), *New media, old news: Journalism and democracy in the digital age* (pp. 71–86). London: Sage.
McManus, J. (1994). *Market driven journalism.* Beverly Hills, CA: Sage.
McManus, J. (1995). A market-based model of news production. *Communication Theory, 5*(4), 301–338.
Micó, J. L., Masip, P., & Domingo, D. (2013). To wish impossible things: Convergence as a process of diffusion of innovations in an actor-network. *The International Communication Gazette, 75*(1), 118–137.
Mitchelstein, E., & Boczkowski, P. (2009). Between tradition and change: A review of recent research on online news production. *Journalism, 10*(5), 562–586.
O'Sullivan, J. (2005). Delivering Ireland: Journalism's search for a role online. *Gazette: The International Journal for Communication Studies, 67*(1), 45–68.

O'Sullivan, J., & Heinonen, A. (2008). Old values, new media. *Journalism Practice, 2*(3), 357–371.
Oblak, T. (2005). The lack of interactivity and hypertextuality in online media. *Gazette: The International Journal for Communication Studies, 67*(1), 87–106.
Örnebring, H. (2013). Anything you can do, I can do better? Professional journalists on citizen journalism in six European countries. *The International Communication Gazette, 75*(1), 35–53.
Örnebring, H. (2009). *The two professionalisms of journalism: Journalism and the changing context of work*. Working paper, Reuters Institute for the Study of Journalism.
Paulussen, S. (2012). Technology and the transformation of news work: Are labor conditions in (online) journalism changing? In E. Siapera & A. Veglis (Eds.), *The handbook of global online journalism* (pp. 192–203). Oxford: Wiley Blackwell.
Paulussen, S., & Ugille, P. (2007). User generated content in the newsroom: Professional and organisational constraints on participatory journalism. *Westminster Papers in Communication and Culture, 5*(2), 24–41.
Pavlik, J. (2001). *Journalism and new media*. New York: Columbia University Press.
Pavlik, J. (2004). A sea change in journalism: Convergence, journalists, their audiences and sources. *Convergence, 10*(4), 21–29.
Pavlik, J. (2013). Innovation and the future of journalism. *Digital Journalism*. doi:10.1080/21670811.2012.756666.
Peebles, J. (2011, March 9). Digital media skills every young journalist needs. *Net Worked*. http://blogs.spjnetwork.org/tech/2011/03/09/digitalmediaskillseveryyoungjournalistneeds
Pew Internet. (2010). *Understanding the participatory news consumer*. http://www.pewinternet.org/~/media/Files/Reports/2010/PIP_Understanding_the_Participatory_News_Consumer.pdf
Pew Project for Excellence in Journalism. (2009). The State of the News Media 2009—citizen based media. http://stateofthemedia.org/2009/specialreportssummaryessay/citizenbasedmedia/
Picard, R. (2011). Certainty and uncertainty of change in journalism. In W. Lowrey & P. Gade (Eds.), *Changing the news* (pp. vii–ix). London: Routledge.
Preston, P. (2009a). *Making the news: Contemporary journalism practices and news cultures in Europe*. London: Taylor & Francis.
Preston, P. (2009b). *Making the news: Journalism and news cultures in Europe*. London: Routledge.
Quandt, T. (2008). (No) News on the World Wide Web? A comparative content analysis of online news in Europe and the United States. *Journalism Studies, 9*(5), 717–738.
Quandt, T., & Singer, J. (2009). Convergence and cross-platform content production. In K. Wahl-Jorgensen & T. Hanitzsch (Eds.), *The handbook of journalism studies* (pp. 130–144). New York: Routledge.
Quinn, S. (2004). An intersection of ideals: Journalism, profits, technology and convergence. *Convergence, 10*(4), 109–123.
Quinn, S. (2005). Convergence's fundamental question. *Journalism Studies, 6*(1), 29–38.
Quinn, S., & Quinn Allan, D. (2005, November 29–December 2). *The worldwide spread of journalism convergence*. Paper presented to the Journalism Education Conference, Griffith University. http://dro.deakin.edu.au/eserv/DU:30005866/quinnworldwidespread2005.pdf
Redden, J., & Witschge, T. (2009). A new order? Online news content examined. In N. Fenton (Ed.), *New media, old news* (pp. 171–186). London: Sage.
Redden, J., & Witschge, T. (2010). A new news order? Online news content examined. In N. Fenton (Ed.), *New media, old news: Journalism and democracy in the digital age* (pp. 171–186). Los Angeles: Sage.
Reese, S. (2001). Understanding the global journalist: A hierarchy of influences approach. *Journalism Studies, 2*(2), 173–187.
Reese, S., Routigliano, L., Hyun, K., & Jeong, J. (2007). Mapping the blogosphere—professional and citizen-based media in the global news arena. *Journalism, 8*(3), 254–280.
Reich, Z. (2006). The process model of news initiative: Sources lead first, reporters thereafter. *Journalism Studies, 7*(4). doi: 10.1080/14616700600757928.
Reinardy, S. (2011). Newspaper journalism in crisis: Burnout on the rise, eroding young journalists' career commitment, *Journalism, 12*(1), 33–50.

Robinson, S. (2010). Traditionalists vs. Convergers. Textual privilege, boundary work, and the journalist–audience relationship in the commenting policies of online news sites. *Convergence: The International Journal of Research into New Media Technologies, 16*(1), 125–143.

Robinson, S. (2011). Convergence crises: News work and news space in the digitally transforming newsroom. *Journal of Communication, 61*, 1122–1141.

Salaverría, R. (2005). An immature medium—strengths and weaknesses of online newspapers on September 11. *Gazette: The International Journal for Communication Studies, 67*(1), 69–86.

Saltzis, K. (2012). Breaking news online. *Journalism Practice, 6*(56), 702–710.

Saltzis, K., & Dickinson, R. (2008). Inside the changing newsroom: Journalists' responses to media convergence. *Aslib Proceedings: New Information Perspectives, 60*(3), 216–228.

Sarrica, M., Fortunati, F., O'Sullivan, J., Balcytiene, A., Macgregor, P., Nuust, V., et al. (2010). The early stages of the integration of the internet in EU newsrooms. *European Journal of Communication, 25*(4), 413–422.

Schultz, B., & Sheffer, M. L. (2009). Newspaper managers report positive attitudes about blogs. *Newspaper Research Journal, 30*(2), 30–42.

Scott, B. (2005). A contemporary history of digital journalism. *Television New Media, 6*(1), 89–126.

Siapera, E., & Spyridou, P. L. (2012). The field of online journalism: A Bourdieusian analysis. In E. Siapera & A. Veglis (Eds.), *The handbook of global online journalism* (pp. 77–97). London: Blackwell.

Singer, J. (2005). The political J-blogger: 'Normalizing' a new media form to fit old norms and practices, journalism: Theory. *Practice and Criticism, 6*(2), 173–198.

Singer, J. (2009a). Convergence and divergence. *Journalism, 10*(3), 375–377.

Singer, J. (2009b). Journalism and digital technologies. In W. Lawrey & P. Gade (Eds.), *Changing the news—the forces shaping journalism in uncertain times* (pp. 213–229). New York: Routledge.

Spyridou, L. P. (2010, June 8–9). *Experimenting with alternative news operations: The tvxs.gr case*. Journalism's next top model: Meeting the cost of journalism tomorrow conference organized by the School of Journalism and Mass Communication, University of Westminster, London.

Spyridou, P., Matsiola, M., Veglis, A., Kalliris, G., & Dimoulas, B. (2013). Journalism in a state of flux: Journalists as agents of technology innovation and emerging news practices. *The International Communication Gazette, 75*(1), 76–98.

Spyridou, L. P., & Veglis, A. (2008, May). Exploring structural interactivity in online newspapers: A look at the Greek web landscape. *First Monday*, 13(5). http://firstmonday.org/htbin/cgiwrap/bin/ojs/index.php/fm/article/view/2164/1960

Steinle, P., & Brown, S. (2012, April/May). Embracing the future. *American Journalism Review*. http://www.ajr.org/Article.asp?id=5295

Thurman, N., & Lupton, B. (2008). Convergence calls: Multimedia storytelling at British news websites. *Convergence: The International Journal of Research into New Media Technologies, 14*(4), 439–455.

Veglis, A. (2010, July). *Teaching data visualization in journalism students*. IEEE Learning Technologies. http://lttf.ieee.org/issues/july2010/IEEELTJul10.pdf

Veglis, A. (2012). Journalism and cross media publishing: The case of Greece. In E. Siapera & A. Veglis (Eds.), *The handbook of global online journalism* (pp. 209–230). Chichester: Blackwell.

Vujnovic, M., Singer, J., Paulussen, S., Heinonen, A., Reich, Z., Quandt, T., et al. (2010). Exploring the political-economic factors of participatory journalism. *Journalism Practice, 4* (3), 285–296.

Wallace, S. (2013). The complexities of convergence: Multi-skilled journalists working in BBC Regional Multimedia Newsrooms. *The International Communication Gazette, 75*(1), 99–117.

Witschge, T., & Nygren, G. (2009). Journalism: A profession under pressure? *Journal of Media Business Studies, 6*(1), 37–59.

Online Presences of Traditional Media vs. Blogs: Redundancy and Unique News Coverage

Kay Hamacher

1 Introduction and Problem Discussion

Media distribute information. As new distribution channels such as the letter-press or the Internet arise and become new paradigms, traditional media have to adjust or will perish. The resulting convergence of media (Lugmayr, Franssila, Safran, & Hammouda, 2011; Zotto, Galichet, & Lugmayr, 2011) is, therefore, and interesting area of research from several points of view. In particular, previous work has shown interesting aspects in specific areas of media convergence such as the technical and content-wise convergence of, e.g., TV and Internet (Bachmayer, Lugmayr, & Kotsis, 2010; Behrendt & Zeppenfeld, 2008).

Now, the capacity of new distribution channels is typically larger and these come with lower entry barriers for competitors. Do these democratizing effects shape the "information landscape" the consumer is able to enjoy? Is there evidence for convergence or divergence from the end-consumer perspective? Typically, practitioners rely on anecdotal evidence, but make strong claims on the necessity to adjust the legal and business framework. This has led—at least in Germany—to a highly emotional debate about the future of the press. More careful assertions on future business model for the global media can be found in academic and industry discussions (Lehtisaari et al., 2012; Picard, Dal Zotto, & Werfel, 2006; van Tuyll, 2010).

K. Hamacher (✉)
Department of Computer Science, Technische Universität Darmstadt, Schnittspahnstr. 10, Darmstadt, Germany

Department of Physics, Technische Universität Darmstadt, Schnittspahnstr. 10, Darmstadt, Germany

Department of Biology, Technische Universität Darmstadt, Schnittspahnstr. 10, Darmstadt, Germany
e-mail: hamacher@bio.tu-darmstadt.de

A. Lugmayr, C. Dal Zotto (eds.), *Media Convergence Handbook - Vol. 1*, Media Business and Innovation, DOI 10.1007/978-3-642-54484-2_7

This chapter is structured as follows: first, we introduce a quantitative framework to cluster and classify (online) texts; we then describe the data sources and the general setting we have chosen to answer the question of (content) divergence or convergence in German speaking media; finally, we show our results and discuss their implications for the discussion of both convergence phenomena and unique business situations for media.

2 State of the Art

Conceptually the information distributed by media can be analyzed by information theory (Cover & Thomas, 2006) and the (Shannon) entropy concepts it is based on. In particular, the used measures are known to be related to so-called type-I errors in statistical testing (Blahut, 1987); thus, our results can be interpreted how likely it is that an otherwise knowledgeable reader would expect a certain news piece to originate from one particular media outlet.

The present study is devoted to develop quantitative measures (adapted from information theory) to bring more reliable, *empirical data* and most of all a substantiated framework to the debate. We will deal with the situation of web presences of traditional media (newspapers) in comparison to new developments such as blogs. Both of these media types (strictly online vs. web-content provided by traditional media) rely heavily on textual representation which is suitable for automated and quantitative analysis as put forward in this work.

2.1 "Bag of Words"

To inspect the contents of a piece of text, such as a web page, it is necessary to clear the text from formatting information such as HTML tags. This was done in the python programming language using the Natural Language ToolKit (NLTK) (Bird, 2002). Furthermore, while focusing on the content in a text, one improves the noise-to-signal ratio by word-stemming, that is by mapping every word that occurs to its respective word stem. This can be done automatically with algorithms well-established in the linguistics community. To this end, we have used the Snow–ball system (Porter, 2006), which is based on the Porter algorithm for stemming (Porter, 1980).

The stream of word stems from a source i can then be converted to a vector $\mathbf{v}_i$ of word occurrence frequencies (a vector space model), which is suitable for information theoretic processing by the "bag of words" (BOW) approach (Harris, 1954; Manning & Schütze, 1999). Here, we simply count the number of times we find a particular word stem in the preprocessed text and finally divide by the overall number of word stems encountered. Thus, the an entry of $\mathbf{v}_i$ is the frequency or ratio of one particular word (stem) in the overall information emitted by the media i.

Focusing on these quantities allows to quantify how important certain phrases are for the content that a media provides to the end-consumer, e.g. the phrase

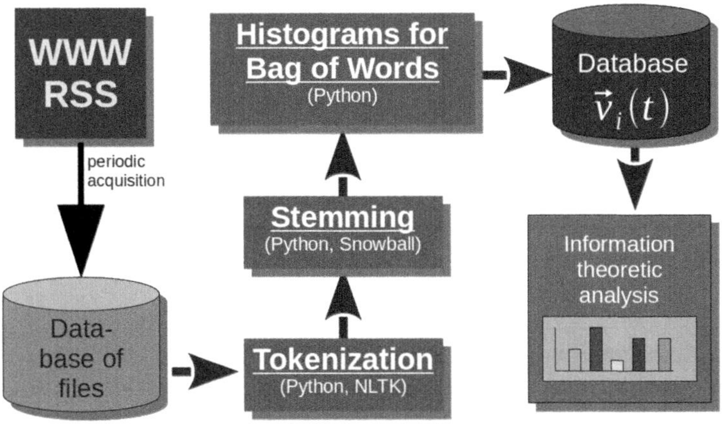

Fig. 1 The pipeline used in this study to automatically: (a) obtain web-content, (b) strip it of meta-information such as HTML tags etc., (c) extract word stems, (d) construct BOW (bag of words) frequency vectors $\mathbf{v}_i(t)$ for source i at time t, and finally analyze these quantitative measures using information theoretic concepts

"Greek budget crisis" is obviously more common in politics and financial news than in sports news. In the following we will therefore refer to this property also as "concepts" that are reported.

We show the data processing pipeline in Fig. 1. The vector $\mathbf{v}_i$ is then a probability function on the word stems and $\mathbf{v}_i(t, w)$ is the probability to obtain word stem w when choosing an arbitrary word from source i at time t.

3 Methodology and Approach

3.1 Information Theoretic Dissection of Textual Sources

In the following, we describe in detail the technicalities of the proposed framework. This section can be skipped if the reader is interested only the results. It is noteworthy, however, that the used approach assigns a "distance" to each pair of (online) media under scrutiny. This "distance" can then be used to cluster closely related media together.

3.1.1 (Shannon) Entropy of Online Presence of Media

The (Shannon) entropy (Shannon, 1948) is a measure to quantify the diversity within a data set. In our setting, we can apply to the BOW vector $\mathbf{v}_i$ for any media i under consideration. Then the entropy S_i for this media reads

$$S_i := -\sum_{w \in \mathscr{W}; t \in \mathscr{T}} \mathbf{v}_i(t, w) \cdot \log_2 \mathbf{v}_i(t, w) \tag{1}$$

where $\mathscr{W}$ is the set of word stems encountered and $\mathscr{T}$ the time interval of observation of the two media. In addition, we use $\lim_{\varepsilon \to 0} \varepsilon \cdot \log_2 \varepsilon = 0$ for those entries in the probability vector $\mathbf{v}_i$ that vanish (a word stem that is not present in media i, but in same other media). This value alone does not necessarily reflect the "information content", but will be used a first measure on the diversity of topics dealt with, see Sect. 4.1.

3.1.2 Relative Entropy to Distinguish Two Sources

The "distance" between the BOW vectors of two media $\mathbf{v}_i$ and $\mathbf{v}_j$ can be quantified by the so-called Kullback-Leibler divergence (Cover & Thomas, 2006) between the BOW vectors. The Kullback-Leibler divergence D_{KL} between such two probability vectors reads (Boba, Weil, Hoffgaard, & Hamacher, 2011; Hamacher, 2007c)

$$D_{\mathrm{KL}}\left(\mathbf{v}_i \middle|\middle| \mathbf{v}_j\right) := \sum_{w \in \mathscr{W}; t \in \mathscr{T}} \mathbf{v}_i(t, w) \cdot \log_2 \frac{\mathbf{v}_i(t, w)}{\mathbf{v}_j(t, w)}$$

Then, D_{KL} is the *amount of information one needs to provide to get from the content of media i to the content of media j*. While this quantity in itself is of uttermost importance in information theory and statistics, it is not suitable as a measure for the divergence or convergence of media content, due to the fact that $D_{\mathrm{KL}}\left(\mathbf{v}_i \middle|\middle| \mathbf{v}_j\right) \neq D_{\mathrm{KL}}\left(\mathbf{v}_j \middle|\middle| \mathbf{v}_i\right)$, that is, the Kullback-Leibler is not symmetric. We can, however, expand its applicability by the construction in the following Sect. 3.1.3.

3.1.3 Clustering by Information Distance

As describe in the previous section, we can measure the "distance" between empirically derived BOW vectors $\mathbf{v}_i(t)$ and $\mathbf{v}_j(t')$ by the Kullback-Leibler divergence D_{KL}, which amounts to a number in bit. Note that this notion is generic: we can compare the same source $(i = j)$ at different times $\left(t \neq t'\right)$, or different sources $(i \neq j)$ at the same time $\left(t = t'\right)$, or even different sources $(i \neq j)$ at different times $\left(t \neq t'\right)$.

As mentioned the Kullback-Leibler divergence is not symmetric. Therefore we use a symmetrized variant namely the Jensen-Shannon divergence H in the subsequent parts of this work:

$$H_{ij} := \frac{1}{2} \cdot D_{\mathrm{KL}}\left(\mathbf{v}_i \middle|\middle| \mathbf{m}_{ij}\right) + \frac{1}{2} \cdot D_{\mathrm{KL}}\left(\mathbf{v}_j \middle|\middle| \mathbf{m}_{ij}\right) \tag{2}$$

$$\text{with} \quad \mathbf{m}_{ij} = \frac{1}{2}(\mathbf{v}_i + \mathbf{v}_j) \quad \text{as an "average distribution".}$$

The Jensen-Shannon-Matrix H_{ij} for all pair of sources (i,j) is still not sufficient as a metric: it does not fulfill the triangle inequality.[1]

In a seminal work for the field of information theory and entropy applications Endres and Schindelin (2003) have shown that the square root of the Jensen-Shannon-Matrix fulfills the triangle inequality. Here we therefore define for our situation $M_{ij} := \sqrt{H_{ij}}$ which is a suitable metric for the vector space of reports in media (as given by the respective BOW).

This approach was previously used to obtain phylogenies of whole genomes via their protein content (Hamacher, 2010) and to derive coarse-grained amino acid alphabets by statistical properties found in structural databases (Pape, Hoffgaard, & Hamacher, 2010).

The resulting matrix $\mathfrak{M} := (M_{ij})$ with M_{ij} as entries can now be used as a distance matrix for well-known clustering algorithms. The result of the clustering is a tree of related sources. We used here the hclust-algorithm of R (R Development Core Team, 2008) implemented in STATLIB (Murtagh, 1985).

By comparing all sources for the same period of time $\mathcal{T}$ we obtain a clustering of all media under consideration. We can find a time-independent consensus clustering by applying the majority-rule to the set of clusterings resulting for various, non-overlapping choices of $\mathcal{T}$. The majority rule we used is implemented in e.g. the well-known PHYLIP-package (Felsenstein, 1989).

This gives insight what sources cluster and which show high redundancy in content among each other. This will allow us to judge on a possible redundancy in traditional media and point to the need for either reduction in market inefficiencies or the need for diversification strategies of traditional media.

3.2 Empirical Data for Media

Complete cross-language comparison is difficult: to judge whether two sources in two different languages report on the same topics, one would need automatic translation methods—a still not sufficiently solved puzzle and therefore beyond the scope of our study. Therefore, we need to focus on media published in a shared language.

There are two cultural zones (German & English speaking countries), which would fulfill the requirements for our study:

1. a common language shared by several countries,
2. sufficiently many media not state-controlled,

[1] Triangle inequality: a generic concept in mathematics to account for consistency of any "distance measure".

3. sufficiently technologically advanced to have those media publishing their content on the Internet, and
4. a large enough population to find a diverse "online community" with blogs etc.

Now, in Germany there is presently a large pressure on legislators to modify copyright and anti-trust laws so as to allow media companies to jointly charge Internet customers for perceived benefits of their news on Internet platforms.

The ongoing discussion is an obvious effect of media convergence and its implications on business models for media conglomerates (Lehtisaari et al., 2012; Picard et al., 2006; van Tuyll, 2010). In the current discussion, the suggested approaches list among others the lifting citation exemptions and the need of copyright protection for text fragments down to individual words (Beckedahl, 2010)—this claim justifies the BOW approximation additionally.

The exchange of arguments in this discussion is based on "anecdotal evidence" at best, but always involves arguments that traditional media have added value in comparison to new, pure online media. Therefore, there is a pressing need to have some objective, quantitative results that either support or falsify those claims. We have chosen German speaking media, as this approach allows for cross-country (Germany, Switzerland, Austria) analysis and the results might be very important for the ongoing discussion.

We have, however, not focused on the WWW alone, but rather used two distinct sources of textual information transferred to the end-consumer via two different channels:

1. *WWW websites:* We have chosen some 40 web presences of media outlets, popular blogs, and tv stations, as well as one news agency. We started the daily (noon) data acquisition on January, 5th 2010 until June, 13th 2010, resulting in more than 6100 entries in our database comprising the "headlines" judged important by the respective content providers.
2. *RSS feeds:* We periodically (every half-hour) stored texts of RSS feeds (Rich Site Summary) from September, 29th 2010 until June, 27th 2012 (637 days). This database consists of 1,042,774 entries overall. Unfortunately, not every website from above provides an RSS feed and not every site with regular RSS updates provides the same information on its (main) website. Therefore, the set of media in the two data sources (WWW and RSS) have only partial overlap.

It is noteworthy, that although the number of RSS entries is much larger by two orders of magnitude than for the WWW, the overall *textual information* is roughly the same, due to the fact, that each individual news makes up one RSS entry, while one WWW data point represents *all* the texts and news deemed relevant at a certain point of time.

4 Results for Empirical Data

4.1 Shannon Entropy

First, we computed the Shannon Entropy on word stems in the WWW-BOW as described in Eq. (1). In Fig. 2 we show the Shannon entropy averaged over the full time period of data acquisition. Obviously, the various sources differ in their overall entropy.

Clearly, there is no pattern or systematic difference between pure online media and "traditional media" with respect to the "disorder" of the content at the level of a BOW. This suggests that the overall variability of new media vs. traditional media is not a global property, but rather depends on the individual media outlet, blog, or newspaper. This is a first hint on content diversification effects and a potential enrichment of the media landscape by new technologies.

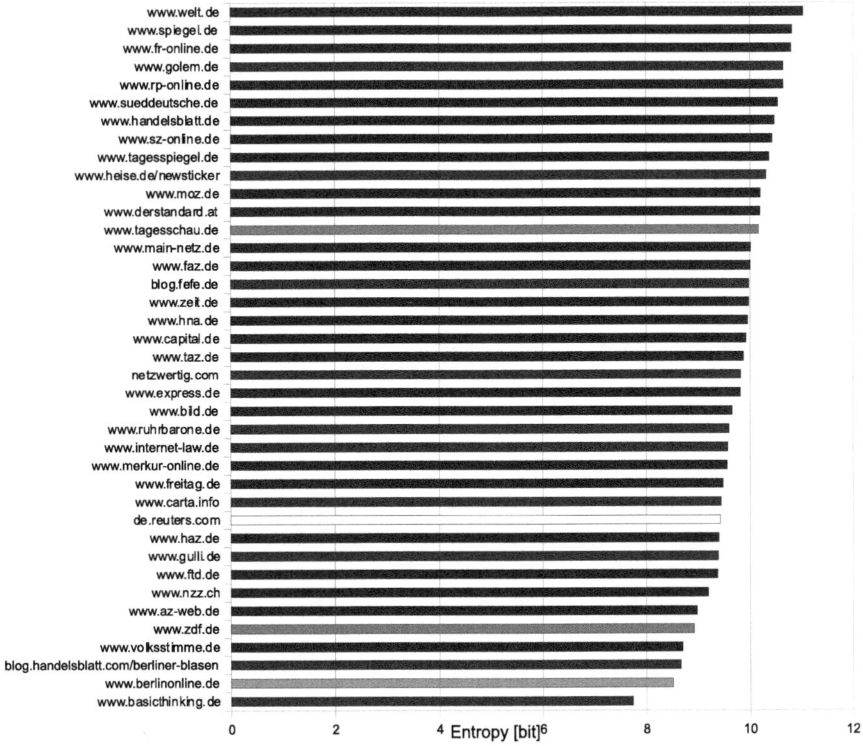

Fig. 2 The average Shannon entropies of all WWW sources of this work. For optical reasons we have sorted them in descending order. *Blue*: online presence of traditional newspaper; *white*: news agency; *green*: tv station; *orange*: city portal with local information; *red*: pure online media, such as blogs and newstickers

4.2 Information Theoretic Dissection

4.2.1 Results for the WWW

Traditional media put obviously a lot of resources into building an information structure, resulting in an ever changing information content. In a competitive environment this can only lead to diversification to obtain a competitive advantage. The differentiation with respect to competitors can be achieved in several dimensions: speed of publication; scope of information reported; brand specific, but not information related details (such as color pictures vs. gray etc.).

For the concepts reported in the sources on a given day we can apply the clustering algorithm proposed in Sect. 3.1.3 and find then—again as laid out in Sect. 3.1.3—an "average" clustering of sources. Neighborhood of sources in such a dissection tree indicates then non-differentiation in the information domain.

In Fig. 3 we show the results obtained by our relative entropy-based protocol for the WWW. Clearly, the results agree with expectations, e.g. tabloids are close to each other; sources that focus on the economy (Capital and Handelsblatt) are related; sources with both politics and business (FAZ and FTD) are related, too.

By including two media from Austria and Switzerland we are able to get a first sensitivity analysis: "Der Standard" and "Neue Zürcher Zeitung" (NZZ) are located in very distinct places in the tree.

This suggests that local news are not too important for the classification. This hypothesis is further supported by the fact that typical local newspapers (SZ, HNA, HAZ, MOZ, AZ) are also distributed throughout the tree and do not a form a sub-cluster to distinguish them from pure nationwide sources such as TAZ, Zeit, Spiegel, or Süddeutsche. The same argument can be made based on the city portal for Berlin with local event news and similar content.

The position of the news agency (Reuters) is interesting, too. Some of the other sources (indicated by a bracket in Fig. 3) are close to this news agency, implying that those sources convey to a large extent the same words and therefore news as the news agency itself. This is further confirmed by the observation that branch lengths in this cluster are smaller than anywhere else, eventually indicating the details *inside* the cluster to be rather variable, but the sources in this particular cluster to be *relative to other* major clusters more distinct.

The online presences of the two TV stations are far away along the dissection tree and therefore different in their content. Interestingly, though, that the most popular news show "Tagesschau" is very close to the most popular news website of the weekly "Der Spiegel". Based on our analysis these two sources report more or less the same concepts and focus on the same issues.

The most striking feature is the grouping of pure online media into one cluster. Although focusing on different issues, they are nevertheless more close to each other than to any traditional media presence. Therefore, there is a distinct quality of new, pure online media. This indicates a higher diversification of those from the traditional ones than among the traditional media themselves—under direct comparison of their respective BOW.

The only two exceptions to this rule are Carta.Info and Golem.De, which come close to traditional newspapers. These two sources deal with the economy of new

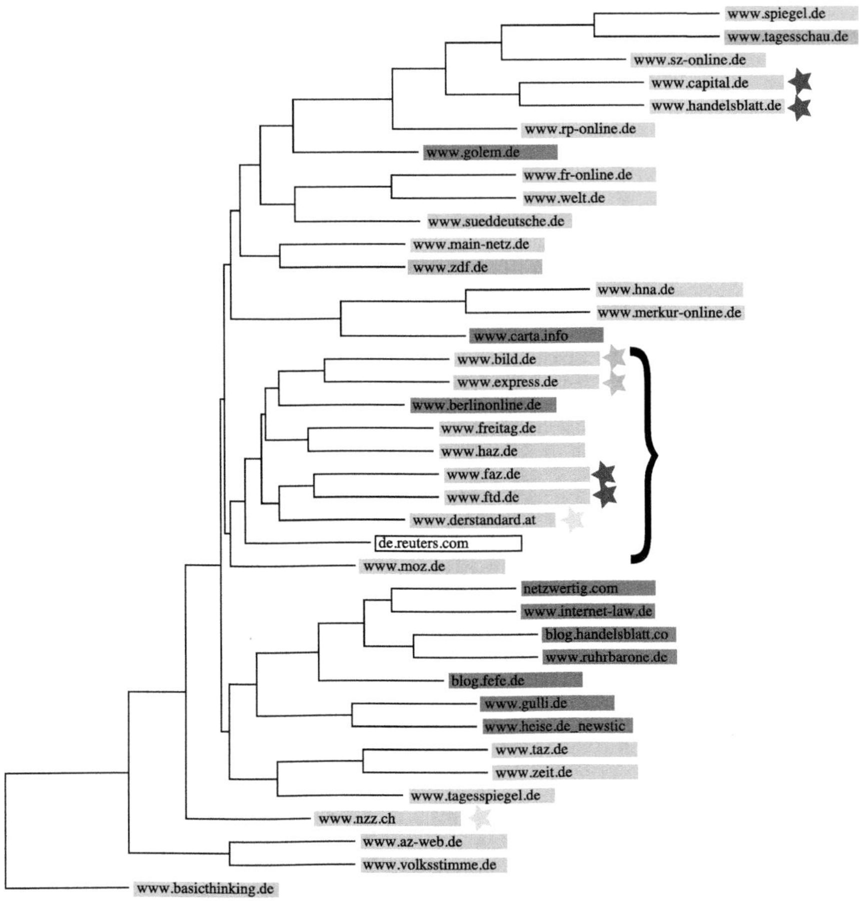

Fig. 3 The consensus tree of sources decomposed by their daily information content differences as measured by the Jensen-Shannon divergence of Eq. (2) for the last 91 days of the period of analysis. *Color* coding of sources as in Fig. 2. *Green stars* indicate tabloids; *blue stars*: exclusive business related magazines (*top* two) or strongly on business focusing newspapers (*bottom* two); *yellow stars*: the two foreign newspapers with German as a publishing language. Note that this tree is unrooted, thus there are no "ancestors" or no temporal order of news, etc. is suggested. Lengths of the branches indicate statistical support for the particular (sub) clustering

media (Carta) and the IT business (Golem); at present, the reason for their separation remains unknown due to the restricted statistics we obtained.

4.2.2 Results for RSS

Now, the data in Sect. 4.2.1 focus on the home- or entry-page of the respective media—thus, on what the editor(s) deem of particular interest for their respective readership and visitors.

The RSS data, however, is much more broad as it contains *all* the information that a source distributes without any priorization as is implicitly achieved by putting content on the entry-page in the WWW. Therefore, a distinct pattern or clustering might arise.

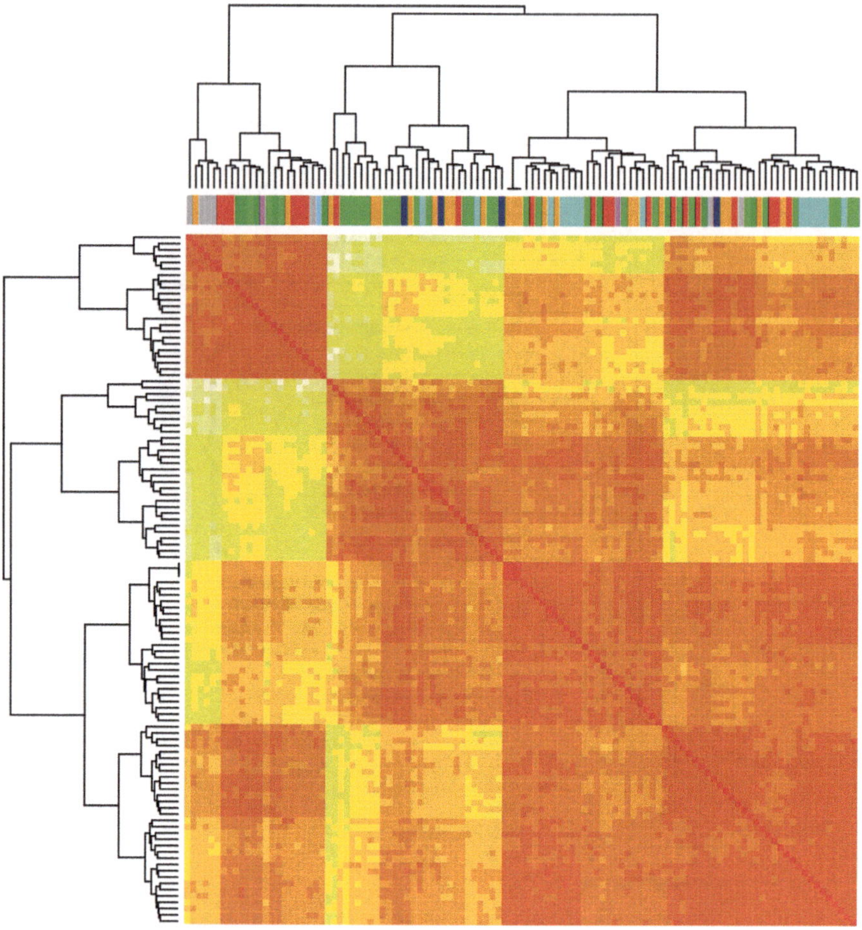

Fig. 4 The Jensen-Shannon distances between all 114 RSS providers for the full time period (*centered, colored box*; *red*: small values, *bright yellow*: high values) and the clustering (*left* and *top*). On the *top* we also show a *color-bar* that indicates classes of sources—in different shades we colored the classification of the sources; *green*: newspaper content, *cyan*: newspaper politics, *orange*: online-only news-services, *red*: blogs and online-only news-services, *gray*: parties represented in the German parliament, *blue*: TV

First, in Fig. 4 we show the results for the full time period. The first observation is the large cluster (cyan in Fig. 4) that are RSS feeds on politics from newspapers only. This results suggests, that the reporting on political events is more or less identical[2] among the online presence of print media.

[2] We could not get data from news-agencies via RSS to investigate whether the finding is caused by a direct routing of news agency data by newspapers—a hypothesis that, however, seems to be a reasonable one.

Secondly, the impact of political parties and the dissemination of their content seems to get more traction on online-only media. This might indicate an editorial role of traditional newspapers while online sources just repeat the press releases of parties.

In addition, online-only news-services mix well with the politics and news RSS feeds of traditional media. The content provided by TV mixes in between—depending on the particular area (e.g., economy, computer related news, politics) and always lies close to newspaper in the respective area.

To investigate the time evolution of the divergence or convergence of content, we split the RSS data in four consecutive time intervals and repeated the clustering procedure from above.

We show the results in Fig. 5. We first observe, that in each period distinct patterns emerge (as is evident from the red-yellow colored distance matrices). Some

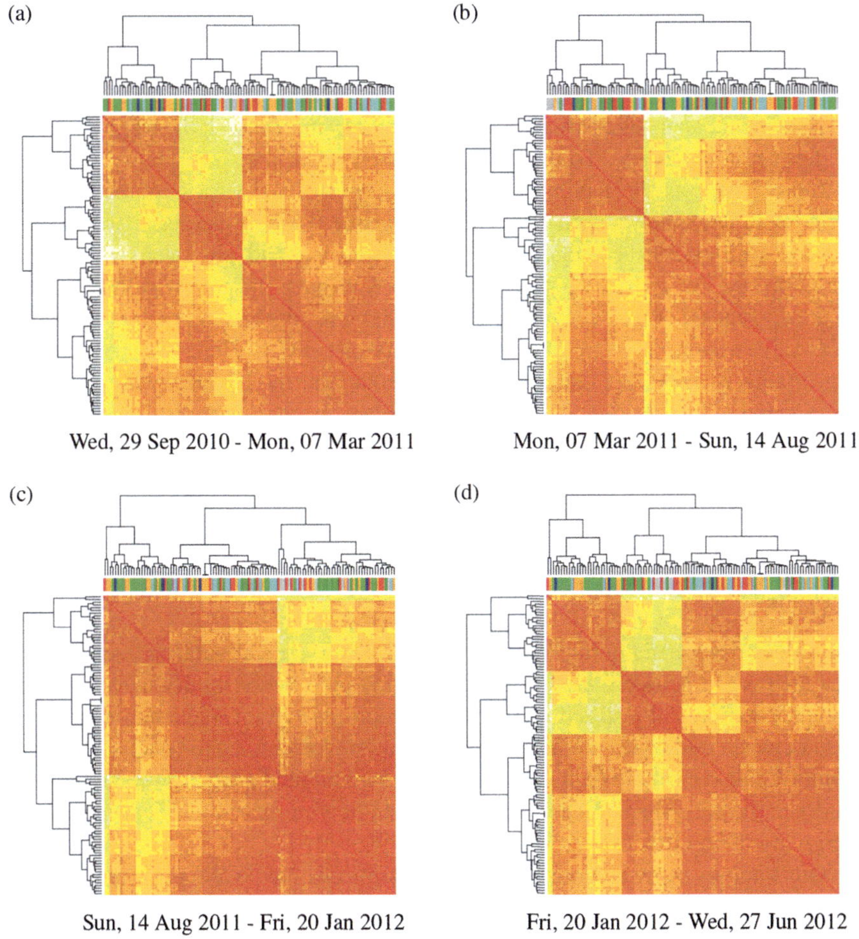

Fig. 5 The Jensen-Shannon distances between RSS feeds for consecutive time periods. *Coloring* as in Fig. 4

of the findings from Fig. 4 are still observable: party feeds are close to blogs etc., newspapers and their politics sections cluster together, only-online news corresponds to respective section in newspapers, TV information follows the lead of other media.

Although, the pattern evolves over time, these distinct features remain a constant and show—content-wise—a rather stable picture and not neither divergence or convergence. Rather a multitude of information is transferred by an information ecosystem, that—nevertheless—is more diversified by the *existence* of online media alone, rather than by any noticeable diversification process.

5 Conclusion

In this work, we have introduced a new way to apply entropic concepts to media analysis. To this end we have used the well understood bag-of-words approach, which is justified here, as we contribute to an ongoing debate that tries to atomize intellectual property down to the level of words.

As a media analysis approach we introduced an information theoretic dissection of the media landscape. To this end, we argued about the applicability of relative entropies, such as the Kullback-Leibler divergence, as metrics. A mathematical sound solution is the Jensen-Shannon divergence. Using this metric we can then employ well-established algorithms from clustering and phylogeny to obtain a "relation tree" of media. This tree for the WWW gave insight into close relations of the content reported on several websites. Some results were in perfect agreement with naive assumptions, such as the proximity of tabloids. Here, the most interesting finding is, however, that pure online media tend to cluster together, although their scope is rather wide. Nevertheless they are more or less distinct from traditional media. Thus, the editorial decision about the *entry-pages* seem to be unique to these online media.

5.1 Viewpoint on Convergence

The above described finding is open to interpretation: in particular, our results show that claims of traditional publishing houses that pure online media "steal" intellectual property are not justified, as the content reported therein is much more different than between traditional media. Instead, one might be tempted to conclude that pure online media contribute unique content to the overall media landscape and thus help to diversify journalism. Thus, we found no indication that traditional media converge to online ones, or vice versa.

Our findings show unambiguously that traditional media are much more related to each other (in a quantitative sense). Therefore one is tempted to question the existence of such high redundancy in the traditional media-sphere itself, rather than to demand compensation from entities, such as blogs, that can be shown to be rather distinct. Instead our results point to the fact that (some) traditional media fail to diversify and provide unique offers to consumers.

For the more automatized RSS feeds we find also a small diversification (in politics) for print media. The distinction between online vs. print media/TV and the respective sections remained constant over time. During the observed time period we cannot find divergence or convergence—however, taken together with the diverse picture for the WWW above, there is definitely a content-based divergence as online-only media came into existence. But whenever they are established they seem to remain as far apart from existing media as from their first day.

Obvious extensions of this study consist of a) adding semantic understanding of sentence structure and b) cross-language results, which would also allow comparison and clusterings of different cultural spheres on their "proximity" in an information theoretic sense. Due to the tremendous linguistic challenges in both regards, these aspects are beyond the scope of this study. Such increased size of data would also allow a full statistical analysis on statistical significance e.g. in the dissection trees.

For an application of the results: using the proposed methodology media outlets can try to optimize their portfolio a) by adjusting to the offerings of competitors, b) by creating unique topics and sub-areas by algorithms, and c) by analyzing the content of news agencies. As such optimization needs to be done in a high-dimensional space, naive optimization that succeeds in low dimensional cases, will almost certainly fail (Hamacher, 2005), while procedures inspired by theoretical physics and computer science (Hamacher, 2006, 2007a, 2007b) can help to explore the high-dimensional, non-convex distance functions dealt with in e.g. clustering most effectively.

Acknowledgments The author would like to thank Franziska Hoffgaard and Philipp Weil for helpful comments and discussions. KH was supported by the Fonds der chemischen Industrie through a grant for junior faculty during this study.

References

Bachmayer, S., Lugmayr, A., & Kotsis, G. (2010). Convergence of collaborative web approaches and interactive tv program formats. *International Journal of Web Information Systems, 6*, 74–94. doi:10.1108/17440081011034493.

Beckedahl, M. (2010). *Verleger: Leistungsschutzrecht soll sprache monopolisieren*. http://www.netzpolitik.org/2010/verleger-leistungsschutzrecht-soll-sprache-monopolisieren

Behrendt, J., & Zeppenfeld, K. (2008). *Web 2.0* (Informatik im Fokus). Auflage: Springer.

Bird, S. (2002). *Nltk: The natural language toolkit* (pp. 62–69). Proceedings of the ACL Workshop on Effective Tools and Methodologies for Teaching Natural Language Processing and Computational Linguistics, Association for Computational Linguistics, Philadelphia.

Blahut, R. E. (1987). *Principles and practice of information theory*. Reading, MA: Addison-Wesley. ISBN 0201107090.

Boba, P., Weil, P., Hoffgaard, F., & Hamacher, K. (2011). Intra- and inter-molecular co-evolution: the case of HIV1 protease and reverse transcriptase. In A. Fred, J. Filipe, & H. Gamboa (Eds.),

Proceeding of BIOINFORMATICS2010 (Vol. 127, pp. 356–366). Springer Communications in Computer and Information Science.

Cover, T. M., & Thomas, J. A. (2006). *Elements of information theory* (2nd ed.). Hoboken, NJ: Wiley.

Endres, D., & Schindelin, J. (2003). A new metric for probability distributions. *IEEE Trans Info Theo, 49*, 1858–1860.

Felsenstein, J. (1989). PHYLIP – phylogeny inference package (version 3.2). *Cladistics, 5*, 164–166.

Hamacher, K. (2005). On stochastic global optimization of one-dimensional functions. *Physica A, 354*, 547–557.

Hamacher, K. (2006). Adaptation in stochastic tunneling global optimization of complex potential energy landscapes. *Europhys Lett, 74*(6), 944–950.

Hamacher, K. (2007a). Adaptive extremal optimization by detrended fluctuation analysis. *J Comp Phys, 227*(2), 1500–1509.

Hamacher, K. (2007b). Energy landscape paving as a perfect optimization approach under detrended fluctuation analysis. *Physica A, 378*(2), 307–314.

Hamacher, K. (2007c). Information theoretical measures to analyze trajectories in rational molecular design. *J Comp Chem, 28*(16), 2576–2580.

Hamacher, K. (2010). Protein domain phylogenies—Information theory and evolutionary dynamics. In: A. Fred, J. Filipe, & H. Gamboa (Eds.), *Proceeding of BIOINFORMATICS* 2010 (pp. 114–122). ISBN 978-989-674-019-1.

Harris, Z. (1954). Distributional structure. *Word, 10*, 146–162.

Lehtisaari, K., Karppinen, K., Harjuniemi, T., Grnlund, M., Linden, C. G., Nieminen, H., et al. (2012). *Media convergence and business models: responses of finnish daily newspapers*. ISBN 978-952-10-8277-1.

Lugmayr, A., Franssila, H., Safran, C., & Hammouda, I. (Eds.) (2011). *Proceedings of the 15th international academic MindTrek conference: Envisioning future media environments*. Tampere, Finland: MindTrek Ry.

Manning, C., & Schütze, H. (1999). *Foundations of statistical natural language processing*. Cambridge, MA: MIT Press.

Murtagh, F. (1985). Multidimensional clustering algorithms. In *COMPSTAT Lectures 4*. Würzburg: Physica-Verlag.

Pape, S., Hoffgaard, F., & Hamacher, K. (2010). Distance-dependent classification of amino acids by information theory. *Proteins: Structure, Function, and Bioinformatics, 78*(10), 2322–2328.

Picard, R. G., Dal Zotto, C., & Werfel, M. (2006). *Business models of newspaper publishing companies*. 1, Ifra Report.

Porter, M. (1980). An algorithm for suffix stripping. *Program, 14*, 130–137.

Porter, M. (2006). *Stemming algorithms for various European languages*. http://www.snowball.tartarus.org/texts/stemmersoverview.html

R Development Core Team. (2008). *R: A language and environment for statistical computing*. Vienna: R Foundation for Statistical Computing. ISBN 3-900051-07-0, http://www.R-project.org

Shannon, C. E. (1948). A mathematical theory of communication. *Bell System Technical Journal, 27*(3), 379–423.

van Tuyll, D. (2010). The past is prologue, or. *Journalism Studies, 11*(4), 477–486. doi:10.1080/14616701003638343.

Zotto, C. D., Galichet, G., & Lugmayr, A. (2011). Managing media convergence: An economic and editorial project. In *European Media Management Conference 2011*. Moscow: European Media Management Association (EMMA).

Towards Intelligent Cross-Media Publishing: Media Practices and Technology Convergence Perspectives

Andreas Veglis, Charalampos Dimoulas, and George Kalliris

1 Introduction and Problem Discussion

In the last 30 years we have witnessed a tremendous evolution in Information and Communication Technologies (ICTs), with Internet and its services being a major force in this growth. These changes have resulted in the digitization of information and have transformed every human activity. Digitization of production in media organizations has facilitated changes in the organization and practices of journalism (Erdal, 2007). These changes involved also production work and resulted in new ways that news is produced, but also fundamental changes in the final products and their dissemination to the end users (Dal Zotto, 2005; Veglis, 2012).

The digitization of production systems enabled content to travel across-media boundaries. Television footage and radio soundbites can be published on the Web, and television sound is frequently used on radio (Erdal, 2007; Veglis, 2012). That resulted in fundamental changes in the workflow of media organizations (Sabelström, 2001; Veglis & Pomportsis, 2009). The news is now produced once and deployed in various formats for different publication channels, which is widely known as cross-media publishing. Thus, media companies can cover more audience needs and offer channels that complement each other. We must also take into account that trends in the worldwide media industry have clearly shown that in order to guarantee long-term success with audience in the future, it is vital to change from a single product to a multimedia, content and user-oriented approach (Dal Zotto, 2005; Dietmar, 2008).

Although the introduction of cross-media publishing has offered new opportunities to media companies to expand their audience, it created many difficulties, among which those that are related to technological issues. Each

A. Veglis (✉) • C. Dimoulas • G. Kalliris
School of Journalism and Mass Communications, Aristotle University of Thessaloniki, Thessaloniki, Greece
e-mail: veglis@jour.auth.gr

A. Lugmayr, C. Dal Zotto (eds.), *Media Convergence Handbook - Vol. 1*, Media Business and Innovation, DOI 10.1007/978-3-642-54484-2_8

publishing channel has its own, unique in many cases, characteristics, and thus requires special information structures in order for the content to be produced. Even though many publishing channels seem to share common content formats (for example WWW, SMS, and blogs all may include text), in most occasions the content must be adapted to every channel. This issue becomes even more complicated if we move to channels with sound and video, which demand higher bandwidth and bitrates, but also in most cases require re-sampling and trans-coding operations, usually along with meta-processing for easier content accessing, browsing and interfacing in general (Dimoulas, Kalliris, Chatzara, Tsipas, & Papanikolaou, 2013; Dimoulas, Veglis, & Kalliris, 2014).

This chapter discusses technological issues that have arisen in implementing cross-media publishing. During the last decade many new technologies, services and capabilities for disseminating news have appeared. The study takes into consideration media entities and structures, user-interaction modes, content types, quality and formats, content description, semantic-annotation and generally meta-information processing and management. The research question that is associated with the current problem statement is how to use all these resources in order to achieve the optimum result that better serve the needs of the media and the audience. The rest of the chapter is organized as follows. In the next section, literature state of the art on the evolution of cross-media publishing is presented, including historical data, definitions, description of channels' characteristics, technical attributes and functional particularities. Next, in the methodological and approach section content types and media models are next discussed with regard to cross-media publishing issues, both at the production and consumption ends. Related requirements and today's practices that are encountered in contemporary media enterprises are also listed, providing the necessary background for the introduction of intelligent cross-media publishing models, which is the main thrust of the chapter. Summary and conclusive remarks are provided in the last section including a view point on convergence.

2 The Evolution of Cross-Media Publishing: Literature and State of the Art

Cross-media practices did not appear out of nowhere. The early cave painting can be considered as the cross-media attempts, since it was the first endeavor to move the experience from the oral storytelling to representations on cave walls. If we move to more recent practices, we must refer to the Star Wars franchise, which is considered to be a well-conceived and implemented cross-media communication experience (Davidson et al., 2010).

The term cross-media was already used in the early days of electronic publishing. In the print world, cross-media was used to describe database publishing (ACTeN, 2004). The first publishing segment that automated cross-media was directory and reference publishing. In the 1980s, novel cross-media publishing systems were proposed that were focused on "content-driven" publications, like manuals, books, treatises, etc. (Veglis, 2012; Walter, 1999). Publishing industries have incorporated cross-media practices for a long time, mainly for marketing and market penetration (Dal Zotto, 2005). In the case of book printing and publishing, cross-media practices appeared as media that were complementary to each other (for example a book and a CD-ROM). In magazine and newspaper publishing, cross-media were used more extensively in order to bridge the time between publications by providing information on the Internet, and to build an online community. In the simplest form of cross-media, newspapers and magazines had an electronic counterpart. They did not really add to the content of the magazine or the newspaper, but reproduced the content of the magazine or the newspaper in a different channel (ACTeN, 2004; Veglis, 2012).

Cross-media can be defined as the production of any content (news, music, text, images, etc.) for more than one media platform/channel (for example print, Web, and TV) within the same media organization (Veglis, 2005). The term refers to integrated experiences across multiple media, including the Internet, video and film, broadcast and cable TV, mobile devices, DVD, print, and radio (Davidson et al., 2010). Cross-media usually involves some high level of audience interactivity, which inter-connects the audience's experience across the various channels.

Another term that is widely used is "Multiple media", which specifies the inter-platform or inter-device possibilities. The term multiple media indicates that the same content is delivered to end users in more than one medium. A medium can be defined as a mean of communication or, more precisely, a system where transmission media are using more than one type of transmission path (e.g., optical fiber, radio, and copper wire) to deliver information (Veglis, 2007).

An alternative term for multiple media is multi-channel publishing. The same content is published on various channels or media. The term multiple media is broader than cross-media, as it expands the concept from devices to content (Antikainen, Kangas, & Vainikainen, 2004).

Cross-media publishing in newspaper organizations has been investigated by some researchers (Davidson et al., 2010; Sabelström, 2000, 2001; Veglis, 2008a, 2012). However, the more specific issue of cross-media publishing in broadcast media has drawn very little attention, although it is worth mentioning that there is a small number of important contributions (Cottle & Ashton, 1999; Duhe, Mortimer, & MChow, 2004).

Cross—media publishing usually employs a significant number of channels. It is worth noting that these channels do not represent different categories of technology but simply distinctive methods for publishing news. These channels are the means for disseminating news by media companies around the world. The channels are WWW, Webcasting, Smartphones, Tablets, e-mail, SMS, PDF, RSS, Twitter, social networks and broadcasting (Veglis, 2008a, 2008b, 2012).

WWW is believed to be the most widely used Internet service. The main advantages of WWW are the transportation of information over great distances (compared to terrestrial broadcasting channels), the theoretically infinite space of information it can facilitate and the possibility of continuous updating (Negroponte, 1995; Veglis, 2012). Usually it is the first alternative publishing channel that media companies utilize.

Webcasting can be broadly defined as the delivery of audiovisual media content on the WWW (Veglis, 2007). Websites can be used for Webcasting audio and video content. This content is available on demand and it is used in order to add value to the services offered by other publishing channels of media companies.

Smartphones are high-end mobile phones that combine the functions of personal digital assistants (PDAs) and mobile phones. These devices support the functions of portable media players, cameras, high-resolution touch screen interfaces, GPS sensors and services, etc. They include Wi-Fi connection and mobile broadband access, enabling users to browse regular websites. But the majority of these devices have small screens (in comparison with the PCs) and thus web access cannot be characterized as user friendly. That is the main reason why many media companies are offering portals prepared for such mobile devices that include limited graphics (with basic navigation functions) but all the necessary text and pictures for each article, as well as tailored made applications (apps) than deliver news content adapted to smartphone user interface . This increases access speed and makes navigation more effortless for the mobile readers (Veglis, 2012).

Tablets are portable PCs that include touch screens and wireless connection to the Internet (through Wi-Fi networks and in some cases mobile phone networks). These devices offer medium size high resolution multi touch displays and an extensive storage capacity (usually several GBytes). The majority of these devices employ the same operating systems with smartphones (with the advantage of the larger screen) thus allowing them to use the previous mentioned (or even improved) apps for receiving news content from media companies (Veglis, 2012; Wearden & Fidler, 2001).

E-mail is considered to be the oldest and one of the most widely used internet services. Media companies use this channel in order to alert their audience about breaking news, relay them the headlines of the main stories (with links to the entire articles included in an online version), or send them the entire edition—in the case of a newspaper—in a PDF file (Schiff, 2003; Veglis, 2012).

Portable Document Format (PDF) is a file format that is capable of presenting documents, featuring interoperability and platform independence in terms of hardware, operating systems and used applications. PDF files are highly compressed, searchable and can include features for interactive document use. PDF is employed by many newspapers and magazines in order to deliver exact copies of their printed edition (Schiff, 2003). This was considered to be a natural choice since the PDF format in widely used in stages of the prepress process.

Short Message Service (SMS) is a service offered by network providers that allows customers to send text messages over their mobile phones. SMS is the most widely used data application in the world (3.6 billion users—78 % of the mobile phone subscribers) (Kelly, 2012). Many media companies employ SMS in order to send their target group the main headlines or to alert them about breaking news (Gillmor, 2004; Veglis, 2012).

Really Simple Syndication (RSS) is a method of describing news or other Web content that is available for feeding from an online publisher to Web users. An RSS document is called a feed and it includes a summarized text, and various metadata.

RSS feeds can be read by users with the help of web-based, desktop-based, or mobile-device-based software called RSS reader, feed reader, or aggregator. Today many media companies are employing RSS in order to alert their readers about the news headlines (Veglis, 2007). An RSS feed usually employs text and often small pictures.

A blog is a website where entries are written in chronological order and displayed in reverse chronological order. An important feature of the blogs is the ability for the users to leave comments. That is the reason why media companies (most often newspapers) have included blogs as a supplement to their web editions, thus giving their journalists the opportunity to comment onto current events and to the audience the ability to interact with them (Veglis, 2012).

Social networks are web-based services that allow individuals to construct public or semi-public profiles within a bounded system, articulate lists of other users with whom they share connections, view and traverse their connection lists and those made by others within the system (Boyd & Ellison, 2008). Many media companies have established a presence in the most popular social networks in order to publish their news articles and attract other members of the social network to their web site. They have also integrated social media links in their web articles in order for the users to link to them through their social network profiles (Veglis, 2012).

Twitter is a social networking and micro-blogging service that enables its users to send and read other users' updates known as tweets. It is worth noting that although one may argue that Twitter belongs to the Social Network category, its distinct characteristics allow us to categorize it as a separate channel. Twitter is often described as the "SMS of Internet", in that the site provides the back-end functionality to other desktop and web-based applications to send and receive short text messages, often obscuring the actual website itself. Tweets are text-based posts of up to 140 characters in length. Updates are displayed on the user's profile page and delivered to other users who have signed up to receive them. Users can send and receive updates via the Twitter website, SMS, RSS (receive only), or through applications. Many media companies are using twitter in order to alert their readers about breaking news (Veglis, 2012).

Last but not least, the oldest electronic mass communication media channel is broadcasting, whereas audio and video content is transmitted to a dispersed audience that utilizes the appropriate access terminals (i.e. radio and TV sets). The main advantage of the broadcasting channel is laid on its wide spreading in most of the social groups, the availability of massive content, media producers and broadcasters, and mostly the low cost demands of the involved equipment. Equally important is considered the fact that audiovisual information is easily and user friendly attended, offering informing and entertainment services (infotainment) that can be launched at various situations and environments (i.e. home, work-office, while travelling, etc.). On the other hand, limited interactivity and bandwidth demands are among the listed difficulties that contemporary digital broadcasting platforms attempt to overcome (Chapman & Chapman, 2009; Dimoulas et al.,

2014a; Kalliris & Dimoulas, 2009; Simpson, 2008; Vegiris, Avdelidis, Dimoulas, & Papanikolaou, 2008).

3 Methodology and Approach: Modeling Intelligent Cross-Media Publishing

3.1 Cross-Media Production and Consumption Issues: Content Types, Media Models and Publishing Channels

In general, there are quite a few classification schemes that content elements and media types can be categorized into. First of all, there are two principal models according to which elements can be combined in different media types: page-based and time-based (Chapman & Chapman, 2009). In the page-based model (PBM) text and images are combined and spatially arranged, so that their two dimensional layout resembles the way that text and images are placed in books and magazines. The time-based model (TBM) includes video clips, sounds, animations and general audiovisual time-sequences that resemble the natural way that human beings are used to communicate with each other (Chapman & Chapman, 2009; Kalliris & Dimoulas, 2009). In this context, TBM offer the advantage of easy attendance with archetypal and vivid presentation of information that is very familiar to the modes that people have learned to be taught, to be informed or even to discuss and debate (Dimoulas, Tsourvakas, Kalliris, & Papakis, 2012; Dimoulas et al. 2014a; Kalliris & Dimoulas, 2009). This also explains the fact that TBMs dominated the early days of analog mass media broadcasting (Chapman & Chapman, 2009; Dimoulas et al., 2012, 2014a; Kalliris & Dimoulas, 2009). On the other hand, audiovisual (AV) information that is involved on TBM cross-media publishing has additional storage and broadcasting bandwidth demands that creates further processing and compression necessities. Considering the high paces that digital AV content is produced and the expansion of the corresponding AV data-bases, proper content description and meta-information management is required in order for the TBM media to be easily accessed, browsed, searched and retrieved (Dimoulas et al., 2012, 2014a; Kalliris & Dimoulas, 2009; Kotsakis, Kalliris, & Dimoulas, 2012).

Extending the above, hybrid models combining both PBM and TBM elements are regularly implemented in digital multimedia broadcasting and in particular in cross-media publishing (Kalliris, Dimoulas, Veglis, & Matsiola, 2011; Kalliris & Dimoulas, 2009; Vegiris et al., 2008). These Multimodal Media Assets (MMA) may contain both the multimedia content (PBM, TBM and their combinations) but also related meta-data, including authoring and semantic annotation, content organization and interactivity structures. In fact, all the publishing channels consist of the above content elements that can be categorized into static and dynamic elements. Textual matter, still images and graphics are usually considered to be static content elements that can be created and edited independently of each other and later compiled and logically connected in an article. Nevertheless, dynamic elements can be also incorporated in these PBM media (i.e. dynamic web-pages) offering the ability of dynamically created content that is driven from user interaction (Chapman & Chapman, 2009; Dimoulas et al., 2014a, 2014b). Similarly, video and sound are usually characterized as dynamic content elements that are sequentially built up, even in cases that static content browsing and reproduction is involved (Chapman & Chapman, 2009). However, the true dynamic TBM and MMA media offer the ability of AV authoring with enhanced user interaction, possibly incorporating semantic ambient media experience contexts in augmented reality and interaction environments (Domingue, Fensel, & Hendler, 2011; Kotsakis et al., 2012; Lugmayr, Stockleben, Risse, Kaario, & Pogorelc, 2013; Veglis, 2012). In any case, the majority of publishing channels employ a combination of all the previously mentioned content elements (Sabelström, 1998; Veglis, 2012). Table 1 presents an overview of the content elements and their classification attributes that are incorporated in each channel.

Based on the previous categorization, we can distinguish the publishing channels into the ones that are considered to be static (E-mail, SMS PDF, RSS, and Twitter), and those that can include both static and dynamic content (WWW, Blogs, Webcasting, Smartphone, Social Networks, Tablets, and Broadcasting). Similarly, channels can be categorized according to the involved media models (TBM, PBM and MMA). It is worth noting that preparing content for a dynamic channel is more time consuming than for a static channel. That is why the fastest channels are always static. It is also worth mentioning that considerable overlapping normally exists in such categorization of content types, media models and publishing channels that is presented in Table 1. For instance, e-mail is not considered as an applicable channel for audio and video distribution, however, low size audiovisual content can be conditionally attached to e-mail, issue that is indicated with the "~" symbol in Table 1. A similar case applies for radio and TV broadcasting channels (analog and especially digital), where textual information is incorporated (i.e. Radio Data System—RDS, teletext, Electronic Program Guides—EPG, etc.), while user interaction can be enhanced via dynamic feedback channels.

One other parameter that must been taken into account is time. The content of each channel requires different time spent in order to be produced. The first channels that relay the headlines news to the readers are SMS, Twitter and RSS. Also short announcements in radio or running text headlines in TV can be facilitated by the broadcasting channels. These channels can be characterized as info-alerts (Antikainen et al., 2004). The idea of the info-alerts is to make the audience aware of content available in different publishing channels. The RSS feeds and tweets link directly to the media company's web site and the SMSs tempt the receiver to seek another publication channel in order to obtain more information. The next set of channels involves e-mails and WWW (running headlines). E-mails can also be categorized as info-alerts since they can be received by mobile devices (Smartphones) or they can attract readers' attention when special software is employed (e-mail notifiers). In the case of WWW we refer only to running headlines that just announce the news, without giving more details. Next short story descriptions are available via voice or video webcasting. This is the equivalent

Table 1 Content types, media models and channels' attributes

Channel	Content				Model			Interaction	
	Text	Picture	Sound	Video	PBM	TBM	MMA	Static	Dynamic
WWW	✓	✓	✓	✓	✓		✓	✓	✓
Blogs	✓	✓	✓	✓	✓		✓	✓	✓
Webcasting	✓	✓	✓	✓		✓	✓	✓	✓
Smartphone	✓	✓	✓	✓[a]	✓		✓	✓	✓
Tablets	✓	✓	✓	✓[a]	✓		✓	✓	✓
E-mail	✓	✓	~	~	✓			✓	
SMS	✓				✓			✓	
PDF	✓	✓			✓			✓	
RSS	✓	✓			✓			✓	
Twitter	✓				✓			✓	
Social networks	✓	✓	✓	✓	✓		✓	✓	✓
Broadcasting	~		✓	✓		✓	✓	✓	~

[a]Probably low quality video

of the voice or video correspondence of the radio and TV channels, which still remain an alternative broadcasting channel (Veglis, 2012).

Subsequent to info-alerts the full story is initially available in webpages (WWW for PCs and Tablets) and later in the form of PDF files sent via e-mails. Usually this story is edited in more detail and presents the facts with more accuracy. This is because there is more time to prepare this content. Webpages can include both static and dynamic elements. In the same category we have included blogs. Blogs are usually updated later than the WWW edition of the newspaper, since they include journalists' comments on the main news.

As far as the broadcast channel is concerned the story is included in the news bulletins that are scheduled to air at specific time periods within the day. The depth of coverage depends on the importance of the news story but also on the predefined time length of the news bulletin. In many cases if the story is very important special live coverage may be provided or shows dedicated to the story at specific time periods. In the later case there is adequate time to prepare the detail presentation of the story, to present the facts with additional accuracy, but also enhance the story with additional multimedia features.

The rhythm of the publishing channels differs and it depends on the type of information they include (Sabelström, 2000). The lead-time from the collection of news by a reporter until the point in time when the audience is able to receive information varies between channels. A reporter writing for the print edition of a newspaper often has several hours to do research and write the report, whereas the same reporter has less than half an hour to write a news item for the WWW. On the other hand, webcasting and broadcasting publish updated news bulletins on schedule several times a day; that is to say the time for research and content creation is shorter than for the printed daily newspaper, which is published only once a day. The same thing stands for all other publishing channels, always depending on the publication schedule of each channel (Veglis, 2012).

Extending the above analysis, some content elements seem very favorable in some channels, while they seem completely useless in some others, requiring specific adaptation treatment in order to become usable. For instance, High Definition (HD) audio and video content is very useful in related broadband digital broadcasting or webcasting channels, while they cannot be regularly incorporated in most of the remaining channels. Hence, AV transcoding might be needed in channels that utilize AV content but with bandwidth, resolution and generally computational resources limitations (i.e. e-mail, Web, smart-phones, tablets, etc.), while textual transcription and abstraction via summarization and highlighting techniques is required in text-oriented channels (i.e. SMS, RSS, blogs, PDF, etc.). Similarly, elongated text is unfavorable in mobile terminals (smart phones, tablets) and short messaging techniques (i.e. SMS, twitter, etc.), so that text classification—visualization and cloud-tag approaches might be quite useful (Dimoulas et al., 2014a; Papapetrou, Papadakis, Ioannou, & Skoutas, 2010; Raj & Mala, 2012).

In fact, any full media story (news, infotainment, entertainment, etc.) that is about to be published is initially prepared for a main target channel, issue that is principally related with the kind of the involved content, its thematic classification,

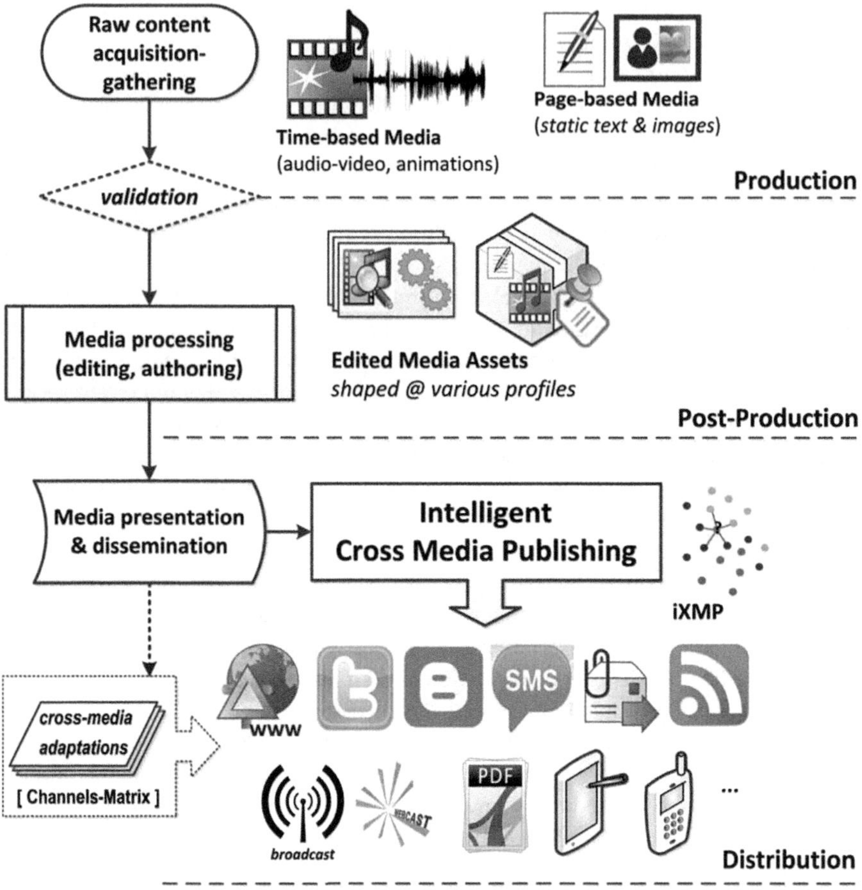

Fig. 1 Information flow-diagram modeling: production, post-production and cross-media channel distribution

the targeted users, the type of the publishing media organization and many other aspects. Then, cross-media publishing examines the suitability of the remaining channels to be exploited along with the necessary content transcoding and adaptation that has to be done. Figure 1, presents the basic information production, post-production and distribution flow diagram (Veglis & Pomportsis, 2014; Dimoulas et al., 2014b), taking into consideration cross-media publishing aspects. It is obvious that multiple time and processing consuming adaptations is needed proportionally to the number (N) of the involved cross-media channels. The current chapter attempts to provide an intelligent cross-media publishing (iXMP) model by clustering content and channel similarities into different profiles. This profiling allows for clustered content and its meta-information to be effectively combined and exploited according to the specifications and the demands of all the users, the content and the involved channels, thus minimizing the necessary adaptation time and processing load.

3.2 Cross-Media Publishing Requirements and Practices in Todays' Media Enterprises

Following Fig. 1, it is clear that the main phases of media production, post-production and distribution remain in the case of the cross-media publishing, as it happens with typical single-channel media communication models. Hence, regardless the involved cross-channels and the primary channel that each story is prepared for, the process involves the same basic production phases whereas content entities have to be created, processed (and meta-processed), edited-assembled and delivered. However, the main channel plays an important role regarding the most likely content types and media models that have to be utilized, along with the necessary equipment and the human personnel to operate them. In addition, as it will be presented and analyzed later on, the main channel affects the way and the form of the content resources and services that are delivered in the remaining cross-media channels. Figure 2 presents an end-to-end content production and distribution flow diagram that incorporates contemporary trends and services, including the intelligent cross-media publishing concept that is discussed in the current chapter.

According to Fig. 2, there are three main modes that media content is initially produced: (a) inside dedicated studios, (b) in outdoor recordings (live correspondences, reportages, etc.), and (c) via Web 2.0 services and user generated content (UGC) models (Kalliris & Dimoulas, 2009; Dimoulas et al., 2012, 2014a; Kotsakis et al., 2012). The degree that each one of the above models is participating and contributing in a specific media coverage scenario is primarily related to the nature of the story itself, the main targeted channel and the involved media organization(s). Hence, along with the evolution of mass media (both press and electronic ones) and prior to the full scale launching of cross-media publishing, each media organization adopted channel and media-type specialization. Although most contemporary media companies currently deploy integrated cross-media publishing services, they still retain a channel-adapted expertise and focusing (Dal Zotto, 2005; Spyridou, Matsiola, Veglis, Kalliris, & Dimoulas, 2013). This is closely related to the associated professional expertise, the corresponding crewing and their knowhow, the available equipment and infrastructures in general. In addition, target group focusing, heritage and brand management considerations seem to play an important role to the main media type and channel selection preferences and concentration that modern media organizations still exhibit (Dal Zotto, 2005; Lowe, 2011). Normally, each media company has at her disposal different studios and equipment that better serve its specialization towards the preparation of the main Channel Adapted Media (CAM) products, at all phases.

Following initial production, content entities are normally combined and concatenated via multimodal processing and editing procedures (Fig. 2). The nature of the covered story also plays a very important role here. For instance, there is need

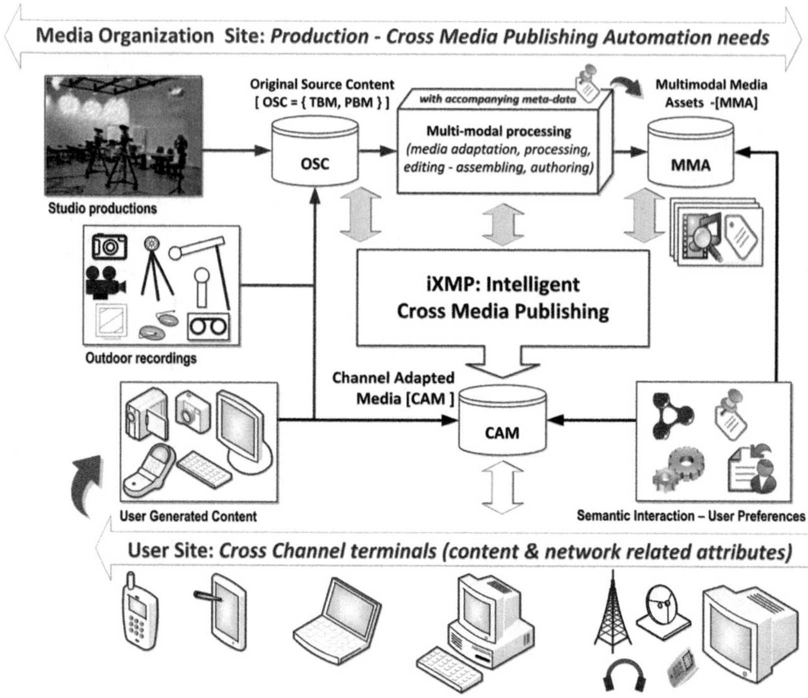

Fig. 2 End-to-end content production—distribution flow incorporating intelligent cross-media publishing

to be different treatment considering content quality and editing formalisms in a breaking news story whereas unedited shots are favorable, in contrast to documentary content and shows. In addition, use of meta-data along with dedicated meta-information management is considered as very important towards efficient content description, searching and retrieval, especially in audiovisual TBM and MMA types. Seeing their usefulness in the preparation of the delivered CAM products, meta-data may include user related feedback and preferences that are associated with content access and browsing, so that they can be very constructive and favorable in cross-media publishing preferences. In any case, media adaptation is necessary in order to be able to incorporate most of the enclosed information of the original CAM_0 to all the remaining CAM_i assets that are associated with the involved—available publishing channels.

Extending the above analysis, the simplest scenario that is mostly applied in today's cross-media publishing practices is to start with the content-preparation of the original channel (CAM_0) and then proceed with the successive extraction of the remaining CAM_i profiles, following the importance and the production time of the involved channels. Hence, simple TBM transcoding (i.e. re-sampling updates, resolution and dynamic range treatment, use of different compression—encoding, etc.) and PBM shaping/re-arranging might represent the easy case, while entirely new content processing-editing, assembling and authoring might be needed in more demanding adaptations. These adaptation processes are usually supervised and manually controlled, while semi-automated operations might also be encountered (especially in cases that fixed publishing channels are predetermined). As Fig. 2 implies, new CAM_i products can be extracted by applying adaptations to all the available CAM_j ($j = 0, 1, \ldots i-1$) assets, but also to the initial Original Source Contents (OSC) and their multi-modal processed versions (MMA), if they are available. In this context, quality losses might be introduced, issue that may be also related to the fact that still cross-channel releases are usually treated as inferior versions—publications of the original channel.

3.3 Intelligent Content Documentation, Selection and Management

As already stated and it is also implied in Figs. 1 and 2, standard cross-media publishing approaches usually require multiple content adaptation and processes according to the number of the involved N channels and the corresponding CAM, OSC and MMA content versions. Figure 3 presents a block diagram of the proposed iXMP model, whereas pruning of unnecessary content adaptation connections-routes is targeted, inspired by neurons-pruning in artificial neural networks (Urolagin, Prema, & Reddy, 2012) and/or path-pruning in multi-hop wireless broadcasting (Rieck & Dhar, 2010). Hence, the main objective of the proposed iXMP concept is to reduce the processing time and the necessary adaptations, while maintaining acceptable content and publishing quality in accordance to the user-terminals and the network-related attributes. In this context, a variety of content-based, user-adapted and channel-related parameters are considered for system configuration, thus offering modular cross-media publishing intelligence (Fig. 3).

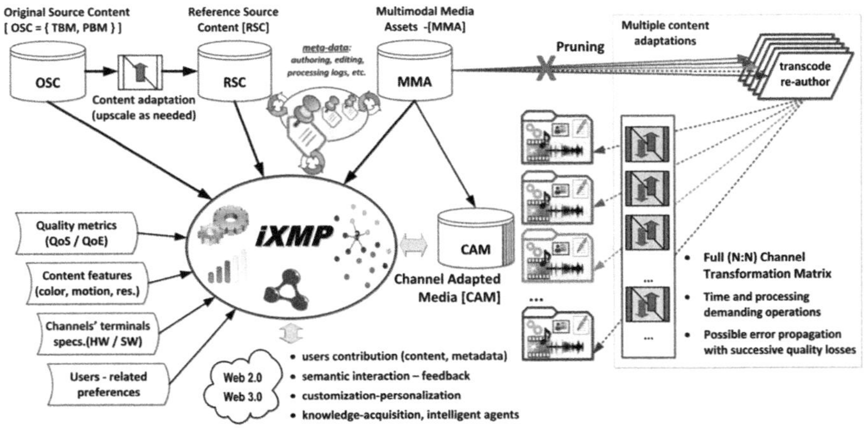

Fig. 3 The proposed intelligent cross-media publishing (iXMP) model

Specifically, the goal of the proposed approach is to take full advantage of all these input parameters, so that the various versions of the involved content (OSC, CAM, MMA, etc.) could be self-organized in similarity clustered profiles. This content profiling would be very helpful for easier adaptation and specific channel extraction during cross-media publishing. Therefore, agent intelligence, data mining and upcoming Web 3.0 semantic interaction technologies are put into the question of intelligent cross media publishing (Dimoulas et al., 2014a, 2014b; Domingue et al., 2011; Kotsakis et al., 2012; Symeonidis & Mitkas, 2005).

Based on the previous discussions, cross-media publishing requires multiple processes of content transcoding and adaptation. In this context, successive error propagation might lead to severe quality losses. This unpleasant continuous quality degradation effect should be avoided, especially in cases that higher resolution versions need to be extracted or rather estimated from sources that naturally exhibit inferior quality (e.g. smartphone audiovisual captures have to be inserted in high definition broadcasting and webcasting services). In any case, source content should be carefully selected during cross-channel adaptation and transcoding, not only for quality maintenance issues, but also in order to lighten up the associated computational load, or even to fully avoid unnecessary transformations. On the other hand, use of highest quality encoding and overdue digitization parameters may also have similar negative impact on the finally achieved Quality of Service (QoS) and the perceived Quality of Experience (QoE), due to losses and inconveniences that can be elsewhere introduced throughout the communication chain (i.e. network, users' terminal, etc.) (Kalliris et al., 2011; Ries & Gardlo, 2010; Vegiris et al., 2008). For instance, in case that content is encoded in higher bitrates than those the involved channels can support, then, packet losses, delays and delay variations can cause even bigger problems, thus further worsening the delivered QoE that it would have been achieved if compromised quality had been initially selected. In other words, there is no need to impose high quality levels that they cannot be supported throughout the end-to-end nodes of the involved publishing channels. Hence, there has to be an optimum balance between the initially selected QoS and QoE that need to comply with the related attributes in all the involved stages (i.e. content encoding, channel transmission—distribution, end-terminal reproduction, etc.), whereas content-based features and channel-related metrics can be utilized. Furthermore, application related standards and prerequisites need to be taken into account. For instance, different QoS/QoE levels are required in news-informing services than in mobile communications, high-definition broadcasting, cultural and e-learning applications, etc., an issue that should also be considered towards optimum resources exploitation and optimal cross publishing planning (Dimoulas et al., 2014a, 2014b; Kalliris et al., 2011; Ries & Gardlo, 2010; Vegiris et al., 2008).

In order to meet the above demands, the proposed model incorporates the Reference Source Content (RSC) concept (Dimoulas et al., 2014a). This content version is the highest quality reference that can be used for all the future content adaptations, eliminating at the same time the multiple-copies problem. For instance, it is likely that high definition (HD) content created at the production phase could serve to the construction of the RSC, even in case that none of the involved cross publishing channels has HD attributes. Besides the quality maintenance issues during content transcoding that have been already stated, this strategy could also serve future use purposes in case that upcoming HD channels would be involved. Furthermore, the insertion of the optional content upscale modality is related to the fact that "small-scale up-sampling (both in resolution and quantization) is feasible, without significant appearance of quality deterioration artifacts" (Dimoulas et al., 2014a). Considering PBM and especially textual entities, the corresponding RSC version may include additional and higher resolution images, as well as additional-elongated text and documentation related information. Again, this information might not be used in any of the involved channels, however, it can be reserved for future use, but also in meta-information processing and management related purposes. In this context, meta-data are expected to play a very important role by keeping all of the involved processing, editing-assembling and authoring meta-information in applicable forms. This settlement can expedite the content adaptation and management process, not only in content transcoding related tasks, but also in media assets rearrangement and channel-related shaping processes that can be employed in the favor of both producers and end-users.

Extending the above analysis, channel interfacing limitations (i.e. smart phone, tablet, etc.) can be faced with the implementation of dedicated meta-processing and multimodal semantic interaction tools (i.e. imaging-based sequential and hierarchical summarization and highlighting, cloud-tag text visualization, text to speech services, content-recognition-based automations, etc.) (Dimoulas et al., 2012, 2014a; Domingue et al., 2011; Kalliris & Dimoulas, 2009; Papapetrou et al., 2010; Raj & Mala, 2012). This intelligent content documentation can be deployed and/or complemented invisibly from users, both at the production and consumption ends, taking advantage of Web 2.0 services and even recent Web 3.0 trends (i.e. UGC contribution models, "invisible" user feedback based on browsing options, on purpose semantic tagging-interaction, etc.). In addition to the above, content and interfacing-related used customization and personalization can be elaborated in combination with novel ubiquitous computing services, such as context and location aware tools.

Figure 4 provides a functional flow diagram of the proposed iXMP model, whereas the main classes of the involved parameters are presented. Hence, QoS/QoE metrics can be considered for all three content-application demands (QoS_A/QoE_A), broadcasting issues (QoS_B/QoE_B) and access requirements related to terminals and interfacing particularities (QoS_C/QoE_C), serving end-to-end ($QoS_{e\text{-}e}/QoE_{e\text{-}e}$) quality modeling, control and assessment (Kalliris et al., 2011; Vegiris et al., 2008). Similarly, content extracted audiovisual features (i.e. spectral, bandwidth and dynamic range audio attributes, color histograms, image-edges and other structural indexes, motion activity, etc.) can be used for the estimation of the applicable sampling parameters that are required for proper content presentation (i.e. sampling-rate/resolution, bit-length quantization, frame rate, interlaced or progressive scanning modes, etc.). All these parameters are assembled along with the remaining channels' and users' related preferences, forming a somewhat

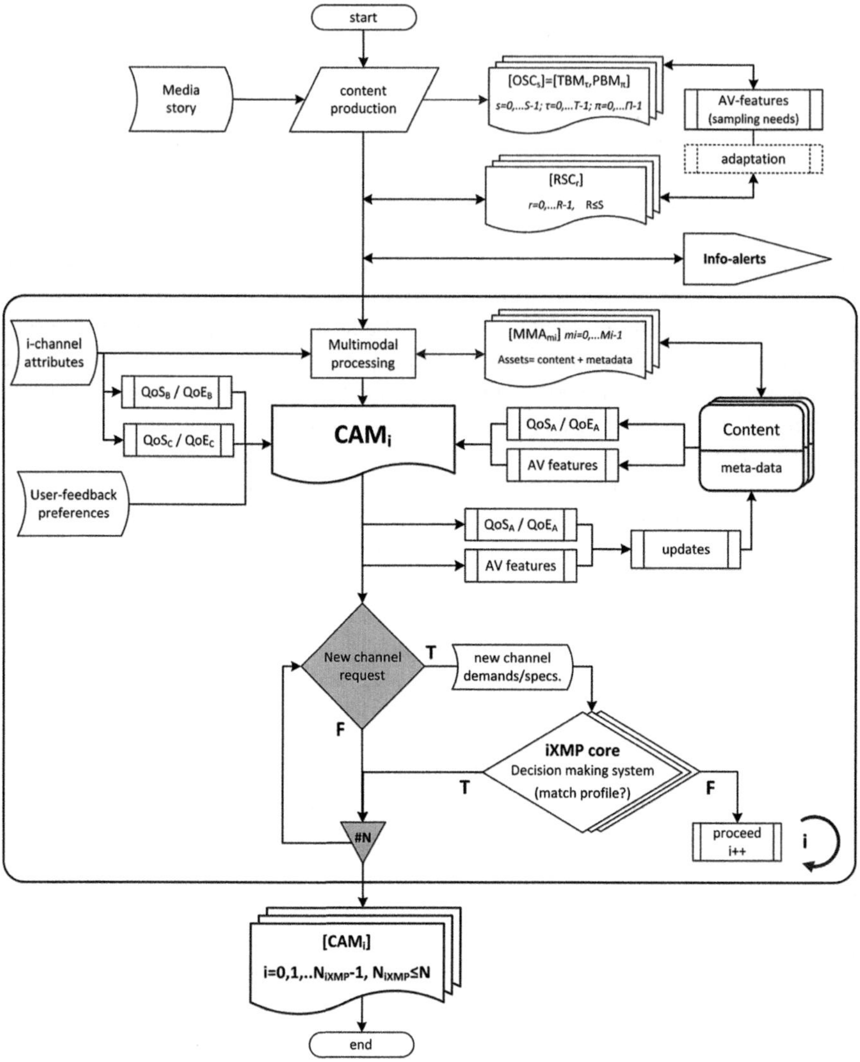

Fig. 4 Functional flow diagram of the proposed intelligent cross-media publishing (iXMP) model and its internal modalities

generic UML activity flow diagram that can be adopted for the materialization of a specific iXMP solution. It is obvious that the role of the iterative-loop procedure is to restrict the involved iterations to the smallest possible number, thus speeding up the cross-publishing process while preserving content quality at the same time. The iXMP core is a multidimensional decision making system, whereas a variety of the available machine learning algorithms and generally artificial intelligence technologies can be utilized. This module can be implemented as a stand-alone system operating locally at the media organization end, or even in Web 3.0 oriented contexts, using remotely-distributed processing nodes and collaborative multi-agents.

There are many advantages that could be pointed out while analyzing the above generic model. First of all, it allows for a better organization and management of the produced content and the media production procedure as a whole, eliminating the time and the human resources that are required for the multiple channels' profiles to be created. Management improvements are also considered in terms of superior storage-units' exploitation and their associated benefits, expediting searching and retrieval processes at the same time. In addition, quality losses due to unnecessary transcoding are avoided, so that optimum format selection and quality maintenance are guaranteed adaptively to each channel's demands. On the other hand, there are many difficulties and problems that have to be resolved before the above model could be massively deployed in practical, real world, scenarios. Hence, technical knowhow and adoption of strict standardization rules might create difficulties that would be discouraging to most traditional media organizations. Furthermore, the configuration of the audiovisual features, the associated content quality metrics and the involved intelligent decision systems is not such straightforward in the implementation of iXMP modules, and a lot of research is still required in these fields. However, once pilot developments and initial trainings of smart cross media publishing systems become available, it will be much easier for the proposed model to be massively adopted from more and more media organizations, professional freelancers and individual users. Summing up, the proposed model is not limited in content-related selection, adaptation and profiling-clustering automation, but also on more complicated media assets structural organization and user interaction-interfacing modalities that are engaged based on the similarities, individualities, limitations and constrains of all the involved cross-media publishing channels. In this context, optimum media assets exploitation and re-use scenarios can be launched, favoring world-wide cross-collaboration between various media organizations and individuals. Likewise, semantic feedback acquisition promotes the gradual construction of related knowledge bases, so that agent-driven machine learning can fully serve ambient media environments and semantic interaction automation.

4 Conclusions and Viewpoints on Convergence

This chapter discusses media convergence issues regarding cross media publishing. Traditionally there seems to be a specialization of media companies concerning their end products. For example traditionally newspapers' product was the print edition of the newspaper, radio stations' product was their radio program and TV stations' product was their TV program. But the convergence of ICTs has blurred the boundaries between the different media organizations, allowing them to incorporate various publishing channels in order to disseminate news. This was also fueled by the increased use of ICTs by the potential audience of the media. Thus the tendency has been for the larger media organizations and companies to have several publishing channels at their disposal. This trend was based mainly on the wide popularity of the Internet as well as 3G mobile networks and beyond. That resulted in a rapid expansion of the news output of many broadcasters that covers a wide range of media platforms. The specific topic is considered to be of great interest within the discussed framework, since cross media publishing inherently appeared as a very distinctive technological convergence outcome. Indeed, cross media publishing became a reality along with the expansion of the digitization process into the media enterprises and throughout all the phases of the involved media production-consumption processes. In this context, introductory information regarding cross-media publishing evolution was presented, providing historical data, definitions, channel and user-related specifications. Hence, media production and consumption issues were analyzed along with the involved content entities, media models and cross-media publishing channels. Consequently, insights regarding cross-media publishing requirements and practices in todays' media enterprises were thoroughly discussed. As a result, a new model was introduced ordering to bring forward and implement intelligent cross media publishing strategies. The proposed model takes advantage of all the recent media convergence advances, including state of the art web 2.0 services and even more upcoming web 3.0 semantic interaction technologies and trends, thus aiming to provide innovative channel and user oriented customization and personalization services along with automated cross media publishing. Beyond this, the proposed model is expected to affect most aspects of current media practices and trends that are encountered in contemporary media enterprises, thus raising new technological, social and economic research directions.

References

ACTeN. (2004). Cross-media E-Content. Report 8.

Antikainen, H., Kangas, S., & Vainikainen, S. (2004). *Three views on mobile cross-media entertainment*. VTT Information Technology, Research report TTE4-2004-17, 22.3.

Boyd, D. M., & Ellison, N. B. (2008). Social network sites: Definition, history, and scholarship. *Journal of Computer-Mediated Communication, 13*(1), 210–230.

Chapman, N., & Chapman, J. (2009). *Digital multimedia* (3rd ed.). Chichester: Wiley. ISBN 978-0-470-51216-6.

Cottle, S., & Ashton, M. (1999). From BBC Newsroom to BBC Newscentre: On changing technology and journalist practices. *Convergence: The International Journal of Research into New Media Technologies, 5*, 22–43.

Dal Zotto, C. (Ed.) (2005). *Growth and dynamics of maturing new media companies*. JIBS Research reports series, No. 2005:2, Jönköping International Business School.

Davidson, D. (Ed.) (2010). *Cross-media communications: An introduction to the art of creating integrated media experiences*. ETC Press.

Dietmar, S. (2008, November). My view on cross-media communication for newspapers. *NEXTNEWSROOM*. Accessed January 22, 2013, from http://schantin.wordpress.com/2007/12/27/my-view-on-cross-media-communication-for-newspapers/

Dimoulas, C., Tsourvakas, G., Kalliris, G., & Papakis, N. (2012). *Audiovisual content management for the new media environment and economic aspects.* 10th World media economics and management conference, Thessaloniki.

Dimoulas, C., Kalliris, G., Chatzara, E., Tsipas, N., & Papanikolaou, G. (2014a). Audiovisual production, restoration-archiving and content management methods to preserve local tradition and folkloric heritage. *Journal of Cultural Heritage, 15*(3), 234–241.

Dimoulas, C., Veglis, A., & Kalliris, G. (2014b). Application of mobile cloud based technologies in news reporting: Current trends and future perspectives. In J. Rodrigues, K. Lin, & J. Lloret (Eds.), *Mobile networks and cloud computing convergence for progressive services and applications*. IGI Global, book chapter 17, pp. 320–343.

Domingue, J., Fensel, D., & Hendler, J. (Eds.). (2011). *Handbook of semantic web technologies.* Berlin: Springer.

Duhe, S. F., Mortimer, M. M., & Chow, S. (2004). Convergence in North American TV newsrooms: A nationwide look. *Convergence: The International Journal of Research into New Media Technologies, 10*, 81–104.

Erdal, I. J. (2007). Researching media convergence and crossmedia news production mapping the field. *Nordicom Review, 28*(2), 51–61.

Gillmor, D. (2004). *We the media—grassroots journalism by the people for the people.* Sebastopol, CA: O'Reilly.

Kalliris, G., & Dimoulas, C. (2009). *Audiovisual content management issues for the new media environment.* International conference on new media and information: Convergences and divergences, Athens.

Kalliris, G., Dimoulas, C., Veglis, A., & Matsiola, M. (2011). *Investigating quality of experience and learning (QoE & QoL) of audiovisual content broadcasting to learners over IP networks* (pp. 836–841). IEEE Symposium on Computers and Communications (ISCC 2011).

Kelly, H. (2012). OMG, the text message turns 20. But has SMS peaked? *CNN International.* Accessed January 19, 2013, from http://edition.cnn.com/2012/12/03/tech/mobile/sms-text-message-20/

Kotsakis, R., Kalliris, G., & Dimoulas, C. (2012). Investigation of broadcast-audio semantic analysis scenarios employing radio-programme-adaptive pattern classification. *Speech Communication, 54*(6), 743–762.

Lowe, G. F. (2011). Respecting the PSB heritage in the PSM brand. *JMM International Journal on Media Management, 13*(1), 21–35.

Lugmayr, A., Stockleben, B., Risse, T., Kaario, J., & Pogorelc, B. (Eds.) (2013). New business, design and models to create semantic ambient media experiences. *Multimedia Tools and Applications, 63*, 1–159, special issue.

Negroponte, N. (1995). *Being digital.* Athens: Kastaniotis.

Papapetrou, O., Papadakis, G., Ioannou, E., & Skoutas, D. (2010). Efficient term cloud generation for streaming web content. *Lecture Notes in Computer Science, 6189*, 385–399. doi:10.1007/978-3-642-13911-6_26.

Raj, D. A. A., & Mala, T. (2012). *Cloud press: A next generation news retrieval system on the cloud.* Paper presented at the 2012 International Conference on Recent Advances in Computing and Software Systems, RACSS 2012. New York, NY

Rieck, M. Q., & Dhar, S. (2010). Efficient multi-hop broadcasting in wireless networks using k-shortest path pruning. In *Distributed computing and networking* (pp. 283–294). Berlin: Springer.

Ries, M., & Gardlo, B. (2010). Audiovisual quality estimation for mobile video services. *IEEE Journal on Selected Areas in Communications, 28*, 501–509.

Sabelström, K. (1998, June). *Consumer needs and competing products in printing and publishing.* 1998 Conference Beyond Convergence, Stockholm, Sweden.

Sabelström, K. (2000, April). *The multimedia news reporter: Technology and work processes* (pp. 53–66). TAGA'S 52nd Annual Technical Conference, Colorado Springs, Colorado, USA, published in TAGA Proceedings, TAGA Office, Rochester, New York, USA.

Sabelström, K. (2001). *Information categories and editorial process in multiple channel publishing*. PhD thesis, Department of NADA, Division of Media Technology and Graphics Arts, Royal Institute of Technology, Stockholm, Sweden.

Schiff, F. (2003, June). Business models of news Web sites: A survey of empirical trends and expert opinion. *First Monday, 8*(6). Accessed January 28, 2013, from http://firstmonday.org/htbin/cgiwrap/bin/ojs/index.php/fm/article/view/1061/981

Simpson, W. (2008). *Video over IP: IPTV, internet video, H.264, P2P, web TV, and streaming: A complete guide to understanding the technology*. Oxford: Focal Press/Elsevier.

Spyridou, L. P., Matsiola, M., Veglis, A., Kalliris, G., & Dimoulas, C. (2013). Journalism in a state of flux: Journalists as agents of technology innovation and emerging news practices. *International Communication Gazette, 75*(1), 76–98.

Symeonidis, A. L., & Mitkas, P. A. (2005). *Agent intelligence through data mining* (Multiagent systems, artificial societies, and simulated organizations series, Vol. 14). Berlin: Springer.

Urolagin, S., Prema, K. V., & Reddy, N. S. (2012). Generalization capability of artificial neural network incorporated with pruning method. In *Advanced computing, networking and security* (pp. 171–178). Berlin: Springer.

Vegiris, C., Avdelidis, K., Dimoulas, C., & Papanikolaou, G. (2008). Live broadcasting of high definition audiovisual content using HDTV over broadband IP networks. *International Journal of Digital Multimedia Broadcasting*. doi: 10.1155/2008/250654

Veglis, A. (2005). Implementation of a computer supported collaborative work system in a newspaper. *WSEAS Transactions on Information Science and Applications, 2*, 891–902.

Veglis, A. (2007, Spring). Cross-media publishing by U.S. newspapers. *Journal of Electronic Publishing*. http://hdl.handle.net/2027/spo.3336451.0010.211.G. O

Veglis, A. (2008a, June). Comparison of alternative channels in cross-media publishing. *Publishing Research Quarterly Journal, 24*(2), 111–123.

Veglis, A. (2008b). Modeling cross-media publishing. In *Third international conference on internet and web applications and services*.

Veglis, A. (2012). Journalism and cross-media publishing: The case of Greece. In A. Veglis & E. Siapera (Eds.), *The Wiley-Blackwell handbook of online journalism*. Oxford: Blackwell.

Veglis, A., & Pomportsis, A. (2009). Workflow management for cross-media publishing. In *Lecture notes in electrical engineering: Proceeding of the European computing conference* (Vol. 1, pp. 533–542). Berlin: Springer.

Veglis, A., & Pomportsis, A. (2014). Journalists in the age of ICTs work demands and educational needs. *Journalism & Mass Communication Educator, 69*(1), 61–75.

Walter, M. (1999, October). *Cross-media publishing systems*. The Seybold Report on Internet Publishing.

Wearden, S., & Fidler, R. (2001, Spring). Crain's Cleveland business: Evaluating an e-Newspaper concept for tablet PCs. *Future of Print Media Journal, Spring*.

Influence of Convergent Media on the Perception of Information Credibility

Daria Mustić and Mario Plenković

1 Introduction and Problem Discussion

Research of media information credibility has become particularly interesting topic since the advent of the Internet and the possibility of every individual to become an author and source of information in the digital space. Internet has enabled access to more information, but because of the information overload, the user also feels uncertainty, mistrust and doubt in the credibility of information. Credibility is a complex concept that is in the same time related to all the components of the communication process—actual veracity of the message quality, communicators, the nature and reputation of the media and, in the end, to all the factors that determine the communication behavior of the communicant (Kurtić, 2007).

Because of the fact that web pages often contain inaccurate, unverified information, designers must also participate in the creation of information credibility (Fogg & Tseng, 1999; Morkes & Nielsen, 1997). According to Müller (2008), digitization has led to major changes in the media visual communication and they are reflected in the three main phenomena: the rise of amateur visual production, global dissemination of visual messages (messages that were originally intended for local or regional level are becoming available to everyone) and de-contextualization of visual messages (this process leads to a different decoding process because of the impact of cultural context). Digital visual media production allows efficient and rapid organization of media mediation of visual messages which in this process are becoming a permanent value that does not disappear—in this way we get a digital asset as a value without an expiration date. Information user today partly assumes the role of creator of visual aspects of information, so there is a repositioning in relation graphic designer—recipient. Today, they are partially sharing the role in

D. Mustić (✉) • M. Plenković
Department of Communication Science, Faculty of Graphic Arts, University of Zagreb, Zagreb, Croatia
e-mail: dmustic@grf.hr

A. Lugmayr, C. Dal Zotto (eds.), *Media Convergence Handbook - Vol. 1*, Media Business and Innovation, DOI 10.1007/978-3-642-54484-2_9

the information chain. Information and communication processes and the dynamic development of graphic technology affect on the information as a cognitive category with increased credibility. Graphic design is today powerful communication tool for the management of media content. Good design enables recipient to have the ability to decode the message content from its own semantic and semiotic perspective, and in its own socio-economic environment.

In this study, analysis and comparison of the two models for presenting information was conducted, to determine whether the credibility of the information is greater, same or lower when we use convergent media content opposed to using content with a smaller share of the convergent content. Research hypothesis was: by using different levels of media convergence in the digital space, the assessment of the media messages can be changed.

The research presented here seeks to obtain a more comprehensive understanding of what the user is doing and reading on the web page and does the convergent presentation of the information influence on this process. Ocular indices enable us to determine what the user is indeed viewing and reading, for how long, and in what order. Throughout the history of eye tracking research, several key variables have emerged as significant indicators of ocular behaviors, including fixations, saccades, pupil dilation, and scan paths.

Research using eye tracking technology provide us with valuable and objective insight into the functioning and characteristics of visual attention, which in turn allows more effective design of media content. While traditional research techniques of computer-mediated communication are sufficient to obtain information about navigating on page, eye tracking technology is successful in analyzing the use of the user interface between clicks and mouse navigation. Such studies gave insight into what most attracts attention. Eye tracking technology has proved to be very effective method in the study of human-computer interaction (Zhu & Ji, 2004).

2 Literature Review

Olaisen (1990) conducted one of the first empirical studies on electronic information credibility. Electronic information in the study were estimated as important, valuable and accessible but also as content with low credibility, poor formatting and user unfriendly. For the assessment of the credibility, participants were often checking and looking for the information about the authorship or information source.

Importance of the source has been confirmed in the research conducted by Rieh (2002) in which subjects also emphasized a source as an element which is important for credibility. Respondents had more confidence in articles with recognizable names of authors, articles written by professionals—doctors, professors, experts. Also, a higher degree of credibility is demonstrated by specifying institutional sources than individuals. Respondents also considered "pay per view" information to be more credible.

Liu and Huang (2005) found a difference in the assessment of the credibility of information on the web between undergraduate and graduate students. Undergraduate students assessed credibility by source, author name or institutions, while graduate students assessed the credibility through the accuracy and quality of information.

Comparative studies of perception of credibility of traditional media and the Internet, have not yielded consistent results (Flanagin & Metzger, 2000). For example, Johnson and Kaye (1998) have shown that people who seek political themes rather believe the Internet than traditional media. The authors replicated similar research in 2002, examining the credibility of online newspapers during the presidential campaign. They have obtained results that confirm the findings from 1998 research.

Wathen and Burkell (2002) divided the evaluation process on the web into three levels: superficial credibility, credibility of the message and the evaluation of the content. For the superficial credibility, users highlighted presentation and organizational elements of the page for credibility assessment. For the credibility of the message they highlighted the importance of sources and for the evaluation of content they combined the evaluation of the sources and personal knowledge and the need for information. Individuals have different expectations and have different levels of knowledge, therefore, credibility never depends solely on the message, but always characteristic of the individual must be included. Believability, honesty, impartiality, reliability, professionalism, accuracy and comprehensiveness are the measures of information credibility perception that have been used in studies of credibility perception (Gaziano & McGrath, 1986; Johnson & Kaye, 1998, 2002; Meyer, 1988; Newhagen & Mass, 1989). Fogg and Tseng (1999) defined four sources of credibility: presumed credibility comes exclusively from the recipient, known credibility is based on the labeling of sources, superficial credibility is based on a first impression and credibility based on experience is based on the experience of using the same source over time.

Previous studies have focused on the role of the sources of information (expertise, professionalism, recognition) on credibility creation, to the role of individual characteristics in the perception of the message, while the influence of the characteristics of the transmission media of the message has not been examined in a large range (Johnson & Kaye, 1998). Research aimed in this direction, mainly compared message credibility in different media, while convergent medium that contains elements of all traditional media has not been studied.

Holmqvist, Holsanova, Barthelson, and Lundqvist (2003) argue that the visual information in the media have largely decorative role and serve to direct the eyes, often not providing any additional information. They think that readers learned that images are not a serious source of information, but only a source of entertainment and leisure.

Researches of visuality are primarily focused on the areas of art, culture and design studies. "Technology driven" researches are focused, among other things, to the development and improvement of technologies for successful visual communication. In this group can be enumerated researches in which system for measuring the movement of the human eye are used. Results can be used to provide information for successful and effective forming of visual information.

It is also important to mention the importance of design in construction of information credibility. The term 'user-centered design' originated in Donald Norman's research laboratory at the University of California San Diego (UCSD) in the 1980s and became widely used after the publication of a co-authored book entitled: *User-Centered System Design: New Perspectives on Human-Computer Interaction* (Norman & Draper, 1986).

Abras, Maloney-Krichmar, and Preece (2004) emphasize that the ways in which users participate can vary. The involvement may be relatively light; they may be consulted about their needs, observed and participate in usability testing or the involvement can be intensive with users participating throughout the design process as partners in the design. A variety of methods have been developed to support UCD including usability testing, usability engineering, heuristic evaluation, discount evaluation and participatory design.

Aesthetic design looks easier to use and has a higher probability of being used, whether or not it is actually easier to use. For example, in a study how people use computers, researchers found that early impressions influenced long-term attitudes about their quality and use (Lidwell, Holden, & Butler, 2003). Because of this phenomenon, the design of the used models in this research was identical.

When users' expectations are not met, they may get frustrated or even angry. The major disadvantage to user-centered design is that is can be quite costly. It takes time to gather data from and about users especially if you seek to understand the environment in which they will be using the products. The process requires resources, both financial and human (Abras et al., 2004). The need to involve actual users, often in the environment in which they would use the product being designed, was a natural evolution in the field of user-centered design (Abras et al., 2004).

Rosenbaum, Rohn, and Humburg (2000) surveyed 134 CHI professionals with a focus on the contribution of organizational approaches and UCD methods to strategic usability. It was found that major obstacles to creating greater strategic impact include resource constraints, which were mentioned by 28.6 % of the respondents, resistance to user-centered design or usability, lack of knowledge about usability.

According to Kurtić (2007), research of media content credibility, among other things, includes components of the experience of authenticity, which is partly related to the actual quality of the news, but also on communicators experience and their intentions, simply put, on general communicative experience in general. Credibility is the issue of experienced quality of the ways in which reality is presented through the media content.

3 Methodology and Approach

There are two basic approaches to measuring visual attention and perception of Internet users: subjective and objective. Subjective evaluation of web pages is usually done using questionnaires before and after the use of the site. In this way we can get feedback from users, which can then be used in shaping the future content. The questionnaire is appropriate because it sets a deliberate, focused questions about what interests us, but for this type of research there is the problem of "human factor" because it is difficult to establish which parameters of the content affect on the answers of participants, and the fact that people often unwittingly give answers that are considered socially acceptable. The advantage of the questionnaire is it can cover a large population. Questionnaire is the most widely used method of collecting data about web design, and this method is often used independently. It provides information on the subjective views of respondents, therefore, in order to obtain more objective data, we also use objective reactive and proactive methods.

These methods have a great advantage in analyzing the usage of the content, but they do not allow real-time analysis of use of any content. Without objectivity of data in research, the producers of media content depend on the experience and judgment of individuals. Creating objective criteria for the evaluation of the content and techniques of production, would enabled the media production with more quality.

3.1 Used Visual Stimuli and Research Procedure Description

Used visual stimuli consisted of an Internet news site, which was made in two versions. One model of the site contained multiple channels for presenting information (text, images, video, sound), while the second model contained a small amount of converged media content (text and photos). The site was comprised of a total of 21 news/articles and two commercials. News were divided into the following sections: General news, Analysis, Croatia, Highlights, Sports, Columns and Fun. When selecting news (information), it was considered that they represented substantially all of the common forms of news: "hard news"—the recent in-formation, "up-to-the-minute" information, events which are interpreted as possible problem in social environment, then "soft news"—events that are often related to the private sphere where speed of the delivery is not crucial.

For each of the participants in both models, time using each individual article (areas of interest) was calculated, and also the number of fixations at the article to determine whether the article was observed, scanned or read. Tobii X60 was used for the experimental part of the work in which eye movements were monitored and measured.

Before testing with the eye-tracking device, participants filled out a brief questionnaire about their gender, age, experience in using the Internet, preference to specific media and media genre preference. Testing was done individually. It was held in a quiet room with a comfortable temperature and enough light. All respondents had the same external conditions.

Before opening the site, once more, on the screen they were briefly introduced to the basic guidelines and instructions. Research began by answering general questions and questions about the use of media and media preferences, and then they could open a model site for testing. Respondents were not time-limited in the use of the content, they could independently decide how they will navigate through the site. The procedure of reading and using the site was not determined for them in advance, because we wanted to a natural behavior and personal motivation and preferences in choosing content.

Gained data were analyzed with the SPSS statistical package (17.0 Version). Results are expressed as mean, standard deviation (SD), median. The Kolmogorov–Smirnov test was used to analyze the normal distribution of the variables ($P > .05$). Quantitative data have had a non-normal distribution, so they were analyzed with non-parametric test. Differences in the number of fixations and visit duration between two models were assessed with the non-parametric Mann–Whitney U test. The statistical analysis was conducted at 95 % confidence level. A P value less than .05 was considered statistically significant.

3.2 Sample Description

The 140 examinees were divided into two groups, and clustering was done randomly. One group used the prototype A (web page with high convergent content), while the other group used the prototype B (web site with low levels of converged content).

Convenient sample was used. Use of this sample derives some methodological problems. First of all, these samples are not representative, since they include only a part of the population available at the time of testing. Therefore, the interpretation and conclusions arising from this study cannot be generalized over the entire population. Due to the nature of research, this kind of sample is the optimal choice because it covers a smaller portion of the population that can be considered as defined by a common trait (in this case, all were active users of the Internet and they use it for information gathering), which makes it homogeneous. This choice of sample is optimal also because of the use of specific equipment for research, because use of the same space for the research, allows the control of the external variables and the same exposure of the participants to the potential impact of external factors (e.g., light, heat, silence, etc.). Another advantage of this sample is its cost-effectiveness and ease of implementation.

Regardless of the fact that it is a homogeneous population, the generalization in relation to the complete population would not be justified, since the sample is not representative. However, it is possible to draw certain conclusions regarding the use of converged media content based on statistical analysis.

Study sample was represented by 70 females and 70 males. While conducting the research, it was took into account that both models were used by the same

Table 1 Gender structure of respondents

Gender	LCC	HCC
Female	35	35
Male	35	35
Total	70	70

Table 2 Age structure of respondents

Age	N(f)	%	Cumulative (%)
18–24	41	29.3	29.3
25–32	41	29.3	58.6
33–38	25	17.9	76.4
39–45	13	9.3	85.7
46–52	10	7.1	92.9
53+	10	7.1	100.0
Total	140	100.0	
Chi square test			
χ^2	46.686		
p	.000		
df	5		

number of females and males, so both groups had 35 female and 35 male participants (Table 1). All measurements were made on the model with low convergent content (LCC) and on the model with a high converged content (HCC).

To better explain and further confirm the hypothesis, it was necessary to take into account other indicators. Some of them are the age structure and completed education of the respondents. The respondents were divided into six age groups (Table 2).

The younger age group is more represented in the research, which is acceptable considering that this age group is still largely using new technologies. Almost 60 % of respondents are between 18 and 32 years old. The fact that many researchers point out that age is not a key element in testing the use of new technologies is taken into account. Jenkins (2007) examines the linguistic compounds "*digital natives*" and "*digital immigrants*" who had originally indicated the generation gap in information and communication literacy. He notes that the age cannot be the sole criterion for classification into one of these groups, since this type of literacy is affected by the socio-economic status, personal interests, etc.

It should be noted that there was no statistically significant differences between the groups themselves ($\chi^2(5) = 1.571, p > .05$) in the age structure, therefore we can conclude that any differences in results between the groups in the following research are not conditioned by the age structure (Table 3).

According to the educational structure, the sample consisted of 32 respondents with secondary school education, 41 respondents with associate degrees and 67 respondents with a university degree (Table 4).

Chi-square test showed that there were statistically significant differences in the frequencies $\chi^2(3) = 65.543$, $p < .01$. This means that we can talk about dominant

Table 3 Age structure of respondents by models

Model	Age	N(f)	%	Cumulative (%)
LCC	18–24	19	27.1	27.1
	25–32	19	27.1	54.3
	33–38	13	18.6	72.9
	39–45	8	11.4	84.3
	46–52	6	8.6	92.9
	53+	5	7.1	100.0
	Total	70	100.0	
HCC	18–24	22	31.4	31.4
	25–32	22	31.4	62.9
	33–38	12	17.1	80.0
	39–45	5	7.1	87.1
	46–52	4	5.7	92.9
	53 +	5	7.1	100.0
	Total	70	100.0	
Chi square test				
χ^2	1.571			
p	.905			
df	5			

Table 4 Education level of respondents

Education level	N(f)	%	Cumulative (%)
Elementary school	0	0	0
Secondary school	32	22.9	22.9
Associate degree	41	29.3	52.1
University degree	67	47.9	100.00
Total	140	100.0	
Chi square test			
χ^2	65.543		
p	.000		
df	3		

educational group in the sample examined, and in our case it is acceptable, because previous researches have shown that this group uses the Internet more than other educational groups. One more reason for this kind of distribution is that the research was conducted at the university, therefore the available population often had higher education.

Education levels by groups is shown in Table 5.

Chi-square test showed that there were no statistically significant differences in the frequencies of both groups $\chi^2(2) = .164$, $p > .05$, suggesting that possible differences in the results obtained by two research groups cannot be caused by the structure of respondents by groups.

Feature of Internet portals is the ability to reach a global audience. Each produced message has the potential to be transferred to the audience in different cultural environments and then a message can be put in a different context and

Table 5 Education levels by models

Model	Education level	N(f)	%	Cumulative (%)
LCC	Elementary school	0	0	0
	Secondary school	15	21.4	21.4
	Associate degree	21	30.0	51.4
	University degree	34	48.6	100.0
	Total	70	100.0	
HCC	Elementary school	0	0	0
	Secondary school	17	24.3	24.3
	Associate degree	20	28.6	52.9
	University degree	33	47.1	100.0
	Total	70	100.0	
Chi square test				
χ^2	0.164			
p	.921			
df	2			

interpreted in a different way. In this study, the audience is still restricted to a limited population of the same cultural environment. Cultural environment can therefore be ruled out as a factor affecting the eventual different responses of survey respondents.

4 Results

Measurements were made on a total of 28 articles, 14 in the model with low convergent content (LCC) and 14 in the model with a highly converged content (HCC). Since participants have had complete freedom in how to use the web page, there are differences in the opening of individual articles. Those articles that have not been popular subjects, those that are open less than 10 times, were not included in the analysis.

Reading involves scanning and reading. Content which had one or more fixation has been noticed, content which had 20 or more fixations has been read and deeply read content has 200 or more fixations.

Analysis of Article no. 1 (Table 6) showed that number of fixations in the model with highly converged content (Mdn = 111) was significantly greater than the number of fixations in the model with low converged content (Mdn = 86), U = 299.500, $p < .05$. Mann-Whitney test showed that the time of the visit in the model with highly converged content (Mdn = 27.52) is significantly longer than the article visit in the model with low converged content (Mdn = 20.28), U = 249.000, $p < .01$. The data indicate that there is a statistically significant difference in the measured two factors: the number of fixation and duration of use of the article. In this case, use of the converged media content positively influenced on the use of the article, because it led to increased interest for the content.

Table 6 Analysis of Article no. 1

	Model	N	Average	SD	Median	Mann Whitney (U)	P
Fixation number	HCC	24	127.62	84.49	111	299.500	.033
	LCC	37	87.7	50.91	86		
Visit duration (s)	HCC	24	32.15	20.57	27.52	249.000	.004
	LCC	37	20	14.4	20.28		

Table 7 Analysis of Article no. 2

	Model	N	Average	SD	Median	Mann Whitney (U)	P
Fixation number	HCC	14	75.5	44.18	76	133.500	.169
	LCC	26	57.65	26.83	75.5		
Visit duration (s)	HCC	14	20.33	13.57	19.85	119.000	.074
	LCC	26	14.04	6.15	15.4		

Second article is the second-largest visually, and it is expected that it will be spotted. Article was located below the main article and it presents important news from the economy. An article in the model with a low converged content was opened by 26 participants, while an article in the model with highly converged media content was opened by 14 participants. Differences in the number of opening are caused exclusively by individual differences in the interest of the respondents for this topic, because the content of both models on the home page was visually formatted in the same way, and was located at the same place. As presented in Table 7, Mann-Whitney test showed no statistically significant differences in the number of fixations in the whole article between the two models, HCC (Mdn = 76), LCC (Mdn = 75.5), U = 133.500, $p > .05$. Mann-Whitney test showed no statistically significant differences in the duration of viewing article between the two models, HCC (Mdn = 19.85), LCC (Mdn = 15.4), U = 133.500, $p > .05$ (Table 7).

Article no. 3, entitled "How to save money on fuel?" was opened on both models by 33 participants, a total of 66. Although it was recorded greater number of fixation on the article in HCC model, Mann-Whitney test shows that this difference is not statistically significant, HCC (Mdn = 92), LCC (Mdn = 61), U = 508.500, $p > .05$. Mann-Whitney test shows that there is a statistically significant difference in the total duration of visits to the article, HCC (Mdn = 26.13) and LCC (Mdn = 15.36), U = 287.000, $p < .01$ (Table 8).

The fourth article in both models was located at the top of the page under the heading "Highlights". The HCC model article was opened by 28 participants and in the LCC model article was opened by an equal number of respondents, which makes a total of 56 subjects who used the above article.

Mann-Whitney test shows that there is a statistically significant difference in the number of fixations on the article, HCC (Mdn = 76) and LCC (Mdn = 47.5), U = 221.000, $p < .01$. Mann-Whitney test shows that there is a statistically significant difference in the total duration of visits to the article, HCC (Mdn = 19.77) and

Table 8 Analysis of Article no. 3

	Model	N	Mean	SD	Median	Mann Whitney (U)	P
Fixation number	HCC	33	101.27	53.24	92	90.000	.492
	LCC	33	71.21	31.14	61		
Visit duration (s)	HCC	33	27.71	16.41	26.13	287.000	.001
	LCC	33	16.45	7.19	15.36		

Table 9 Analysis of Article no. 4

	Model	N	Average	SD	Median	Mann Whitney (U)	P
Fixation number	HCC	28	78.68	45.23	76	221.000	.005
	LCC	28	46.82	24.08	47.5		
Visit duration (s)	HCC	28	23.23	17.25	19.77	201.000	.002
	LCC	28	10.79	5.96	10.67		

LCC (Mdn = 10.67), U = 201.000, $p < .01$ (Table 9). Analysis of article no. 4 showed that there were statistically significant differences in the number of fixations and the duration of use of the article, if we used the converged media content. It increased the time spent on the article.

Article no. 5 was located in the sports section. Mann-Whitney test shows that there is a statistically significant difference in the number of fixations on the article, HCC (Mdn = 83.5) and LCC (Mdn = 49.35), U = 90.500, $p < .05$.

Mann-Whitney test shows that there is a statistically significant difference in the total duration of visits to the article, HCC (Mdn = 24.66) and LCC (Mdn = 10.76), U = 69.000, $p < .01$. Looking at the overall results, the presence of video caused the longer duration of the visit and bigger number of fixation on the content (Table 10).

Article no. 6 was located in the sports section, and in the HCC model it had videos, while LCC model contained only a photograph. Text was opened by 15 participants in HCC model and 16 participants in the LCC model, a total of 31 respondents. Mann-Whitney test shows that there is no statistically significant difference in the number of fixations to the article, HCC (Mdn = 80) and LCC (Mdn = 42.5), U = 82.000, $p > .05$. Mann-Whitney test shows that there is a statistically significant difference in the duration of visits to the article, HCC (Mdn = 24.85) and LCC (Mdn = 9.76), U = 69.000, $p < .05$ (Table 11).

Analysis of article no. 6 suggests that the use of the converged media content affected the visit duration. Use of the article was significantly longer when using converged media content.

Article no. 7 theme falls under the section "Fun" and this is so called light news. In both models the article was placed in the bottom of the page, and was opened by 43 participants, 26 in the model with a highly converged content and 17 in the model with low converged content (Table 12).

Mann-Whitney test shows that there is a statistically significant difference in the number of fixations on the article in two models, HCC (Mdn = 60.5), LCC (31), $p < .001$. Mann-Whitney test shows that there is a statistically significant difference

Table 10 Analysis of Article no. 5

	Model	N	Average	SD	Median	Mann Whitney (U)	P
Fixation number	HCC	18	83.5	54.51	84	90.500	.039
	LCC	17	49.35	21.82	51		
Visit duration (s)	HCC	18	29.01	20.25	24.66	69.000	.006
	LCC	17	11.91	5.7	10.76		

Table 11 Analysis of Article no. 6

	Model	N	Average	SD	Median	Mann Whitney (U)	P
Fixation number	HCC	15	79.33	54.64	80	82.000	.133
	LCC	16	47.19	21.33	42.5		
Visit duration (s)	HCC	15	26.91	22.39	24.85	69.000	.044
	LCC	16	11.07	5.36	9.76		

Table 12 Analysis of Article no. 7

	Model	N	Average	SD	Median	Mann Whitney (U)	P
Fixation number	HCC	26	79.58	65.14	60.5	53.500	.000
	LCC	17	31.82	17.22	31		
Visit duration (s)	HCC	26	25.91	29.7	18.25	49.000	.000
	LCC	17	7.61	4.18	8.04		

in the duration of the use of the article, the respondents devoted more time to an article that had the text with also visual media content, HCC (Mdn = 18.25), LCC (8.04), $p < .001$.

In article no. 8, the average number of fixations is larger in model with highly converged media content, Mann-Whitney test shows that this difference is not statistically significant, HCC (Mdn = 113), LCC (Mdn = 87), U = 79.000, $p > .05$.

Mann-Whitney test showed no statistically significant difference in the visit duration on the article in two models, HCC (Mdn = 30.76), LCC (Mdn = 21.02), U = 69.000, $p > .05$ (Table 13).

Article no. 9 in both models was located right in the central part of the home page. The article was opened a total of 73 times, making 38 in a model with highly converged media content and 35 times in the model with low converged media content. Mann-Whitney test shows that there is a statistically significant difference in the number of fixations on the article, HCC (Mdn = 95.5), LCC (Mdn = 72), U = 376.500, $p < .01$. Mann-Whitney test shows that there is a statistically significant difference in the time of the visit to the article, HCC (Mdn = 27.75), LCC (Mdn = 16.84), U = 323.000, $p < .01$ (Table 14).

Article no. 10 was located in the bottom right of the home page, in the section "Analysis". Article was opened by a total of 28 subjects, of which 16 on the site with highly converged media content and 12 on the page with low converged media content. Articles differ only in one photograph—author of the article, which was to see if there is an influence of a visual presentation of author on opening the article

Table 13 Analysis of Article no. 8

	Model	N	Average	SD	Median	Mann Whitney (U)	P
Fixation number	HCC	12	110.75	45.81	113	79.000	.308
	LCC	17	92.24	41.26	30.76		
Visit duration (s)	HCC	12	28.03	10.01	5.51	69.000	.144
	LCC	17	22.68	11.71	21.02		

Table 14 Analysis of Article no. 9

	Model	N	Average	SD	Median	Mann Whitney (U)	P
Fixation number	HCC	38	118.63	75.83	95.5	376.500	.001
	LCC	35	69	35.87	72		
Visit duration (s)	HCC	38	31.55	20.03	27.75	323.000	.000
	LCC	35	17.27	12.55	16.84		

Table 15 Analysis of Article no. 10

	Model	N	Average	SD	Median	Mann Whitney (U)	P
Visit duration (s)	HCC	16	24.51	25.89	14.35	86.000	.642
	LCC	12	15.91	10.8	14.57		

and reading it. In the version with the photography, the text was opened a few times more, and the analysis of visits duration, indicated whether there were differences in the use of the article, due to the presence of author photography. Mann-Whitney test shows that there is no statistically significant difference in the duration of visits to the article, HCC (Mdn = 14.35) and LCC (Mdn = 14.57), U = 86.000, $p > .05$ (Table 15).

Article no. 11 was located in the lower part of the home page, in the section "Fun". In total, it was opened by 20 respondents, 10 respondents in each model. Mann-Whitney test shows that there is no statistically significant difference in the number of fixations on the article, HCC (Mdn = 40) and LCC (Mdn = 27), U = 39.500, $p > .05$. Mann-Whitney test shows that there is no statistically significant difference in the duration of visits to the article, HCC (Mdn = 10.39) and LCC (Mdn = 6.55), U = 31.000, $p > .05$ (Table 16).

Article no. 12 was located in the central part of the site, on the right side in the "Highlights" section. Article was opened by a total of 33 subjects, 16 in the HCC model and 17 in the LCC model. Mann-Whitney test shows that there is a statistically significant difference in the number of fixations on the article HCC (Mdn = 17.5) and the LCC (Mdn = 65), U = 28.000, $p < .001$. Mann-Whitney test shows that there is a statistically significant difference in the duration of visits to the article HCC (Mdn = 4.68) and the LCC (Mdn = 15.9), U = 33.000, $p < .001$ (Table 17).

Article no. 13 was located in the lower part of the home page in the section "Fun". In the model with low converged media content, the article was

Table 16 Analysis of Article no. 11

	Model	N	Average	SD	Median	Mann Whitney (U)	P
Fixation number	HCC	10	43.3	25.75	40	39.500	.427
	LCC	10	37.6	22.46	27		
Visit duration (s)	HCC	10	11.2	6.21	10.39	31.000	.151
	LCC	10	10.24	9.77	6.55		

Table 17 Analysis of Article no. 12

	Model	N	Average	SD	Median	Mann Whitney (U)	P
Fixation number	HCC	16	28.31	36.43	17.5	28.000	.000
	LCC	17	73.47	35.16	65		
Visit duration (s)	HCC	16	7.9	11.61	4.68	33.000	.000
	LCC	17	17.3	8.05	15.9		

accompanied by a photograph, while in the model with a highly converged media content article has had a photo gallery, that offered more visual information. Article was opened by a total of 43 subjects, including 20 in the model with low converged content and 23 in the model with highly converged media content. Mann-Whitney test shows that there is a statistically significant difference in the number of fixations to the article in HCC (Mdn = 52) and LCC (Mdn = 38), U = 139.500, $p < .05$.

Mann-Whitney test shows that there is no statistically significant difference in the duration of visits to the article in HCC (Mdn = 6.42) and LCC (Mdn = 6.59), U = 167.000, $p > .05$ (Table 18).

Article no. 14 was located in the upper right corner of the home page, in the section "Analysis". Article was opened by a total of 45 participants, 23 in the version with high convergent content and 22 in the version with low convergent content. Mann-Whitney test shows that there is no statistically significant difference in the number of fixations on the article, HCC (Mdn = 55) and LCC (Mdn = 65.5), U = 171.000, $p > .05$. Mann-Whitney test shows that there is no statistically significant difference in the duration of visits to the article, HCC (Mdn = 14.12) and LCC (Mdn = 17.44), U = 199.000, $p > .05$ (Table 19).

Credibility was expressed through four dimensions: impartiality, reliability, expertise and stringency. The intensity of the response was measured using a Likert scale (1 = not at all impartial, reliable, competent, persuasive and 7 = completely impartial, reliable, competent, persuasive). Results from each dimension are summed to obtain the index of credibility (Table 20).

Mann Whitney test shows the differences in the assessment of the credibility dimensions we tested: impartiality (HCC Mdn = 5.50, LCC Mdn = 5.0, U = 1944.500, $p < .05$), reliability (HCC Mdn = 5.00, LCC Mdn = 5.0, U = 1631.000, $p < .01$), competence (HCC Mdn = 5.00, LCC Mdn = 4.0, U = 1650.500, $p < .01$) and persuasiveness (stringency) (HCC Mdn = 5.00, LCC

Table 18 Analysis of Article no. 13

	Model	N	Average	SD	Median	Mann Whitney (U)	P
Fixation number	HCC	23	58.04	28.17	52	139.500	.027
	LCC	20	40.05	22.8	38		
Visit duration (s)	HCC	23	13.28	12.37	6.42	167.000	.125
	LCC	20	10.57	10.01	6.59		

Table 19 Analysis of Article no. 14

	Model	N	Average	SD	Median	Mann Whitney (U)	P
Fixation number	HCC	23	49.52	26.13	55	171.000	.063
	LCC	22	67.05	29.32	65.5		
Visit duration (s)	HCC	23	13.46	7.06	14.12	199.000	.220
	LCC	22	16.85	7.98	17.44		

Table 20 Index of credibility in high convergence media content and low convergence media content

Dimensions of credibility	High convergence content			Low convergence content			Mann Whitney (U)	Sig.
	%	SD	Median	%	SD	Median		
Impartiality	5.19	1.53	5.50	4.97	1.07	5.0	1944.500	.027
Reliability	5.00	1.46	5.00	4.43	1.12	5.0	1631.000	.000
Competence	4.66	1.24	5.00	4.26	0.94	4.0	1650.000	.000
Persuasiveness	5.14	1.40	5.00	4.41	1.03	5.00	1336.500	.000
Index of credibility	4.99	0.24	5.07	4.52	0.31	4.42		

Mdn = 5.0, U = 1336.500, $p < .01$). In all four dimensions credibility index increased in the model with highly converged media content.

5 Discussion

Studies of digital media, including this one, are faced with many challenges and dilemmas in the interpretation of research results. Possible problems are:

- The need to control variables which would be impossible using real Internet portals. Readers have patterns of reading and habitual eye movements when they use familiar content. Behavior then depends on the extent in using specific pages and how they are skilled in finding the desired content.
- With using real Internet portals there would be a great possibility of influence made by respondents media preferences. Since the transmission channel (in this case, web portal) has an impact on the message itself (Barthes, 1977), using models prevents the social positioning of the message source and excludes the

possibility of estimating messages under the influence of user preferences and worldviews.

- Extended period of research on actual sites would lead to usage of a different media content. Information on the web portals are exposed to constant change. Substantially, some information do not change, but it can happen, for example, that an article on the website changes its position on the page several times during the day, thereby changing the experimental conditions and the data obtained in the study would thus become mutually incomparable.
- Inability to control exposure of subjects to other media. If we record the content and present it to participants at various intervals, there would be a possibility to influence the answers in the questionnaire, due to the "freshness" of information and influence to the perception by other media sources. Information content of such sites through a research time would be "eaten". Using models provides an assortment of information and news that will not have such a short expiration date. Cognitive paradigm defines the selective effect of the message because the recipient already has prior knowledge, values, attitudes, needs, etc. which affect the interpretation of objects (Kress & Van Leeuwen, 1996), therefore exposure of subjects to other media, which consequently affect the knowledge, values, attitudes, needs, cannot be completely eliminated from the study.
- Information used in research models must be selected so that they are not tied to any particular moment in time, so the user cannot "experience" the information as "old." This method of selection of information, necessarily leads to a narrowing of choices in selecting information. The value of information depends on the assessment of its timeliness.

The results show that respondents who were offered content in more visual forms, having more choice in what manner to use the content, assessed information to be more reliable, more professional, more persuasive and more impartial. Analysis with eye tracking device has shown that the mere presence of visual information does not necessarily mean that the person will use it or even notice it, but the results of assessment of credibility of information by the respondents confirmed that the presence of visual information and possibility of choice of receptive message channel has a positive effect on the valuation of media content.

Increased number of fixations and longer retention time on content indicates increased interest, which should definitely be a goal of any media content on the Internet, especially if we think about user habits regarding short retention on individual content. Analysis of a total of 28 articles showed that the visual changes in shaping the content lead to changes in user behavior, and thus in the process of decoding information. According to the results we conclude that the increased use of the media converged content has a positive impact on the number of fixations (which points to increased interest) and the retention time of the article. Statistically significant increase in the number of fixations appeared using a highly convergent media content for the 7 articles of the 14 analyzed. Using highly converged media, eight articles had significantly increased retention time. The use of video has been increased in the articles in which the video brought new information, something meaningful, that is, in cases where the visual information enriches the text with the new content, and contains no mere repetition of text information.

6 Viewpoint on Convergence

Media convergence has changed the way citizens access to information, the way they handle them and how they react to them—on a local, national and global level. Technological convergence of media is not reflected in the transition to a new information distribution medium, but in the creation of information from multiple sources and their presentation through all available forms of presentation—text, video and audio, while leaving to the user ability to choose the way they will meet their communication needs. Although media is converging with each other, they cannot simply merge. Visual design must be adapted to the new medium, to meet the needs of users.

This research has gave results in guidelines for the optimal design of visual media in a convergent digital environment:

- The use of converged media content allows the user a greater degree of selectivity in the use of digital media content;
- The use of converged media content has a positive effect on the assessment of the credibility of the message;
- The existence of multiple modes of media presentation of content, does not necessarily mean that users will use all kinds of information, but the existence of choice has a positive effect on the assessment of the credibility of information.

Based on the theoretical analysis and empirical study we can conclude that visual communication, in the new convergent media environment, is experiencing a redefinition. Visual mode of information is not only a decoration and complement to the written word, but also a real content carrier. The use of converged media content can have a positive impact on the assessment of the credibility of the media, and encourage the use of semantic and epistemological visual elements (graphic elements are in function of a fuller understanding of the message).

7 Conclusions

New graphic media discourse involves multiple media channels synergies, sharing of information content in multiple graphic presentations, allowing the user to select the information he wants to use, for how long and in what way. The research results indicate that the use of media synergy in graphic communication has a positive effect on the assessment of the credibility of the content, and thus the evaluation and use value of media content.

The research in this study showed that the use of appropriate and meaningful visual design, through the use of a high degree of media converged content, can shape the visual information as the cognitive content with targeted mediated meaning, which is tailored to maximize decoding capabilities of visual messages recipient, what increases the possibility of acquiring disseminated media messages.

Due to the dynamics of the development of information and communication technologies and social changes that they cause, it is difficult to shape one model that could encompass a broad range of technological and social changes that could affect the valuation of media content, but in this study the characteristics of impartiality, reliability, competence and persuasiveness emerged as essential elements in shaping the credibility of media information. Today, many media information cannot be checked in direct human interaction, so people often have the possibility to believe or reject information. However, graphic design should be the thread that will lead to acceptance rather than rejection of media information. Graphic design is characterized by the creative participation of several professions that seek to shape the information in the best possible way in order to focus fully on the needs of the user. The new rhetoric of graphic communication implies the need for synergy of existing graphics media discourse through the introduction of choices in the way the recipient will receive the information. This kind of media convergence fulfills the need of user divergence.

References

Abras, C., Maloney-Krichmar, D., & Preece, J. (2004). User-centered design. In W. Bainbridge (Ed.), *Encyclopedia of human-computer interaction*. London: Sage.

Barthes, R. (1977). *Image music text*. London: Fontana Press.

Flanagin, A. J., & Metzger, M. J. (2000). Perceptions of internet information credibility. *Journalism and Mass Communication Quarterly, 77*(3), 515–540. doi:10.1177/107769900007700304.

Fogg, B. J., & Tseng, H. (1999). The elements of computer credibility. In *Proceedings of the CHI 99 Conference on human factors in computing systems* (pp. 80–87). New York: ACM Press. doi:10.1145/302979.303001.

Gaziano, C., & McGrath, K. (1986). Measuring the concept of credibility. *Journalism Quarterly, 63*(3), 451–462.

Holmqvist, K., Holsanova, J., Barthelson, M., & Lundqvist, D. (2003). Reading or scanning? A study of newspaper and net paper reading. In J. Hyönä, R. Radach, & H. Deubel (Eds.), *The mind's eye: Cognitive and applied aspects of eye movement research* (Section 5: Eye movements in media applications and communication, pp. 657–703). Amsterdam: Elsevier Science.

Jenkins, H. (2007). *Reconsidering digital immigrants…* Accessed June 26, 2012, from http://henryjenkins.org/2007/12/reconsidering_digital_immigran.html

Johnson, T. J., & Kaye, B. K. (1998). Cruising is believing? Comparing Internet and traditional sources on media credibility measures. *Journalism and Mass Communication Quarterly, 75*(2), 325–340. doi:10.1177/107769909807500208.

Johnson, T. J., & Kaye, B. K. (2002). Webelievability: A path model examining how convenience and reliance predict online credibility. *Journalism and Mass Communication Quarterly, 79*(3), 619–642. doi:10.1177/107769900207900306.

Kress, G., & Van Leeuwen, T. (1996). *Reading images: The grammar of visual design*. London: Routledge.

Kurtić, N. (2007). Konceptualizacija istraživanja. In S. Malovic (Ed.), *Vjerodostojnost novina* (pp. 39–65). Zagreb: ICEJ.

Lidwell, W., Holden, K., & Butler, J. (2003). *Universal principles of design*. Gloucester, MA: Rockport Publishers.

Liu, Z., & Huang, X. (2005). Evaluating the credibility of scholarly information on the Web: A cross cultural study. *International Information & Library Review, 37*(2), 99–106.

Meyer, P. (1988). Defining and measuring credibility of newspapers: Developing an index. *Journalism Quarterly, 65*(3), 567–574. doi:10.1177/107769908806500301.

Morkes, J., & Nielsen, J. (1997). *Concise, scannable and objective: How to write for the Web*. Accessed June 26, 2012, from www.useit.com/papers/webwriting/writing.html

Müller, M. D. (2008). Visual competence: A new paradigm for studying visuals in the social sciences? *Visual Studies, 23*(2), 101–112.

Newhagen, J., & Mass, C. (1989). Differential criteria for evaluating credibility of newspaper and TV news. *Journalism Quarterly, 66*(2), 277–284.

Norman, D. A., & Draper, S. W. (1986). *User centered system design: New perspectives on human-computer interaction*. Hillsdale, NJ: Lawrence Erlbaum Associates.

Olaisen, J. (1990). Information quality factors and the cognitive authority of electronic information. In I. Wormell (Ed.), *Information quality: Definitions and dimensions* (pp. 91–121). Los Angeles: Taylor Graham.

Rieh, S. Y. (2002). Judgment of information quality and cognitive authority in the Web. *Journal of the American Society for Information Science and Technology, 53*(2), 145–161. doi:10.1002/asi.10017.

Rosenbaum, S., Rohn, J. A., & Humburg, J. (2000). A toolkit for strategic usability: Results from workshops, panels, and surveys. In T. Turner, G. Szwillus, M. Czerwinski, F. Peterno, & S. Pemberton (Eds.), *Proceedings of the ACM CHI 2000 human factors in computing systems conference* (pp. 337–344). Hague: ACM Press.

Wathen, C. N., & Burkell, J. (2002). Believe it or not: Factors influencing credibility on the Web. *Journal of the American Society for Information Science and Technology, 53*(2), 134–144. doi:10.1002/asi.10016.

Zhu, Z., & Ji, Q. (2004). Eye and gaze tracking for interactive graphic design. *Machine Vision and Applications, 15*, 139–148.

Print Your Personal Book: A New Service and Revenue Model for Open Access Journals

Martin Ebner, Sandra Schön, and Aldi Alimucaj

1 Introduction

Nowadays we are living in a world with big changes in the field of publishing and media. On the one side, new technology is growing with an amazing speed and new devices are available within a very short time frame. Smartphones and tablets are becoming ubiquitous in everyone's daily life, additionally to personal computers and laptops. Even the traditional TV stations are more and more merging with the Internet. The so-called "second screen" brings social media in our living rooms. Furthermore, Amazon and other companies are offering e-readers to deliver books and journals as e-books and e-journals directly to the device. The famous Kindle is just one of them. It can be stated that the world becomes more and more digitalized assisted by different surrounding technologies.

Furthermore the so-called Web 2.0 changed the way people deal with the Internet (O'Reilly, 2004). Nowadays anyone can easily write and therefore publish on the Worldwide Web by using different kind of media, like Weblogs, Wikis or even Video- and Audio-podcasts. Finally many different applications appeared allowing collaboration through technology and writing together on documents in real-time (Schaffert & Ebner, 2010).

Through these developments many different questions arise, concerning the traditional media market. For example: Should our daily newspaper still be delivered to our home in a printed form? Is a printed book maybe a thing of the past?

M. Ebner (✉) • A. Alimucaj
Social Learning, Information Technology Services, Graz University of Technology, Münzgrabenstraße 35/A, 8010, Graz, Austria
e-mail: martin.ebner@tugraz.at; aldi@outlook.com

S. Schön
ILAB–Innovation Lab, Salzburg New Media Lab, Salzburg Research Forschungsgesellschaft, 5020, Salzburg, Austria
e-mail: sandra.schoen@salzburgresearch.at

A. Lugmayr, C. Dal Zotto (eds.), *Media Convergence Handbook - Vol. 1*, Media Business and Innovation, DOI 10.1007/978-3-642-54484-2_10

Will humanity read only on digital devices in future? Even more questions arise, when we think about the entire process and business models. What will the reader pay for and in which ways? Considering the application markets of Apple (iTunes) and Google (Play Store) today's digital business model is individualized as much as possible. Finally, the World Wide Web has brought us openness, summarized by the open movement, where different kind of resources can be accessed and also used by anyone for free. So why should we pay for information, when we can get it for free on our device on any time?—Currently, traditional media and digital media are diverging more and more. There is a need for concepts, developments, and implementations to bring them together, to use synergies, and to develop innovative cross-medial solutions.

Within our contribution we will describe and discuss a new service and revenue model for open access journals: the personal book printing service for the Open Journal System (OJS). Our question was: How can open access journals be assisted by traditional (print) media to strengthen convergence, to build a new service and a new revenue model?

To start with, we will give a short introduction to open access journals and their revenue models, to the new print-on-demand services, and to the most often used open journal system, OJS. Building upon this state of the art, we will describe the technical prototype of our new "personal book printing service" for OJS as well as our experiences with it.

2 Literature State of the Art

2.1 Open Access Journals as Part of the Open Movement and Their Revenue Models

Open access journals are part of the "open movement". Sharing intellectual property with other people is one of the most important movements of the last two decades. The World Wide Web allows anyone to offer his/her work, documents, programming code, or any other resources online to anyone. The OECD (2007) defined the open movement goal and ambitions as *offering everybody the possibility to access intellectual resources freely for (non-) commercial purposes*. The open movement is a general term for many different developments:

- Open source software includes software published under certain licenses that allows the end user to use it without any charge.
- Open source hardware is the necessary resource needed to understand and rebuild a project that describes a hardware design.
- Open standards are a set of rules that assist the exchange of open data.
- Open design is mainly free design templates and resources for web design.
- Open knowledge is any content, information or data that is free accessible and reusable without any restrictions.

- Open data refers to an initiative that makes public information available and reusable for everyone.
- Open content is any kind of resource, freely available for reuse, revise, remix, and redistribute.
- Open educational resources are free online teaching and learning materials. "Open courseware" are whole courses and lectures series provided online.
- Last but not least, open access concentrates mainly on scientific journals and allows the download and free access to articles.

Hence, the primary goal of the open access initiative is that results of science research must be available for free for the public. Several arguments strengthen and support this open access movement: First of all, very often the authors and reviewers of scientific publications and content are paid or co-funded with public money. Naturally, the scientific community and the public have to get free access. Further arguments for open access are: Scientific research will be improved, if any scientific institution has access to any result. Organizations (like startups or companies) can use free scientific publications to bring innovations faster to the public market.

In general, there are different ways for open access, tagged with "colors" (Harnad et al., 2004): The "green road" to open access are pre- or re-publications of printed texts, e.g. in institutional repositories. The "golden road" and best way to support the movement is the publication in an open access journal. Open access journals are per definition free available, i.e. every article and issue is for everybody freely accessible within the Web. Concerning to the Budapest Open Access Initiative users should have the right to "read, download, copy, distribute, print, search, or link to the full texts of these articles" (BOAI FAQ, 2012).

By January 2013, the Directory of Open Access Journals (DOAJ, 2013) lists more than 8600 journals that are available as defined, use a peer review system and have an ISSN. The story of open access is a success story, but nevertheless it still tackles with traditional reputation arguments of impact factors bound to traditional (printed or electronic fee-affordable) journals. Nevertheless, there are already disciplines such as computer sciences and physics where it seems to be obligatory to publish open access to be recognized by the scientific community. Additionally, a majority of studies shows that publications with open access are more often cited than non-free publications (Wagner, 2010)—an important argument for future publications strategies for researchers.

Offering something for free does not mean it is produced and developed without costs. Open access journals are often managed and offered by researchers or institutions without sufficient knowledge in journal publishing. Not surprisingly, they sometimes struggle with the monetary component. These new editors provide us with enthusiastic and very often inspiring work, but with a lack of financial interest and therefore tend to be naïve in a mid-term perspective of the journals. In addition, public funded open access journals (e.g. by research funds) very often needs to look for funding alternatives after the first kick-off funding phase. As a result, open access journals has developed new business models to (re-) fund and co-finance their costs of setup, hosting, editorial work and management.

Some of the revenue models for open access journals are listed below (OpenAccess-business-models, 2013):

- Author pays: The authors of the article pay the full publication fees. This is also known as "article processing charge".
- Research foundation: The publishing of the article is paid from research project funds.
- Institutional funding: The open access journal is part of the institutional work, e.g. publication strategies, and funded with internal money.
- Institutional subsidies: For example, universities take care about the repository using their existing infrastructure and know-how.
- Publishing support fund: The open access article is paid by the institution or library of the researchers.
- Hybrid business model: This model is a mixture of subscription and publication fee. For example, the editor Springer offers to publish an article online for a defined fee.
- Community-fee model: The members of a community gather funds to finance open access.
- Other financing options: Newer funding options using the idea of donations, sponsoring, or crowd funding. In the case of crowd funding, publishing is only possible if many people provide the necessary money beforehand.
- Combined model: Very often, open access journals use a combination of different models mentioned above.

But "open access" does not mean that it is not allowed to offer additional services where users (readers) have to pay for, especially printed version of its issues. Some open access journals use printed versions as an additional service and (additional) way to get some (small) earnings. For example, the open access journal J.UCS also appears in an annual printed archive edition (J.UCS, 2013). As we will show in the following paragraphs, the availability of new print-on-demand services can be seen as the base for new services and also revenue model for open access journals.

2.2 Print-on-Demand as a Possible Base for a New Service for Open Access Journals

Print-on-Demand (PoD) denotes mostly the technology based on digital printing that enables printing different kinds of documents even in a small editions. With other words, the digital document is sent to a service provider who does the print job.

Mostly PoD is directly connected to self-publishing which allows to bring a book directly to the market without a traditional publisher. Due to the fact that such

	29,90	Gross price (set from the publisher)
-	1,96	Value added tax (VAT)
=	27,94	Price after tax
-	1,40	5% from marketing campaign
-	13,97	50% PoD's share
-	4,20	Print costs
-	4,00	Stock, shipping, digital administration
=	4,37	Author/Publisher share

Fig. 1 Amazon PoD costs and revenue example

providers offer also an appropriate online portal, publishing a book for anyone was never easier. The main steps are uploading a document and paying a valid ISBN number; afterwards the book is offered on the platform for any interested readers. Amazon, the world greatest bookstore, also follows this trend and is offering "Amazon Print on Demand"; other enterprises offer similar services. So any private author finds a convenient way of selling his/her books. The revenue model is rather simple: Figure 1 shows an example of the price of a PoD book calculated with Amazon's service. The author defines the price of and finally gets revenue for each sold book.

PoD services allow anyone to publish his/her document as book available in the bookstore without the need of a traditional publisher. Usually, the book is printed when ordered or printed in small editions when it is on demand. On the one side, this leads to a high flexibility, on the other side, the book price is more expensive due to higher print costs. Book on demand as a printing technology was the base for a new publishing process: Authors are now able to print, publish, and sell books, even with a valid ISBN, with relatively very low cost and low risk. The main advantages of self-publishing are that it is easy to publish, it is easy to update the content and it is seen as a step to a more democratic publication business (Kirschner, 2003).

Not surprisingly, open access journals and other open content initiatives, which develop text materials, use such new PoD and self-publishing possibilities to offer additional printed versions. It has to be seen as a service for readers and an additional way to find readers that do not want to search for free texts but for a book on a portal. Additionally, it is an additional (small) revenue for the authors and providers.

The open access journal "bildungsforschung" (bildungsforschung, 2013) is such a scientific interdisciplinary journal (on education and learning in German language) that already uses print on demand services for 4 years. Several issues are available as printed books within online and also traditional book services with the help of a German PoD publisher called "Book On Demand" (BOD) (Norderstedt, Germany). Another example for scientific texts is the German e-learning textbook project "L3T" (Ebner & Schön, 2011a). It is available for free, via mobile devices (Ebner & Schön, 2011b) and in the iTunesU store, as well as with cost as e-book app and in two different printed editions (as softcover and hardcover)—and with two different German PoD-publisher (BOD from Norderstedt and "epubli" from Berlin).

From our own experiences as editors of these two open access initiatives we have seen that it is just a limited extra workload to publish such texts as book (e.g. we have to develop a cover and to order it, about 2 days of extra work) and a manageable sum (about US$300, excluding copies for authors) for the printing. These costs cover the cost for the contract, the ISBN, and the monthly fee for hosting of the book.

In general, we see that the print service means additional effort for us but that this extra effort is more or less refunded by the author margins; in the case of our textbook project we even have a more positive result. Nevertheless, we have seen two limitations: (a) The print of issues is not very convincing, as we have additional other articles with similar focus in our journal/textbook that are not included in the issue and (b) the print of a whole issue is sometimes not needed and therefore too much/expensive. To sum up: the print-on demand service is not flexible enough to meet the interests of our readers. We guess that they would like to collect and select articles from our whole journal and textbook collection for their "personal printed issue".

Building on this, we sketched the idea of a "personal printing service". As it has to be included and part of our running system, the Open Journal System, we will now describe this before we will go on with our new service.

2.3 The Open Journal System as Major Technical System of Open Access Journals

The Open Journal System (OJS) is *a research and development initiative directed toward improving the scholarly and public quality of academic research through the development of innovative online publishing and knowledge-sharing environments* (Willinsky, Stranack, Smecher, & MacGregor, 2010). The open source project was started in 1998 to be able to manage, to publish, and to index online journals. The system has designed to reduce the time and energy devoted to all the tasks associated with editing a journal and all editorial processed. Due to the fact that the whole process is highly complex the software assists and provides following roles:

- Administrators manage the whole system as such as well as the user management. They are responsible for all preferences of the journal and they have to install additional plugins to extend the core system.
- Journal managers are responsible for a specific journal. He/she is able to publish a new volume and allocates editors.
- Editors can change and configure requirements, sections, review process etc. for one volume of a journal. Furthermore they are responsible for the whole publishing process—for the call for papers, the correct submissions, the allocation of

reviewers, the delivering of the reviews results and the acceptance of articles. Finally he/she is doing the final edits before publishing.
- Authors are people submitting articles.
- Reviewers are responsible for doing a review of allocated articles in predefined forms.
- Readers are the public who is interested in the final journal.

As mentioned, the software is open source and can be downloaded for free, currently in its version 2.4.1. It is written in object-oriented language PHP using Smarty template framework for the user interface abstraction. The backend supports most free SQL database management systems. OJS offers a huge variety of features due to its extendable architecture and the possibility of additional plugins.

The introduction of the OJS can be seen as a milestone for the open access initiative and movement. On the webpage it is visualized that in January 2012 more than 10,000 installations worldwide were in use (PKP, 2013): This number is even larger than the current number (8600) of registered open access journals in the (DOAJ, 2013). Consequently, we guess that the OJS is the most often used system used for open access journals. Therefore, every attempt to add new facilities and extra features for this open journal system can be seen as a support for open access journals and the whole open access movement.

Building on this insight, our department, Social Learning at Graz University of Technology (AT) already developed some add-ons for the OJS:

- Mobile access: There are already three mobile apps existing for reading any article on each OJS that is using our mobile access plugin. This plugin provides an API allowing mobile devices to download and read any content. In the appropriate stores are applications for iPad, iPhone, and Android phones.
- Social media: Another plugin concerns the embedding of different social media networks like Facebook, Twitter, or Google+. Posts done on one of these platforms will be directly shown in the sidebar of the OJS.
- Recommender system: This plugin recommends directly at the article page another similar articles, based on content similarity or user reading paths (Taraghi, Grossegger, Ebner, & Holzinger, 2013).

The following add-on is our newest development.

3 Methodology and Approach

The methodology of our work is strongly following a user requirement analyses and a follow-up prototyping. According to the field of software engineering prototyping is implementing an early and very first software product to test and trial a new approach. Before starting the prototype careful system requirements, user roles and work flows are designed. After the implementation a first evaluation has been done to improve the first prototype.

4 Implementation of a Prototype for a Personal Printing Service as Add-on for the Open Journal System

Within a master thesis project at the University of Technologies in Graz (Alimucaj, 2011) we sketched and implemented a new service called "personal printing service" as add-on for the OJS.

4.1 Focus and General Assumptions

The main idea of our research work is to combine the idea of open access with the emerging technology of PoD using the OJS to create a new printing service and revenue model. Due to the fact that each OJS holds a huge number of articles, it should be possible to combine different articles of different volumes in order to meet the expectations of readers. Afterwards the chosen articles are converted to just one document with additional overall journal information and sent to the PoD-provider who is responsible for the production and shipping of the personal book.

4.2 User Roles

As mentioned before, the OJS has a flexible and extendable architecture. With the help of so-called "plugins" the system can be enhanced in different ways. In our particular case a block-plugin has been developed, due to the fact that those plugins react on user inputs and display a different user interface.

First of all, use cases were designed for our plugin to generate the necessary workflows. Figure 2 shows the general overview. It must be considered that the implementation will work with two different systems—the OJS as well as the PoD-system. There are four different user roles:

- Author: The author writes an article, submit it to the system, and get feedback according to the review process.
- Publisher: The publishers are responsible for the whole publishing process. This group includes editors, sections editors, layout editors, copyeditors, and proofreaders.
- Reader: The readers link both systems, because he/she decided to choose different articles for print and customize the book on the PoD-provider site.
- PoD-Provider: The provider is responsible for the book print, its delivering, and finally also for the financial part of the process.

4.3 Workflow

Figure 3 points out the workflow from the user's perspective. He/she decides to print articles from the OJS. Afterwards he/she has to select articles, to rearrange it and to start the PoD-process. If the automated process is accepted from the

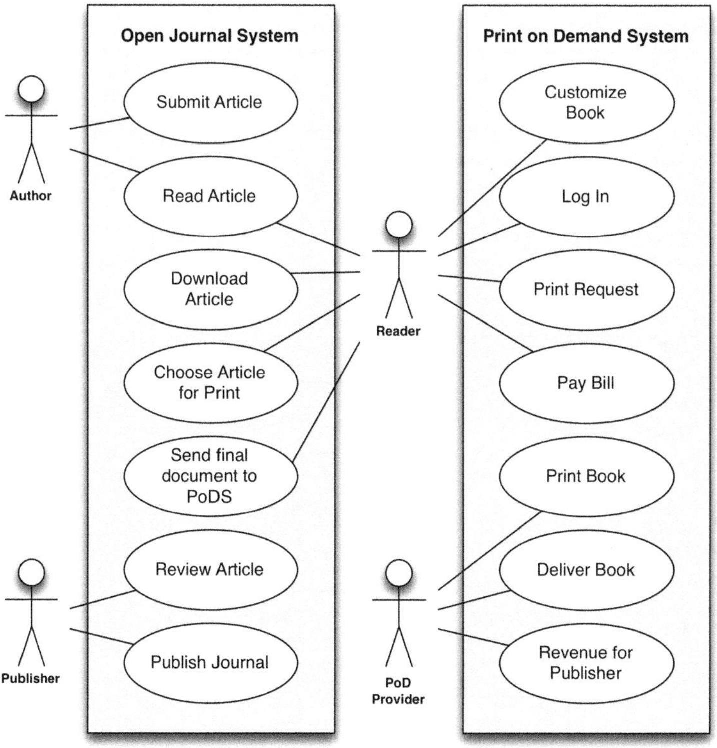

Fig. 2 OJS and PoD system use case

PoD-provider the user will be asked to make the final customizations of the book as well as accepting the order. The order will be finished as soon as the user pays and the production will be started from the PoD-provider.

It must be noticed that such a workflow is only possible in case the PoD-provider offers a web-based Application Programming Interface (API) to send the final document and the additional parameters. The technical communication of the OJS with the PoD system was a precondition for developing this plugin.

4.3.1 Configuration

In this section the mandatory parameters of the developed plugin are discussed. For example, the partner ID, a security key, and the repository must be provided by the PoD-provider. The plugin itself is programmed independently to give the possibility to cooperate with any PoD-provider. In our particular case the company epubli joined us for this research work.

Two further parameters "cover" and "format" are needed for generating the final document. After the selecting of articles by the user and pressing the button

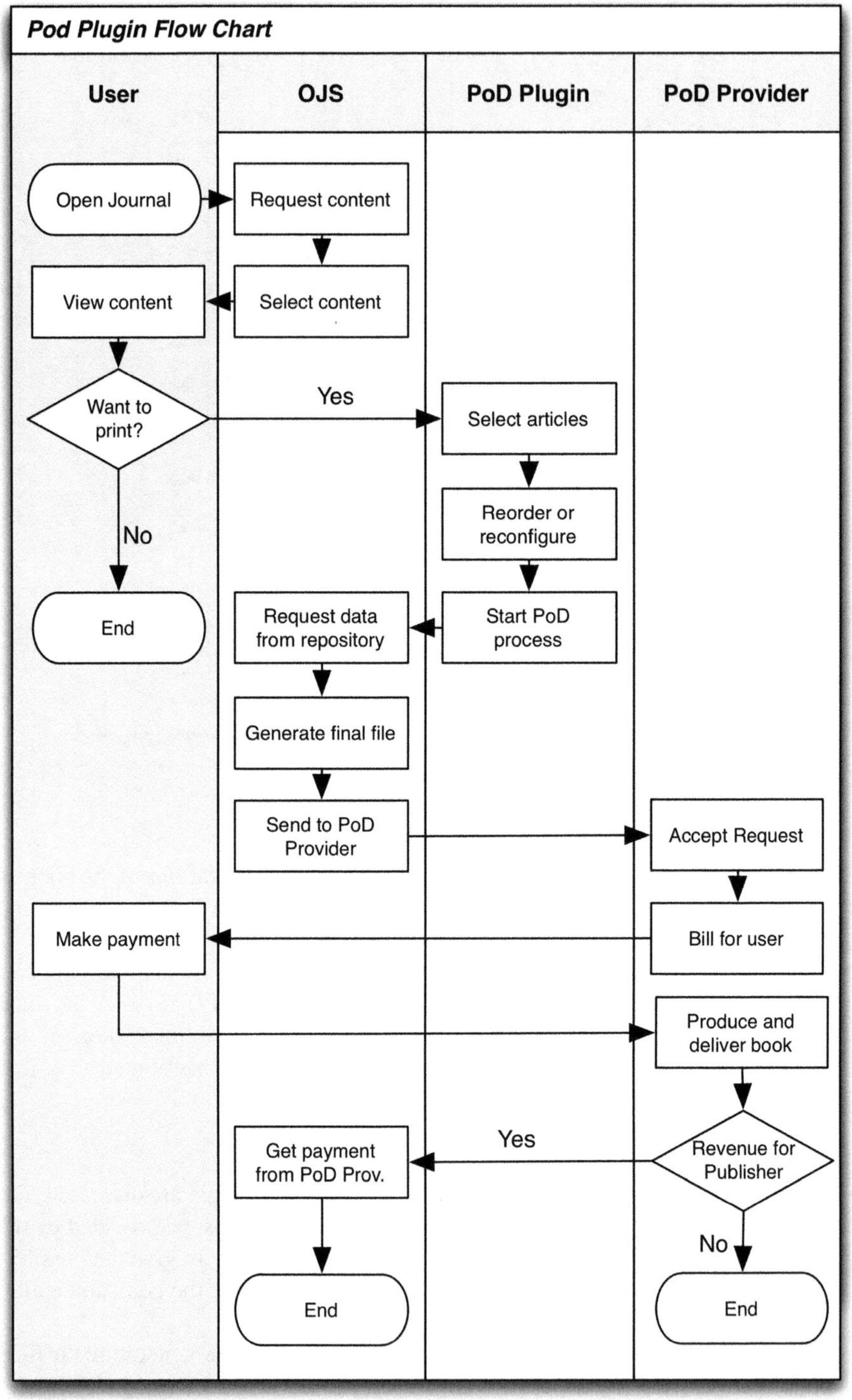

Fig. 3 PoD flow chart

"generate" the cover is automatically added to the pdf-file, which will be delivered to the PoD-provider in the predefined format.

Finally, there is also a parameter called "price". There are two different ways to define the amount of money that should be transferred back to the publisher (revenue):

- Static price: The publisher defines a fixed price that will be added automatically to the PoD-providers one.
- Dynamic price: The publisher defines a ratio. With other words the price depends on the number of printed sites; for example: 1:100 means that 1 € (if € is the defined currency of the PoD-publisher) is paid for each 100 sites; or 0.1:10 means that 10 sites cost 10 cents.

4.3.2 The Prototype Online

For our real-life test we implemented the prototype within the OJS of our textbook project L3T (Ebner & Schön, 2011a, 2011b) in cooperation with the German PoD publisher (epubli, 2013). After installation of the plugin and activation on the plugin setup site the parameters has been defined. Afterwards, the "Select Article Box" appeared in the sidebar of the OJS. Figure 4 shows the box, where different articles can be easily chosen by the reader and be added to the shopping cart. Before the user generates the final document he/she is able to rearrange or delete articles. After pressing the button "Print on Demand" the document is directly sent to the PoD-publisher with all predefined parameters. In the reader's browser the homepage of the PoD-provider appears and all further steps must be done there.

The strength of this application is that just this small box is needed to handle the whole process to maximize usability experiences to the consumers.

Fig. 4 Select article box at http://l3t.tugraz.at

5 First Experiences with the New Personal OJS Printing Service

As a reader of our report you may guess that we were pretty curious, and yes, we were also a little bit enthusiastic about the possible reactions on our new service. Nevertheless, the first reactions did not surprise us: The new service was implemented and it worked: We ordered a personal issue of L3T and got a small softcover book with our selected chapters. We count more than 180,000 downloads of single chapters at L3T homepage, therefore the traffic promised potential customers. But nobody else used our service within the first 2 weeks.

We made a short advertising video for our new service, distributed in our community-network, e.g. at our Facebook fan page, Twitter account and Google+ page. Additionally, we installed the add-on at the open access journal "bildungs-forschung" with more than 1000 visits in a single month.

Same there: Within the next 3 weeks, no one ordered a personal printed version. Nevertheless, we got very positive feedback over our social media channels ("cool service!", "great work"). For example, on Facebook the post of the video attracted 5146 people, but only 84 watched the video (1.6 %) and only 52 of them (1 %) clicked on the link to the system. When a closer look at the API is done, it can be carried out that till today 138 people generated a final document and were passed on to the PoD page, but nobody of them ordered the final book. To clarify the barriers of ordering we asked selected people and did some small updates (e.g. revising spelling error). Our investigations, e.g. at an Austrian e-book conference, did result that the whole idea is hardly understandable for people who never heard about PoD possibilities. Additionally, the service itself was confusing: Why should an open access text be printed? The main idea of our approach seems to be the main hurdle: Is our innovative cross-media solution too new?

6 Discussion

The goal of our research and implementation work was to develop and implement a new concept for the convergence of digital media (in our case: open access journal articles) and traditional media (printed journals) by combining existing systems via a web-interface. With the help of new developed plugin for the OJS, readers now can order their personal book by a PoD-provider who is delivering the book to costumers as well as the defined revenue to the publisher.

The advantages of this business model are manifold. For each party they are:

- Publisher: For the publisher of an open access journal this plugin offers the possibility for a complete new revenue model. For the first time publishers can get revenue by printing their free available articles.
- Reader: The user can get a printed book as flexible and personal as possible. He/she defines which articles in which order are parts of the final product. The possibility to combine articles of different volumes makes it even more

interesting. For example, with the aid of the recommender plugin, articles of the same research field can be easily found and added to the shopping cart. The reader only gets what he/she really wants to pay for.
- PoD-Provider: The provider benefits from an increased turnover and a new market, due to the fact that open access journals are supposed to have a big future.

On the other side, they're also some critical facts:

- Very often, the open access journal editors did not ask for or simply got the necessary rights to print and sell (!) the text delivered by the authors. This is especially the case when licenses are used e.g. the creative commons licenses CC BY-NC or CC BY-SA.
- The provided revenue model did not take care about the voluntarily work of the authors and reviewers. Nevertheless this is an important task for the future to think about how also these stakeholders can profit from selling their articles (authors) and giving valuable feedback (reviewers).
- Due to the fact that all articles are online available for free the number of prints will be small compared to the number of downloads. Therefore it can be expected that this revenue model will not revolutionize the open access market, its impact is limited to a revenue model for parts of the costs.
- For the reader there seems to be only one real disadvantage: Ordering a personal book means definitely more work than just ordering a predefined volume. Therefore the authors implemented a one-click-order of the current issue, but nevertheless the necessary steps on the PoD-platform still exist. We guess that this aspect is the key pro-argument for the service and in the same way one of the hard challenges.
- In the launch phase of the plugin the idea of printing a personal book was spread by an intensive marketing campaign. Short videos were produced to explain the idea in general (http://youtu.be/JRzr8psJ840) as well as to show the necessary steps that must be done by the customer. Anyway, in personal interviews the authors recognized that the concept is hardly understandable to people who never got in touch with Print-on-Demand possibilities.
- There is an arbitrarily effort for PoD-publishers, because they have to provide a platform where the users (readers) can be managed as well as an innovative API (Application Programming Interface) which allows data exchange between the OJS and the publisher's system. In our particular case the young company epubli cooperated in the research work. Another PoD-publisher could not be convinced so far, mostly because of the missing API and the arbitrarily effort to develop it.

7 Conclusion: Relevance to Convergence

Our research work aims to bring digital and traditional media together in a way that both benefit from each other. Therefore a plugin for the worldwide largest open source platform for publishing open access journals, the Open Journal System, has been developed that allow readers to select a number of articles and sent the selection to a Print-on-Demand publisher for printing. Due to the fact that the OA-publisher can define the revenue added by the plugin automatically to the price of the PoD-provider a possibility to allocate money is given.

It can be summarized that the proposed plugin is able to close the gap between open access journals and printed journals, with a new and an attractive possibility to print a personalized book. From this perspective it can be stated that an innovative cross-medial solution was developed. On the other side, this idea needs time to establish, due to the fact that readers have to see their benefits and recognize the power of PoD solutions in combination with innovative web technologies. The provided implementation can be seen as step towards convergence of media, but much more research effort is needed to help open access to become a serious market competitor.

References

Alimucaj, A. (2011). *Individualization possibilities for books with "Print On Demand" technique*. Unpublished Master thesis, Graz University of Technology.

bildungsforschung. (2013). *bildungsforschung*. Retrieved January 15, 2013, from http://bildungsforschung.org

BOAI FAQ. (2012). *Budapest open access initiative. FAQ*. Retrieved January 10, 2013, from http://www.earlham.edu/~peters/fos/boaifaq.htm#openaccess

DOAJ. (2013). *Directory of open access journals*. Retrieved January 15, 2013, from http://doaj.org

Ebner, M., & Schön, S. (2011a). *Lehrbuch für Lernen und Lehren mit Technologien (L3T)*. Retrieved January 15, 2013, from http://l3t.eu

Ebner, M., & Schön, S. (2011b). L3T assists m-Learning. In *Mobile learning: Crossing boundaries in convergent environments* (S. 69–72).

Epubli. (2013). *Epubli*. Retrieved January 15, 2013, from http://epubli.de

Harnad, S., Brody, T., Vallieres, F., Carr, L., Hitchcock, S., Gingras, Y., et al. (2004). The access/impact problem and the green and gold roads to open access. *Serials Review*, *30*(4), 310–314. doi:10.1016/j.serrev.2004.09.013

J.UCS. (2013). *Journal of Universal Computer Science*. Retrieved January 10, 2013, from http://www.jucs.org/

Kirschner, T. (2003). *Selbst Publizieren mit eBooks und Print-On-Demand*. Norderstedt: Books on Demand GmbH.

O'Reilly, T. (2004). *What is Web 2.0: Design patterns and business models for the next generation of software*. Retrieved July 26, 2013, from http://www.oreillynet.com/pub/a/oreilly/tim/news/2005/09/30/what-is-web-20.html

OECD. (2007). *Giving knowledge for free: The emergence of open educational resources*. Retrieved September 2015, from http://www.oecd.org/edu/ceri/38654317.pdf

OpenAccess-business-models. (2013). *OpenAccess*. Retrieved January 15, 2013, from http://open-access.net/de/allgemeines/geschaeftsmodelle/

PKP. (2013). *Journals using open journal systems by continent*. Retrieved January 15, 2013, from http://pkp.sfu.ca/ojs-geog

Schaffert, S., & Ebner, M. (2010). New forms of and tools for cooperative learning with social software in higher education. In B. A. Morris & G. M. Ferguson (Eds.), *Computer-assisted teaching: New developments* (pp. 151–165). Hauppauge, NY: Nova Science Publishers.

Taraghi, B., Grossegger, M., Ebner, M., & Holzinger, A. (2013). Web analytics of user path tracing and a novel algorithm for generating recommendations in Open Journal Systems. *Open Information Review, 37*(5), 672–691.

Wagner, A. B. (2010). Open access citation advantage: An annotated bibliography. In *Issues in science and technology librarianship*. Retrieved January 15, 2013, from http://www.istl.org/10-winter/article2.html

Willinsky, J., Stranack, K., Smecher, A., & MacGregor, J. (2010, September 29). *Open journal systems: A complete guide to online publishing*. Stanford, CA, USA.

Business Model Convergence and Divergence in Publishing Industries

Seppo Leminen, Juho-Petteri Huhtala, Mervi Rajahonka, and Mika Westerlund

1 Introduction and Problem Discussion

The rapidly progressing digitalization of media content has impacted traditional publishing industries heavily in the last decade (Fetscherin & Knolmayer, 2004). Moreover, there are other momentous paradigm shifts taking place in the media business, such as the transformation from products to services and from efficiency to flexibility (Westerlund, Rajala, & Leminen, 2011). The digital disruption has led to the launch of new devices, such as e-readers and tablets, which further boost the evolution in the electronic publishing industry, and the number of various e-readers is expected to grow remarkably in the future (Huhtala & Sihvonen, 2012). As technology for media industry is developing promptly, media consumption patterns have shifted towards digital format, which has raised publishers' interests to expand their digital content offerings (Flavián & Gurrea, 2006).

New technological innovations together with changes in consumer preferences to use media have stipulated both traditional publishers and industry entrants to

S. Leminen (✉)
Laurea University of Applied Sciences, Espoo, Finland

Department of Marketing, Aalto University School of Business, Aalto, Finland
e-mail: seppo.leminen@laurea.fi; seppo.leminen@aalto.fi

J.-P. Huhtala
Department of Marketing, Aalto University School of Business, Aalto, Finland
e-mail: juho-petteri.huhtala@aalto.fi

M. Rajahonka
Department of Information and Service Economy, Aalto University School of Business, Aalto, Finland
e-mail: mervi.rajahonka@aalto.fi

M. Westerlund
Sprott School of Business, Carleton University, Ottawa, ON, Canada
e-mail: mika.westerlund@carleton.ca

A. Lugmayr, C. Dal Zotto (eds.), *Media Convergence Handbook - Vol. 1*, Media Business and Innovation, DOI 10.1007/978-3-642-54484-2_11

create new business models for digital media services (Fetscherin & Knolmayer, 2004; Huhtala, Tölö, Penttinen, Rajahonka, & Leminen, 2011; Stahl, Schäfer, & Mass, 2004; Tian, Martin, & Deng, 2008). As computing technology and telecommunication networks make it possible to merge content from newspapers and magazines, music, radio, television, films, and entertainment, business models have become convergent in the new media context. Companies share similarities in multiple dimensions of business model, such as the revenue logic and infrastructure (Huhtala et al., 2011). They are also encouraged to develop divergent business models in the new media industry, because both media content and its audiences become fragmentized (Appelgren, 2004).

The understanding of the 'business model' concept has also evolved during the previous decade (Casadesus-Masanell & Ricart, 2010; Osterwalder, Pigneur, & Tucci, 2005; Westerlund, Rajala, & Leminen, 2008; Zott & Amit, 2008). Osterwalder (2004) provides probably the most widely used definition of the concept by defining business model as a firm's logic for creating and commercializing value. Furthermore, this definition suggests that corporate business models consist of various 'pillars'. Recent studies (e.g., Leminen, Huhtala, Rajahonka, & Siuruainen, 2011) suggest that, due to the digitalization of content, publishers face various management challenges that address every pillar of digital business models.

Other topics of interest for business model researchers include business model elements, change factors, and dynamics (Jarzabkowski, 2005; Tikkanen, Lamberg, Parvinen, & Kallunki, 2005). For instance, Tikkanen et al. (2005) introduce a business model framework to explain the relation of material aspects, belief systems, and the evolution of the business. Understanding publishers' belief systems is critical, since there is an ongoing transformation and convergence of media industries and business models that publishers apply. These changes lead to necessity to challenge the existing assumption of business models.

While the focus has been on conceptualizing business models and their change factors, the extant literature provides little empirical evidence of the evolution of business models and their convergence and divergence in both cross-industry and industry levels. Therefore, this study depicts the evolution paths of business models and their underlying belief systems in order to reveal industry-specific and cross-industrial convergence and divergence of new media business models in Finland. In particular, the research problems are:

- What are the underlying belief systems that enable or prevent industry-specific and cross-industrial convergence and divergence of media business models?
- What are the business model evolution paths in the newspaper, magazine, and book sectors of the Finnish publishing industries?
- How are business model convergence and divergence defined in these publishing industries, and what kinds of examples can be found?

The article is divided as follows. First, after a brief introduction to the study, the article reviews the theoretical foundations for business model evolution in publishing industries. Second, it describes the research methodology as well as data collection and analysis. Third, it presents the empirical findings on media business models' material aspects and their underlying belief systems. Finally, it discusses the findings and concludes on the convergence and divergence of media business models in the publishing industry.

2 Literature Review

2.1 Material Aspects, Belief Systems, and Business Model Evolution

During the previous decade, the interest of business model research has shifted from defining business model towards understanding the business model dynamics (Tikkanen et al., 2005). Recent studies have focused on the conceptualization of business model evolution (Tikkanen et al., 2005) and different types of business model change (Cavalcante, Kesting, & Ulhøi, 2011). According to Cavalcante et al. (2011), there are four different types for business model change involving specific challenges and characteristics: (i) business model creation, (ii) business model extension, (iii) business model revision, and (iv) business model termination.

Business model creation is defined as a transition of new ideas into standardized processes of a new business model (Cavalcante et al., 2011), and it occurs, e.g., when a company from one industry starts doing business in another industry. Business model extension takes place when a company adds new processes in its existing business model (Cavalcante et al., 2011). For example, a publisher's current business model is extended when the company increases its product portfolio from print to both print and digital products. Business model revision refers to removing something from the existing business model and replacing it with new processes (Cavalcante et al., 2011). For instance, change in the revenue logic of a business model is counted as business model revision. Finally, business model termination means extinguishing the current business model.

In order to develop business models towards a more effective direction, an organization has to understand and recognize the barriers of business model change (Chesbrough, 2010). Tikkanen et al. (2005) emphasize the role of managerial cognition in the business model evolution and argue that industry-specific belief systems affect managerial decision-making; thus, enabling or preventing business model change. These belief systems include industry recipe, reputational rankings, boundary beliefs, and product ontologies. Industry recipes express the persuasions of the management related to economic, competitive, and institutional logic of the firm (Spender, 1990). Boundary beliefs define the identity of the company with a certain inter-organizational community (Porac, Ventresca, & Mishina, 2002). Product ontologies link product or service attributes, usage conditions, and buyer characteristics into a hypothetically superior offering on the target market (Porac et al., 2002). Reputational ranking denotes firms' own performance related to its socially evaluated competition (Tikkanen et al., 2005). Tikkanen et al. (2005)

maintain that any relevant business-related outcomes consist of the interaction of material aspects of the business model with managerial belief systems.

2.2 Media Convergence

Media convergence interlocks computing and information technology together with telecommunications networks, newspaper and magazines publishers, music, radio, television, films, and entertainment. It brings together the "three Cs"—computing, communications, and content. Convergence takes place at the levels of technologies and industries, i.e. industry-standards that enable content delivery to different devices lead to technological convergence. Industry convergence takes place when companies from media, telecommunications, and technology industries merge or form strategic alliances to develop new business models together (Britannica Online Encyclopedia, 2013). Saltzis and Dickinson (2008) state that convergence means the coming together of different media, industries and products, and that from a technological perspective it functions at three levels: network ("information superhighway"), production ("write once, publish anywhere"), and distribution (convergence towards a single device).

Saltzis and Dickinson (2008) investigated how the working practices of journalists change due to convergence in news production. They argue that convergence is evolution rather than revolution. Traditional news organizations expand from single to multi-media journalism. Multimedia journalism means that journalists produce output for more than one medium at the same time. Organizations integrate their newsrooms to share resources and to manage a multimedia production process. Journalists have to cope with a widening range of responsibilities in their work. However, Saltzis and Dickinson (2008) argue that multi-skilled multimedia journalists are still a minority. In addition, the need for specialized workforce is decreasing as the production technology is becoming easier to handle. Saltzis and Dickinson (2008) conclude that news production is in continuous change and information can be treated more and more as a medium-neutral commodity.

The traditional media business model was based on media owners providing content to consumers, consumers paying for the content, companies delivering commercial messages to the media owners' audience, and regulators determining the parameters of the industry (McPhillips & Merlo, 2008). McPhillips and Merlo (2008) studied business models of the new media age and defined media convergence as the ability to deliver different media channels via one digital platform. According to their view, the industry will experience not a revolution but an evolution, as the old and new models firstly co-exist, and ultimately converge. It is evident that while horizontal integration will merge industries, distinct sectors will experience vertical disintegration because producers do not need to rely on media owners anymore. That is, media owners will cease to dominate the value chain from creative work to production, marketing, and distribution. However, as McPhillips and Merlo (2008) state, media owners are able to focus on the profitable parts of the value chain. The fundamentals of the business model based on the relationship between media owners, consumers, and advertisers are not changing, but the media landscape is becoming more complex.

Appelgren (2004) reviewed literature on convergence and divergence in the media industry and argues that there are as many definitions as there are authors. Moreover, albeit being conceptually opposites, convergence and divergence often appear together or follow each other. As technology behind the services and the production workflow converge, interactivity and niche publishing encourages divergence. A media company experiences divergence as media content becomes fragmentized and a consumer as amount of media channels increases. Appelgren (2004) also states that convergence has no end, but can be seen as continuously ongoing process; if the process ends, another process starts.

The last-mentioned idea of continuous change levels with Fine's (2000) description of supply chain structures. Supply chain structures are not stable, but they continuously cycle between vertical/integral and horizontal/modular. The fluctuation of supply chain structures can be explained with different set of forces that push the changes from vertical/integral to horizontal/modular structures and vice versa. However, the clock-speeds of industries are individual; for the computer industry two decades were enough for a full cycle. According to Fine (2000), the computer industry's supply chain structure experienced a structural shift from a vertical/integral industry structure to a horizontal/modular one in the late 1970s. This shift was generated by a single product/supply chain decision by a dominant producer, IBM. The company had decided to respond to a challenge it had faced from Apple Computer by launching the personal computer (PC). The modular architecture of PC encouraged multiple companies to enter the industry producing sub-systems, e.g., semiconductors, software, peripherals, network services, design, and assembly. These segments then evolved into separate sub-industries.

As a conclusion, we maintain that the publishing industry will experience horizontal integration and vertical disintegration due to media convergence and increased modularization of media content and production. The industry structures and borders will be redefined. The environment will at first become more complex (divergent) as the old and new models co-exist, but in the end these models slowly convergence. However, the changes are a continuously ongoing process, and the development of industry structures and business models can best be described as a pendulum, not a one-way trip.

2.3 Framework for Analysis

After reviewing prior literature on business models and media convergence, we propose a framework for analyzing the barriers and drivers of business model evolution in publishing industries. A company may aim at (1) creating a new business model, (2) extending its current business model, or (3) revising its current business model. It will either succeed or fail in achieving this goal, and we argue that the emphasis of four belief systems has a major impact on a firm's capability to

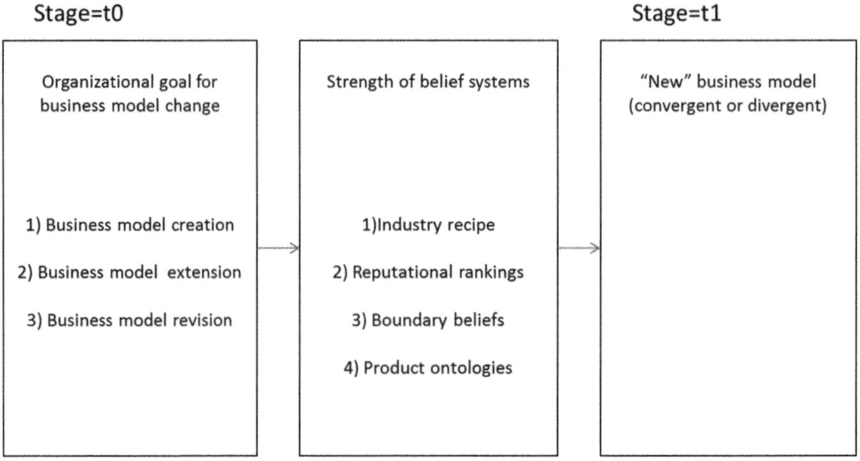

Fig. 1 Framework for analyzing business model evolution in the media industry

achieve its goals in business model change. Specifically, we speculate that strong belief systems engaged to the current business model will prevent business model change, while belief systems characterized by openness are driving the change. We use the proposed framework in analyzing four company cases to identify how belief systems effect on business model change. Additionally, we examine what kind of combinations of organizational goals of business model change and belief systems are driving business model convergence and divergence.

3 Methodology and Approach

We apply a multiple case study design to analyze business model evolution in the publishing industry. The cases were chosen because they were initiated as convergence and divergence of media business models in the publishing industry, which enables us to explore how their underlying belief systems enhance or inhibit changes of business models. The selected cases represent different types of business model evolution; (1) create new business model, (2) extend current business model, and (3) revise current business model. Therefore, we expect that findings from those cases are representative.

3.1 Data Collection and Analysis

The study puts forward four case descriptions of companies operating in the Finnish publishing industries. We obtained preliminary understanding of these industries and their ongoing change through interviews with 15 publishers (4 magazine, 7 newspaper, and 4 book publishers) that were developing digital content services between 2010 and 2012, during the time when tablet devices were launched in Finland. The data were collected as a part of the Next Media research program which focused on developing future concepts and business models for media.

We used multiple researcher participation in creating the case descriptions. They are based on publicly available information including content from web sites, bulletins, magazines, and public case reports. We focused on the four aspects of belief systems in media business models; industry recipe, reputational rankings, boundary beliefs, and product ontologies. The created framework was applied to track the business model evolution during the observed period. It helped to understand the types of belief systems and business model change that enhance convergence within and cross industries.

4 Findings

4.1 Market Disruptions

Market disruptions create business model opportunities for entrants from other industries to the publishing industries. In particular, the launch of Apple's iPad and Android tablets create both market and technology contingency in the market. The shakedown or disruption is a prerequisite for the media/publishing industries convergence to occur. The change of revenue logic, technological platform, and content to be published at diverse channels represent evolution of diverse business model elements. Consequently, factors hampering or enabling business model evolution are critical for understanding convergence and divergence.

The changes of business models are steered by assumptions at belief systems, while both market and technical disruptions challenge underlying belief systems. Factors enabling or hampering changes are related to four elements of the belief system; reputational ranking, industry recipe, boundary belief, and product ontology. We found that these elements forestall convergence of diverse media industries, while traditional publishing is characterized by lock-in to the current belief systems. Lock-in mechanisms create business opportunities for entrants from other industries to enter the market and challenge the existing players by creating new business models.

We illustrate the adaptation to and exploitation of market disruption through four companies operating in the publishing industries: (1) Gummerus, a traditional book publisher, (2) Sanoma, the publisher of Helsingin Sanomat, the leading newspaper in Finland, (3) Elisa, a newcomer to a book publishing industry from the telecommunication industry, and (4) Teknari, an organization converging different media industries. In the following, we will describe these companies' current business models and their underlying belief systems. We will also discuss their goal of business model development, and sketch their business model evolution paths from the material aspects perspective.

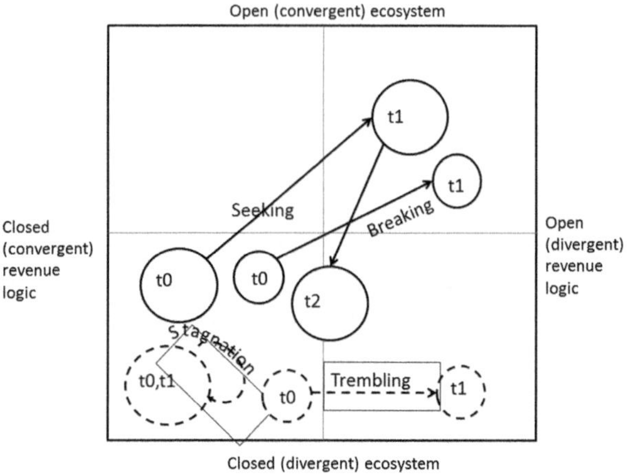

Fig. 2 Diverse media business model transformations

4.2 Evolution Paths of Business Models

According to our analysis, evolution paths of media business models vary among different publishing industries and sectors. After reviewing our findings, we were able to identify several types of media business model transformations (Fig. 2). Each type reflects distinctive characteristics that differentiate it from other types of transformation and affect respective publishers' competitive positions in the industry. The different media business model transformations include:

(a) *Seeking* characterizes companies such as Sanoma (particularly the Helsingin Sanomat newspaper) that pursue lucrative business models by revising the current business model and by seeking a balance between closed (divergent) and open (convergent) ecosystems and old (convergent) and new (divergent) revenue logics. This is especially true in the newspaper industry which has faced fierce competition from the emergence of online news and even citizen journalism.

(b) *Stagnation* is faced by those book publishers (e.g., Gummerus) that apply the traditional model of selling books for readers within closed (divergent) ecosystems also on selling e-books. In other words, they do not change their business model in any relevant way.

(c) *Breaking* enables industry newcomers such as Elisa to extend their current business models to ecosystems that have been opened for them due to convergence. Consequently, they can for example bundle and unbundle books as a part of larger digital media service, in which content may be personalized and modularized and financed with new (divergent) revenue logics.

(d) *Trembling* means that a newcomer from another industry, such as Teknari, enters into the publishing industry and disrupts the established industry logic by creating a genuinely novel business model with their digital offerings converging several products and services offered by traditional actors. The trembler creates intentionally a closed ecosystem based on new divergent earning logics.

Table 1 Belief systems in publishing industries

	Reputational ranking	Industry recipe	Boundary belief	Product ontology
Gummerus	Traditional book publisher	Edit, publish, and print books	Books for readers	Product is a book
Sanoma (Helsingin Sanomat)	Trusted publisher of the leading nationwide newspaper	Foundation for publishing newspaper	Quality content by professional journalists for readers Diversification of the boundary belief while transformation from journalist-driven content product to partly user driven approach	Diversification of the product ontology: newspaper, internet sites, apps for tablets and e-readers (e.g., iPad, Android tablets)
Elisa	No reputational ranking at publishing industries, new player from telecommunication industry	No industry recipe at publishing industry, integration of all media content to create lock-in situations for customers to the existing product portfolio (fixed and mobile broadband)	Diversification of the boundary belief Challenge traditional boundary belief while authors create single storylines, (book chapters) commented by readers Gamification of content	Diversification of the product ontology: rely on product ontology of books (existing ones), and create own modularized product ontology of books chapters
Teknari	No reputational ranking, a newcomer into the publishing industries	No industry recipe Convergence of diverse publishing industries, magazines, TV, and websites	No boundary belief Produce storylines financed by advertisers	Convergence of existing product ontology to new type of product portfolio (divergence)

4.3 Belief Systems Behind the Business Models

To create a profound understanding of business model evolution, as well as convergence and divergence in publishing industries, we need to depict the elements of belief systems in addition to material aspects of business models (Table 1).

Gummerus is a family-owned book publisher established in 1872. The owners respect the original business concept and philosophy of the founder of the company, who wanted to publish books "for the joy and benefit of all Finnish people" (www.gummerus.fi). The company's focus is on traditional book publishing, although it also produces audio books and e-books, as well as different kinds of mobile applications. The company is not the market leader; consequently, it has no reputational pressures, but can follow the industry developments from a safe distance. Therefore, the company has strong belief systems based on a traditional industry recipe ("we are in the book industry"), a boundary belief ("we are a book publisher") and product ontology ("our products are books", and "a book is a book is a book"). The products are books; however, they can be either traditional physical books or digital books. The revenue logic remains the same, i.e. the readers pay, and the ecosystem is closed.

Helsingin Sanomat is a newspaper that is a part of a big media conglomerate, Sanoma Oyj (www.sanoma.com). The company is a market leader and Helsingin Sanomat is the most reputable newspaper in Finland. Therefore, it has a strong reputational ranking that guides its behavior. It has been in the vanguard of industry development, with incremental innovations leading the newspaper industry developments in Finland. This ranking has been beneficial in its newspaper business. As the market leader, Helsingin Sanomat encourages in a limited manner industry convergence; only its reputational ranking and industry recipe are still locked in the newspaper industry. Instead, Helsingin Sanomat has been able to develop its product ontology by moving forward towards digital content, because the products have been modified into different delivery channels in slightly different formats. Boundary beliefs are tumbling down, and the company is reaching towards new customers with its digital offerings.

Previously, the company's journalists prepared quality content for its readers, but now this is changing: customer—created content is being tested and applied by Metro, a part of Sanoma (http://metro.fi/). Metro has tapped into the reader community in the Internet in an expectation that the online community would produce content for their advertise-based print and e-magazines. The same logic is now applied to Helsingin Sanomat and the readers' role is changing, as they act as content creators, submitting pictures and ideas for articles. Helsingin Sanomat has been active in changing its product ontology (also the physical product has been changed into tabloid format), delivering content to Internet pages, tablets, and e-reading devices.

Elisa is a strong actor in the telecommunications industry (www.elisa.fi). It has no reputational ranking in the book publishing industry despite of its new business model in the book industry (https://kirja.elisa.fi). Elisa challenges the traditional thinking in book publishing by entering the industry. The credibility of the company stems from its size and the amount of customers in telecommunications. The industry recipe of the book publishing industry does not limit the company, because its context is digital from inception, and it defines its industry in a broader way. Its industry recipe originates from the ICT sector where it pursues to take over digital context delivery business. By doing this the company wants to ensure customer loyalty and sell more of its traditional services, namely bandwidth.

Elisa is not limited by boundary beliefs or product ontology; it can produce services for any potential customers, because the current business model does not limit its actions. Elisa can break down the logic of the traditional book publishing, e.g., writing of a book can be done by more than one author, and the plot can be commented or advanced also by the readers. Thus, Elisa approves the book publishing industry as its field of competition, but it does not approve the traditional product ontology of the book publishing. Elisa sells and distributes classics and digital content is not sold only as a book, but the format of the book can be changed. New books can be sold also in smaller modules. In this manner, Elisa is diversifying the product ontology.

Teknari is a newcomer in publishing. It describes itself as being the biggest tablet magazine in Finland, and is published by T-Magazines Oy for MTV3 which is a Finnish commercial television station owned by Bonnier (www.teknari.fi). However, Teknari is also a TV program. It has no reputational ranking in the publishing industries. It exploits the shakedown of the industry and seizes new business opportunities. It combines features from different industry sectors, and shakes traditional publishing by acting in several industries. A frequently asked question is what Teknari is and in which industry it is; is it a magazine, a TV programs, or a website? Teknari has no boundary beliefs from the publishing industry since its storylines are financed by advertisers. It challenges the popular subscriber- and advertiser-financed products such as the Tekniikan Maailma magazine (http://tekniikanmaailma.fi/). Teknari promotes divergence by converging existing product ontologies into new types of products.

The current business models of publishing industries were created through established belief systems. The more there are lock-in elements in the belief systems, the longer the industry stays independent, and the longer there are clear border lines between diverse industries. The belief systems also create regional and technological barriers. However, the publishing industries face threat of new entrants from other industries, for example Elisa and Teknari. The business models of these new entrants are not hampered by belief systems of the incumbents in publishing industries. Therefore these companies have been able to develop new business models and trigger change in the industry. However, whereas Elisa's business model is limited by the industry boundaries, Teknari's industry boundaries have been tumbled down and the company has accelerated convergence of industries.

We argue that the change of product ontologies and boundary beliefs do not empower industry-specific nor cross-industrial convergence of media business models, rather changes industry recipes. The studied four cases support the findings that traditional publishing industries are reluctant to foster the convergence of business models. This is concordant to Leminen et al. (2011), who document that publishing industries are not willing to risk their existing profitable businesses by entering into insecure new media business models.

However, our cases show signs of convergence in publishing industries. Value creation logic is changing while not only small editorial local newspapers but also the leading newspaper Helsingin Sanomat, have a pressure to merge the editorial staffs in Finland. Local newspapers utilize and spread the same editorial content such as international news and concentrate on producing only local news. In the case of Helsingin Sanomat, which is a part of the big media conglomerate, journalist staffs of the newspaper and TV4 (significant TV broadcaster in Finland) are merged. That way, content for news, newspaper, web pages, and iPad and Android apps are generated simultaneously although their distinct product ontologies remain.

Modularized content enables the convergence of books, newspapers, and magazines while the storylines will shorten, thus being easier to conflate. However, we were unable to find an example of a business model which would integrate content created from other parties. Books and newspapers are composed by modularized content, but the existing product ontology seems to exist. In addition, users as co-creators of content are progressing in newspapers, magazines, and books. To that end, Nousiainen and Leminen (2012) illustrate three various storylines for creating content.

5 Conclusions

5.1 Convergence

Disruption of media market creates new opportunities. This article discussed the evolution of media business models by revealing their convergence and divergence. It depicted the business model evolution paths in newspaper, magazine, and book publishing industries and revealed the underlying belief systems of diverse publishing industries; newspaper, book, and magazine industries in Finland. The evolution of business models is seen at their material aspects. The study argues that the underlying belief systems either enable or prevent the development of new business models, thus prohibiting development and convergence of business models at the publishing industries. In other words, the more open the belief systems are the more new business model innovations may be found.

5.2 Can We Say If a Particular Publishing Industry or Publishing Industries Converge or Diverge?

Different types of convergence and divergence transpire: the industry specific convergence and cross-industrial convergence occurred in the studied publishing industries. We argue that a strong belief system prevents a company from shifting from one industry to another industry. New entrants from other industries foster cross-industrial convergence, because their belief system boundaries do not prevent the evolution of business models, while traditional publishers were locked in boundaries of their belief systems in the existing industries. The divergence of media business models is increased if the existing business models are not terminated when new models are introduced into market. However, industry specific convergence of different business model elements, such as technology platform, revenue logic, and content production occurs between existing players in specific industries. Players seem to adapt to the same value creation logic. We propose that new entrants foster cross-industrial convergence, not the existing traditional publishing industry.

5.3 Managerial Contribution

From the managerial contributions perspective, this study offers a straightforward tool in terms of a framework to analyze media business model evolution. In addition, the study provides new knowledge on the evolutionary paths of media publishers' business models by contrasting and comparing developments in different publishing industries. The findings can help publishers and other media companies steer their business model development to a chosen direction by focusing on the type of ecosystem and the type of revenue logic at a specific publishing industry.

Each and every research has its limitations. Our analysis is based on limited amount of case companies in publishing industries in one country because of limited research resources. We could not cover all publishers in the industry, which may affect the results on described evolution paths and belief system. We studied the case firm's business model evolution between 2010 and 2012. Therefore, we highlight the need to extend the study period and to include other countries and market areas. The view and framework proposed in the study offer remarkable potential for future research on the evolution of business models and underlying belief systems. While the industries are in a continuous pendulum-like motion, the direction and interplay of convergence and divergence are interesting topics to study. Finally, we call for more research on media convergence and divergence between and within diverse publishing sectors.

References

Appelgren, E. (2004). *Convergence and divergence in media: Different perspectives.* 8th ICCC International Conference on Electronic Publishing, Brasilia, DF, Brazil. http://creativegames.org.uk/modules/Media_and_Cultural_Theory/Appelgren_Convergence%20and%20Divergence%20in%20Media.pdf

Britannica Online Encyclopedia. (2013). Accessed January 12, 2013, from http://www.britannica.com/EBchecked/topic/1425043/media-convergence

Casadesus-Masanell, R., & Ricart, J. E. (2010). From strategy to business model and onto tactics. *Long Range Planning, 43*, 195–215.

Cavalcante, S., Kesting, P., & Ulhøi, J. (2011). Business model dynamics and innovation: (Re) establishing the missing linkages. *Management Decision, 49*(8), 1327–1342.

Chesbrough, H. (2010). Business model innovation: Opportunities and barriers. *Long Range Planning, 43*(2/3), 354–363.

Fetscherin, M., & Knolmayer, G. (2004). Business models for content delivery: An empirical analysis of the newspaper and magazine entrantindustry. *The International Journal on Media Management, 6*(1–2), 4–11.

Fine, C. H. (2000). Clockspeed-based strategies for supply chain design. *Production and Operations Management, 9*(3), 213–221.

Flavián, C., & Gurrea, R. (2006). The choice of digital newspapers: Influence of reader goals and user experience. *Internet Research, 16*(3), 231–247.

Huhtala, J.-P., & Sihvonen, A. (2012). The role of learning in eReading service development. In S. Leminen & J.-P. Huhtala (Eds.), *eReading advertising and value creation best practices – Outlook 2010–2011* (Edita Prima Oy Helsinki 2012, pp. 21–26). Vantaa: Laurea Publications 8.

Huhtala, J.-P., Tölö, M., Penttinen, E., Rajahonka, M., & Leminen, S. (2011). Business models in digital landscape. In S. Leminen & J. Salo (Eds.), *eReading services business models and concepts in media industry* (pp. 40–43). Vantaa: Laurea Publications A73. 49–68.

Jarzabkowski, P. (2005). *Strategy as practice: An activity-based approach.* London: Sage. 216p.

Leminen, S., Huhtala, J.-P., Rajahonka, M., & Siuruainen, R. (2011). Recommendations and managerial implications for media industry. In S. Leminen & J. Salo (Eds.), *eReading services business models and concepts in media industry* (pp. 136–149). Vantaa: Laurea Publications A73.

McPhillips, S., & Merlo, O. (2008). Media convergence and the evolving media business model: an overview and strategic opportunities. *The Marketing Review, 8*(3), 237–253.

Nousiainen, A., & Leminen, S. (2012). Future digital space. In S. Leminen & J.-P. Huhtala (Eds.), *eReading advertising and value creation best practices – Outlook 2010–2011* (Edita Prima Oy, Helsinki 2012, pp. 36–48). Vantaa: Laurea Publications 8.

Osterwalder, A. (2004). *The business model ontology, a proposition in a design science approach.* These Presente'e a' l'Ecole des Hautes Etudes Commerciales de l'Universite' de Lausanne. Accessed December 16, 2011, from www.hec.unil.ch/aosterwa/PhD/Osterwalder_PhD_BM_Ontology.pdf

Osterwalder, A., Pigneur, Y., & Tucci, C. L. (2005). *Clarifying business models: Origins, present, and future of the concept.* Communications of AIS 15 1. 40 p.

Porac, J., Ventresca, M., & Mishina, Y. (2002). Interorganizational cognition and interpretation. In J. Baum (Ed.), *Companion to organizations* (pp. 579–598). Oxford: Blackwell.

Saltzis, K., & Dickinson, R. (2008). Inside the changing newsroom: Journalists' responses to media convergence. *Aslib Proceedings, 60*(3), 216–228.

Spender, J. C. (1990). *Industry recipes.* Oxford: Basil Blackwell.

Stahl, F., Schäfer, M.-F., & Mass, W. (2004). Strategies for selling paid content on newspaper and magazine web sites: An empirical analysis of bundling and splitting of news and magazine articles. *The International Journal on Media Management, 6*(1–2), 59–66.

Tian, X., Martin, B., & Deng, H. (2008). The impact of digitization on business models for publishing some indicators from a research project. *Journal of Systems and Information Technology, 10*(3), 232–250.

Tikkanen, H., Lamberg, J.-A., Parvinen, P., & Kallunki, J.-P. (2005). Managerial cognition, action and the business model of the firm. *Management Decision, 43*(6), 789–809.

Westerlund, M., Rajala, R., & Leminen, S. (2008). SME business models in global competition: A network perspective. *International Journal of Globalisation and Small Business, 2*(3), 342–358.

Westerlund, M., Rajala, R., & Leminen, S. (2011). *Insights into the dynamics of business models in the media industry.* Vantaa: Laurea Publications A74. 42 p.

Zott, C., & Amit, R. (2008). The fit between product market strategy and business model: Implications for firm performance. *Strategic Management Journal, 29*, 1–26.

Part III

Convergence and Broadcast Media

Standards, IPR and Digital TV Convergence: Theories and Empirical Evidence

Nicola Matteucci

1 Introduction

After almost three decades of "digital revolution", worldwide TV remains the most diffused, important and revenue-generating media outlet. By definition, the move towards digital TV (DTV, henceforth) was rooted in powerful convergence dynamics: binary coding of the analogue audio-visual signal into bits of digital information blurred sectoral boundaries and services specificities, enabling lateral entry and more variety—hence competition—in a new convergent "TV sector". At the same time, so far convergence did not seem to imply standardization and homogeneity at each layer of the value chain. Instead, a peculiar mix of expanding diversity of devices, "boxes", contents and services coupled with increasing standardization of core technological components seem to characterized today's DTV. Similarly, while technological drivers brought television closer to information technology/ Internet (henceforth identified as ICT—information and communication technologies) and telecoms, other economic forces and firms' choices led existing TV platforms to evolve along diverging and differentiated trajectories, introducing different contents, services and business models and enabling new original styles of TV consumption (pay-TV, mobile TV, catch-up/web TV, etc.).

All in all, worldwide forces of convergence and divergence seem to coexist in the DTV ecosystem. The sector itself—whose blurred boundaries now span from the former audiovisual value chain to new telecoms and IT-Internet domains—appears characterized by fast dynamics: rapid technological change, increasingly global initiatives of standardization, high market turbulence and business shakeouts, and accompanying socio-institutional change. The resulting regimes of standardization and market competition, so far, seem to contain both Schumpeterian as well as cooperative-collusive features.

N. Matteucci (✉)
DiSES, Marche Polytechnic University, Ancona, Italy
e-mail: n.matteucci@univpm.it

A. Lugmayr, C. Dal Zotto (eds.), *Media Convergence Handbook - Vol. 1*, Media Business and Innovation, DOI 10.1007/978-3-642-54484-2_12

The present work addresses these industry dynamics focusing on the key role played by IPR (such as patents, copyrights, trademarks and industrial secrets), as embedded in technological standards. These IPR do constitute a main ingredient of firms' competitive strategies, being capable of supporting technical and user-level interoperability among different vendors' solutions; alternatively, once appropriately managed, they can also enforce strong proprietary strategies and disrupt product and service compatibility, enabling the market pre-emption of rivals.

The chapter is structured as follows. In Sect. 2, an interdisciplinary literature review surveys the most relevant economic models and legal conceptualizations. First, the basic economic functions of technical standards are illustrated, and the conflicting incentives that different operators may have towards interoperability are explained. Second, we present a compact summary of the most recent IPR phenomena, standardization trends and legal issues, highlighting the polyhedral and ambiguous contribution of standards and patent pools to technology and market convergence. The following empirical analysis (Sect. 3) studies the main standardization initiatives for digital TV[1]: the specifications for signal transmission, those for content and security management, for the applications interfaces and, last, the emerging trends of the convergence between TV and Internet. Together, the main convergence/divergence trade-offs are discussed, and the outcomes so far achieved are evaluated. Section 4 summarizes the main results, mentions a few important points left for the future research agenda, and concludes.

2 Literature Review

2.1 Technical Standards and Strategies in Network Industries[2]

A large literature (among the reviews, David & Greenstein, 1990; Matutes & Regibeau, 1996) explains how technical standards cater for scale and network economies, potentially improving total welfare. Technical standards are consensual, official and descriptive technical norms; this type is known as *de jure* standard and it is usually produced by an official/formal standards developing organization (henceforth, SDO), having a national status—like DIN in Germany—or a supranational one—such as ISO, ITU or ETSI. Alternatively, consensus might often emerge as a spontaneous market outcome, rather than being the product of a specific official procedure; in this case, we have a *de facto* standard,[3] which differs in many

[1] Due to space limitations, this empirical analysis mostly concentrates on the DVB and EU experience, although keeping an eye on global trends.

[2] This section builds on a wide selection of papers and books; for sake of synthesis, we only mention the references most directly connected to our main argument, while skipping other background materials. For a proper presentation of the standardization literature, among others, Blind (2004), Swann (2010).

[3] Here, a single company manages to have its own specification acquiring a leading market share and becoming a *de facto* standard.

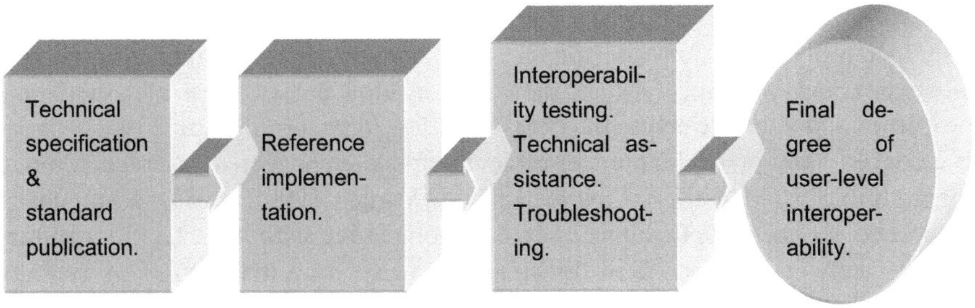

Fig. 1 Factors affecting the final degree of user-end interoperability of a standard. *Source*: Our elaboration

respects from *de jure*, being a privately owned and managed piece of intellectual property.[4]

In this work, we restrict our attention to standards designed for achieving compatibility and technical interoperability. The current "Knowledge economy" era has greatly emphasised the role of interoperability standards and their large scale diffusion, introducing a more flexible type of interoperability, that is increasingly achieved at the software layer of the concerned goods and services, in parallel with the growing dematerialization of manufacturing output. In particular, we say that a particular digital communication system is interoperable with others if it can exchange digital bits of information and process them according to stated purposes.[5] To this end, beside standards, other less frequent instruments are also available: "ad-hoc" converters[6] (or adapters)—either voluntary or mandatory—provide an alternative path to interoperability (for an economic analysis, Choi, 1996, 1997; Farrell & Saloner, 1992).[7]

At the same time, standardization might not be sufficient to ensure effective user-end interoperability (see Fig. 1): in fact, complementary measures are needed, such as producing a good (working) reference implementation,[8] performing interoperability testing and organizing services for technical assistance and users troubleshooting.

---

[4] Also most of *de jure* standards are privately possessed, but the ownership is usually shared by a multiplicity of subjects/licensors. A main example is provided by patent pool members (see *infra*).

[5] These compatibility standards in the literature have been also called "interface standards" (David & Greenstein, 1990).

[6] Generally, a converter is an adapter used to match the physical or functional characteristics of different objects/technologies.

[7] Many examples are available in ICT domains, from the simplest (plug adapters and power transformers for electricity) to more elaborated ones, such as analogue-to digital converters for TV receivers (or set top boxes, STB henceforth) or, in computer science, file format converters.

[8] This is the implementation of the standard used as its official/definitive interpretation. It checks that the specification is implementable, correcting potential errors and disambiguating its elements. The reference implementation also works as a benchmark for the following (derived) implementations.

A main feature of the ICT and digital media worlds is that standardization and interoperability create network structures among the concerned products/services—either physical of simply virtual; networks, in turn, possess powerful functional properties and valuable economic characteristics. *In primis*, the value of a network is a convex function of the number of devices/people connected to it, according to the notorious R. Metcalfe's Law (Shapiro & Varian, 1999); or of the amount of associated services, applications or content compatible with it. In fact, in communication networks is useful to distinguish between two main types of network effects—the direct and the indirect ones (see the surveys of Economides, 1996; Gandal, 2002; Katz & Shapiro, 1994). While the first type of network effect is related to the number of users connected, like in the typical case of the utility deriving from the fax transmission, in the second example, peculiar to media products, the utility is mostly stemming from the number of attractive contents and complementary services jointly available with the specific communication network.

While big firms managing incumbent communications networks want to control the access to their installed base of customers (potentially foreclosing the entry of competitors), small and new entrants have the opposite incentive, being eager to join and interconnect with the larger established base of customers: in fact, interconnection between networks would enable them, at least potentially, to steal customers connected to the incumbent, offering better services or contents. In this respect, Besen and Farrell (1994) define this set of incentives, behaviours and stylised business facts as the property of "time inconsistency of the compatibility choice".

Similarly, firms endowed with a large base of connected customers strive for maintaining them captive, increasing the consumers' "switching costs" to rival alternatives: this strategy, extensively analysed in the literature (for a classic review, Shapiro & Varian, 1999), is called "walled garden". As a matter of fact, it is typically enacted by means of strategic standardization and interoperability management based on proprietary and closed technologies; or, in the case of open standards, with manipulative patent pool strategies (see *infra*). For example, in digital pay-TV, most of the times consumers are forced to buy firm-specific reception equipment/STB (such a situation is called "vertical market"), which cannot be used for watching rival channels or unauthorised third party services, or for accessing other TV platforms; in other words, switching to other commercial alternatives is impossible without costly and cumbersome duplications of equipment, and new training. All these inertial factors increase customers' change costs.

In particular, in digital communications, physical networks interconnection has been the most typical instrument for achieving the interoperability of services, and is based on preliminary technological standardization of the underlying infrastructure, including both signal transmission and signal reception equipments. Then, for a smaller competitor willing to enter the market, the lack of interoperability with a larger incumbent would severely diminish the size of the available network effect—thereby reducing the competitive value of its offer and the likelihood of entry.

The market provision of standards typically leads to complex and uncertainty-intensive coordination games. In fact, in analogy with Katz and Shapiro (1994), standards provision, beside being rooted in the hard laws of physics and chemistry, also depends on agents' expectations and economic incentives, as extensively illustrated in mainstream oligopoly models featuring network externalities. The main conclusion of this literature is that decentralised (*id est*, market-based) systems of choice are rarely optimal in terms of social welfare. Stango (2004), for example, observes that spontaneous market acceptance of standards could result either too slow or too fast. Then, in a dynamic setting of standards choice, multiple equilibria are likely to emerge. Moreover, if there are strong network effects reinforcing the first standard arriving on the market, premature standardization might irreversibly displace future superior alternatives (excess inertia equilibrium, Besen & Farrell, 1994).[9] At the same time, firms' tactics and consumers' expectations might also cause inefficient switches to new (inferior) standards (excess momentum equilibrium), prematurely foreclosing the market for the 'next big thing'. To sum up, in both cases, due to network effects, a suboptimal equilibrium type "the winner takes all" might well occur, due to the 'wrong' winner (Shapiro & Varian, 1999).

However, neither the Government or other policy makers are immune from the risk of standardization failure. On one side, a central coordination is mostly valuable when markets feature both increasing returns (arising from network effects) and strong uncertainty (due to high technological turbulence), to induce a Pareto-superior equilibrium. On the other side, as underlined by Cowan (1991), this is also the context where Government's intervention aimed at picking the best standard faces the highest risk of foresight and error.

Finally, other drivers of complexity, like the presence of multiple market stakeholders, augment the error probability of the policy-maker. This is the case of markets qualifying as "two-sided" (or "multi-sided", by logical extension; Rochet & Tirole, 2003). A main example is free to air (FTA, henceforth) TV broadcasting, supported by commercial advertising, where the revenues earned from the ads collection side of the market finance the operations (content acquisition, transmission, etc.) in the broadcasting (audience generation) side, yielding cross-cumulative effects. Here, the working of the multi-sided system relies on the usage of a common technical platform (family of standards and other components, like those developed by DVB), which fuels the mechanics of the intra-market effects.

Finally, the previous literature review does not imply that standards wars inevitably end up with one unique technical specification, dictating an inexorable trend of technological and market convergence. In fact, long periods of coexistence of

[9] In a similar vein, other authors stressed the path dependent nature of standards choice and technology evolution. In fact, initial fortuitous events might result in permanent suboptimal trajectories, due to network effects and lock-in dynamics engendered by consumers' switching costs (hence the dependence—David, 1985; Arthur, 1989).

rival standards within the same market are also possible, and this outcome is most likely in those cases where network effects are counterbalanced by platform-specific characters or demand preferences: these uniqueness traits may help the weaker alternative to retain a minimal critical mass of loyal customers. In the PC world, a famous text-book example is the long term struggle played by the two competing systems Apple-Macintosh and IBM-Wintel (Windows-Intel): Apple managed to survive in the PC market segment, although with a tiny market share,[10] before rejuvenating with new families of products (tablets, smart-phones, etc.) and becoming a global ICT and media player. We also notice that Apple is expected to become a relevant player even in future hybrid TV, being on the verge of launching innovative Internet-based products and services.

2.2 Recent Trends in Standardization and IPR Management

Since the late-1980s/early 1990s, the previous "command" approach entrusted to European bureaucrats performing top-down detailed standardization activities lost its momentum. Face to the disappointing results of major standardization initiatives led by the EU Commission (for e.g., the failure of the European project for high definition TV), a new system based on delegation to formal and independent SDO was believed to cater more effectively for the technological unification of the European internal market, while ensuring fast, consensual and more market-driven standardization processes. Additionally, in several ICT and digital media sectors a progressive hybridization of standardization activities has taken place (Funk & Methe, 2001; Hawkins, 1999): here, *de jure* standards are initially promoted by private consortia or industrial fora, which eventually submit the developed technical specification for approval to the concerned SDO. A main example, featured in our empirical analysis, is DVB, founded in 1993[11] and aiming at the standardization of digital TV.

Most SDO, implicitly or explicitly, are committed to the development of "open standards" (OS, henceforth). Preliminary, we need to remember that standards embody pieces of knowledge, usually protected by IPR: patents, copyrights, trademarks and industrial secrets. Often, in the most complex cases, a standard may arrive to include hundreds of different IPR, directly or indirectly. Hence, the owner of the IPR embedded in a standard is entrusted by law with powerful legal tools to charge monopoly rents (royalties and fees) for the usage of that standard, to restrain (or even block) the downstream market (product implementation) for third

[10] A valuable analysis of the Apple-Macintosh and IBM-Wintel saga can be found in Bresnahan and Greenstein (1999).

[11] Being DVB a private body, the developed specifications are submitted to official SDO for formal approval. DVB was born as a temporary consortium, and later transformed into a permanent body.

party vendors, or to perpetuate its legal monopoly blocking further related advances (for e.g., by preventing improvements over the original standard).

OS emerged as a sort of compromise, believed to mitigate the inner monopolistic potential possessed by successful standards. Paradoxically, there is no 'standard' definition of OS in the literature, but several different and partly overlapping ones (Cerri & Fuggetta, 2007; West, 2003, 2007). For our purposes, we define an OS that whose detailed technical specification is publicly available for interested third parties, thanks to the publication of its descriptive documents.[12] After this basic common denominator, definitions start to diverge.

From a meta-analysis of the literature, three main profiles of OS can be highlighted. First, the openness of the standardization process. In OS, stakeholders act in a consensus-based process open to a plurality of participants, while proprietary or "closed" standards are dominated by a unique vertical and self-referential managing subject, following autarchic rules. However, openness may be a transitory character[13]: Krechmer (2006) emphasises that ICT standards are increasingly controlled by implementers, and the latter is the crucial layer where to check that an OS does not fall prey of enclosure strategies. In fact, technical change and strategic firms' incentives may transform an initially 'genuine' OS into a series of semi-open or even proprietary implementations, turning it into a *de facto* closed standard[14]; in this respect, OS may suffer from the same phenomenon of "time-inconsistency" of compatibility as network technologies (see Sect. 2.1).

Second, the degree of public (external) availability of an OS may vary; in fact, its actual licensing policy could restrain effective access to the "essential" IPR[15] embedded in the standard for some implementers. For example, West (2003) recalls that some third parties might not qualify for access to IPR (typically, the direct competitors of the standard owner); or the scope of this access could vary (for instance, blocking the right of modification of the standard and its implementations).

Third, the welfare properties of OS (including its purported superiority over proprietary standards) depend on the specific licensing procedure and policy enacted. Generally, the minimal necessary requisite for an OS is to offer FRAND (fair, reasonable and non discriminatory) terms of licensing.[16] However, FRAND terms also conceal areas of ambiguity and indeterminacy:

[12] Generally OS have their technical documentations publicly accessible for free or at a nominal fee.

[13] Moreover, the concept of "openness" is many-sided and touches upon several dimensions and phases: Krechmer (2006) arrives to categorise ten dimensions of openness, and three main types of stakeholders involved: creators, implementers and users.

[14] The software industry provides a high number of examples for enclosure strategies (Shapiro & Varian, 1999).

[15] Those necessary to build a material implementation (a product or a service) of the standard.

[16] In fact, the most credited definition does not assimilate an OS to a free standard (royalty-free, typical of the open source software world, OSS henceforth).

1. FRAND does not mean necessarily a small fee, for prospective licensees. FRAND advocates point out that, for stimulating innovation, the monopoly rent accruing from IPR should be aligned and proportionate to the "importance" of the concerned OS, somehow determined.
2. Remembering that a single ICT/media standard may arrive to contain hundreds of essential IPR, potentially linked to previous ones, this structural complexity originates many intertwined and recursive linkages between IPR. Similar cases are often defined as "patent thickets", due to the "dense web of overlapping intellectual property rights that a company must hack its way through in order to actually commercialize new technology" (Shapiro, 2001, p. 120). As a consequence, IPR licensing terms, also in the case of OS, in practice may get very complicate and involve a high level of transaction costs—especially for prospective third party implementers.
3. A further problem is the occurrence of "royalty stacking" (Lemley & Shapiro, 2007). This is due to the fact that radically new ICT/media products and services involve *per se* risky business plans[17]; then, complex standards might add to it, when a cumulative overload of royalties endogenously depresses the *ex-ante* net value of the project,[18] irrespective of its real market potential.
4. More radically, some scholars and practitioners point to the conflicting plurality of criteria implementing FRAND obligations, and to the practical indeterminacy involved in most of them (Treacy & Lawrance, 2008).[19]

All in all, these licensing issues may easily become causes of lengthy litigations and costly settlements, and in several cases they do seem to be responsible for hold-up dynamics, market uncertainty and lost business opportunities, dampening the diffusion of the concerned OS.

Is there any effective remedy available for solving these IPR licensing shortcomings, increasingly affecting most ICT and media OS? Patent pools have been representing an appropriate institutional answer, at least in some respects.[20] A patent pool is a private entity (association, consortium, firm) formed by qualified IPR owners of the concerned standard. These owners gather and confer their IPR on the pool, to be available for licensing to interested third parties; then, when the patent pool is effectively set up and run, it should be welfare-increasing, since it

[17] In mass network markets the entry phase may require intensive efforts (high sunk costs) to build the installed base and activate network effects.

[18] In other words, in this case the net present value of the project is low mostly because of high licensing costs, rather than negative supply and demand conditions.

[19] A main example is the methodology to be used for calculating the royalty base. When patents are embedded in high value added products—e.g.: Apple's smart-phones and tablets—and percentage royalties schemes are used, they earn high absolute royalties, irrespective of their intrinsic stand alone technical and innovative contribution to the system.

[20] A compact historical analysis of patent pools and antitrust practice in the US case is provided by Gilbert (2010).

acts as a bundling device, lowering transaction costs and reducing "double marginalization" welfare losses (with respect to the individual purchase of IPR).

At the same time, patent pools also display several critical dimensions. First, most of the times the patent pool members exchange reciprocally their IPR licences, instead of paying their full price.[21] So, IPRs cross-licensing automatically confers upon the pool members a superior cost structure, with respect to non-members. This sort of joint market dominance, when pool members are also implementers, is likely to lead to anticompetitive licensing strategies, aimed at "raising rivals' costs" (Salop & Scheffman, 1983) and deter potential downstream competitors (Lerner & Tirole, 2004); similarly, patent pools can become a natural venue for illegal price fixing. On overall, the risk of having abusing and welfare-reducing patent pools is substantial. Second, royalty stacking may come from patent pools, being itself related to the number of essential patents included in the standard, and to their types. This happens because the essentiality character in practice conceals a strong subjective nature and is plagued by asymmetry of information, being self-stated by patent holders: hence, an incentive to overfill the standard pool with one's own IPR arises, for obvious rent-seeking reasons.

In brief, for these and other reasons, an increasing number of scholars, practitioners and operators have started to campaign for a serious reconsideration of the current system of patent-driven standardization.

3 An Empirical Analysis of Standardization and Convergence in Digital TV Sector

3.1 Methodology and Scope

In the following sections we test against the empirical evidence the theories and the models reviewed in Sect. 2, to contribute to a fundamental missing point of the research agenda: the unexplored relation between standardization, IPR and convergence. To this end, we carried out an original and interdisciplinary analysis of major innovative and regulatory activities performed in the TV sector—mostly in Europe; this exam is based on our scrutiny of standardization and policy documents, technological assessment exercises and market analysis.

Before presenting the main results, a first important distinction has to be made between technological convergence and industry convergence. For defining the first process we refer to the landmark contribution of Negroponte (1995), who characterised the digital revolution as the transformation of traditional media from atoms (meant in their material supports) into digital bits of information: hence, the dematerialization of media content and delivery platforms possesses strong technological determinants, being dictated by the 'hard laws' of physics and

[21] Obviously, the bargaining power of each member of the pool depends on the size and importance of its IPR portfolio.

chemistry. The latter spurred the development of micro-electronics as the main technological support for increasingly powerful systems of digital coding and transmission of information, to be used in the telecoms, Internet and television sectors.

With the term of industry convergence, instead, we refer to the industry transformations provoked or stimulated by technological convergence, and to the stakeholders' deliberate strategies or counter responses (either supply or demand side). So to say, we capture with the second concept the economic implications of technological convergence, without assuming an absolute technologically-deterministic stance where, with one-way causation, technology alone would shape markets, social behaviours or institutions. On the contrary, systemic feedbacks are also possible, and technological artefacts themselves are in part modelled and conditioned by economic factors and the socio-institutional sphere.

Having defined the two concepts, we can conclude that, while technological convergence means a trend of progressive similarity (or, at least, interoperability)[22] of the underlying standards and user-level devices, market convergence indicates that, on the supply side, operators along a particular value chain (vertically) and across markets (horizontally) tend to become similar or even coincident (the last case happens with horizontal/vertical mergers and acquisitions). On the demand side, instead, media markets converge when consumers increasingly buy or use different services or devices as if they were similar/identical, and their cross-prices elasticity augments in absolute value. Obviously, these technological and market-mediated processes of media convergence interact with closely related phenomena of social and cultural media convergence, as comprehensively explored in communications studies, sociology and political science (for example, Jenkins, 2006; other chapters in this book).

Hence, our point of departure is the informed belief that 'spontaneous' forces of technological convergence and standardization activities are inextricably linked to IPR strategies and to the overall political and institutional spheres governing contemporaneous societies; these spheres see intellectual property and its governing rights as crucial assets to control investments, production, and national competitiveness, beside empowering new effective strategies of trade policy. As a main consequence, we believe that also the very phenomenology of convergence (or divergence) in digital TV cannot be properly understood without a deep appraisal of firms' IPR strategies, economic incentives and related phenomena such as patent portfolio strategies (including patent pooling), IPR proliferation and litigation, which stand as clear stylised facts of contemporary ICT and media sectors (for a recent assessment of mounting IPR litigation, PWC, 2012).

In the following empirical analysis, the DVB consortium will feature as the main protagonist of the story (for a broad presentation of its experience and operative

[22] In the first case, standards do not proliferate and unique designs prevail. In the second case, standards may proliferate, but services and devices are interoperable, supporting multiple standards.

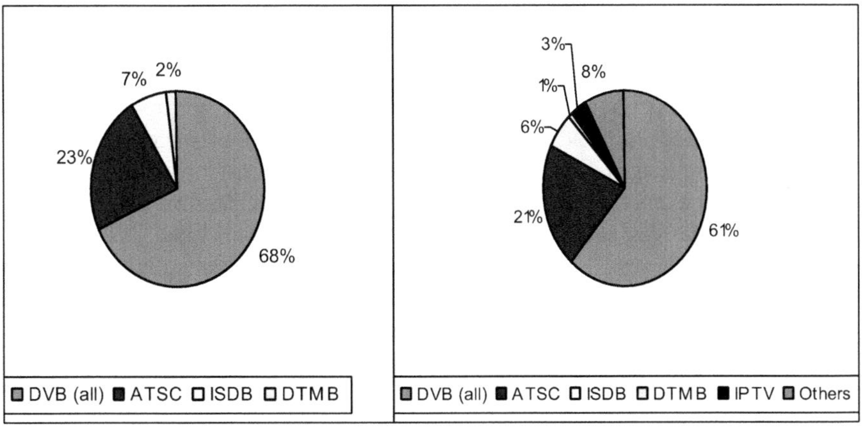

Fig. 2 DVB share of the worldwide installed base of household digital STB (*right*) and household digital broadcast (not IPTV) STB (*left*)—2011. *Source*: DVB and HIS-screen digest

functioning, see Eltzroth, 2008, 2009; Grimme, 2002). We are aware that important standardization activities for digital TV were also carried out by other competing SDO and institutions, based in different geographic areas (primarily, US and Asia), as nicely described by previous works (Farrell & Shapiro, 1992; Gerbarg, 1999; Hart, 2004; Shapiro & Varian, 1999). Despite this, our focus on DVB is principally motivated by its leading role acquired within the global system of ICT and media standardization—particularly for digital TV. The noticeable impact DVB managed to exert on worldwide DTV deployments is indirectly showed by its substantial share of the global installed base of digital set-top-box (henceforth, STB)—this base being estimated at 1.4 billion pieces by 2012. In particular, the left part of Fig. 2 demonstrates that, considering only broadcast transmissions, in 2011 DVB controls 68 % of the world installed base of DTV, while the closest rival (ATSC, the US alternative) commands only 23 %. Then, if we consider the DVB share on the total world installed base of digital STB—including narrowcasting modes (such as IPTV)—its share slightly falls to 61 % (right part).

3.2 Standards for Digital Transmission

The first generation of DVB standards for signal transmission began to be worked out a few months after DVB formal foundation (1993): first, in 1994 was issued that for satellite (DVB-S), shortly after came that for cable (DVB-C) and finally that for terrestrial TV (DVB-T), in 1997.[23] DVB-S, DVB-C and DVB-T are based on a

[23] Other ancillary supporting standards were also developed, such as DVB-SI (for service information), DVB-SUB (for subtitling), and MHP, the open middleware specification analyzed *infra*. The transmission-independent part of MHP has led to the GEM specification, used also in off-line media (Blu-ray Disc).

common signal compression specification (MPEG-2, previously standardized at ISO), a then powerful algorithm able to compress the large video output of TV studios[24] to a datarate of a few Mb/s (so called "source coding"); the transmitted signal is later decompressed once received by the STB. Before transmission, the source signal needs to be adapted to the transmission medium ("channel coding"): this involves the usage of specific algorithms to perform errors correction, modulation, format conversion and filtering of the signal. At that time channel coding operations were designed and developed as platform-specific. For example, DVB-S, DVB-C and DVB-T incorporate different modulation systems, implemented with tailored hardware: hence, they result to be different and not interoperable standards, so that a TV viewer willing to subscribe to all the three platforms' services, transmitted with the first generation, should buy and pile up at least three different STB.

Was this divergent outcome inevitable? On one side is technically dubious if a unique transmission standard for the three platforms would have been feasible with the technology of mid-1990s, given the pioneer stage of digital signalling technologies and the country specificities permeating terrestrial services (and the delays likely to affect their switchover processes). Lacking any counterfactual, we tend to think that its retail cost would have been prohibitive and its niche business model unsustainable. On the other side, a higher degree of interoperability might have been feasible between cable and satellite transmission standards, since the modulation requisites of the two platforms were more similar.[25] More explicitly, the absolute and symmetric incompatibility choice[26] between satellite and cable equipment made while designing the first generation seems to have been dictated firstly by the diverging commercial interests of the incumbents, oriented to stifle the first signs of strong competition between the two platforms. Consequently, the economies of scale actually enjoyed by the CE industries were significantly lower with respect to those originally anticipated during the standardization process, and this, due to the higher unit cost of the STB, might have somehow dampened the DTV diffusion under the first generation.

Residually, one can investigate if the degree of STB interoperability was sensibly higher at least within each transmission platform. Also in this respect, DVB transmission standards did not enable any relevant degree of intra-platform interoperability between devices and services, across Europe and even within single EU members. The case of DVB-T is probably the most paradigmatic, since its STB

[24] At that time, typically, a signal of 166 Mb/s or more.

[25] Historically, satellite and cable have been used complementarily for primary transmission of TV signals. The wide footprint enjoyed by satellite helped to connect separate distant cable networks (Parsons & Frieden, 1998).

[26] Concerning modulation, two different techniques were chosen. For satellite, QPSK (quadrature phase shift keying), for cable QAM (quadrature amplitude modulation). These techniques, through different implementations, are also the modulation core of other platforms such as DSL modems, CDMA, 3G, Wi-Fi and WiMAX.

market across EU remained strictly national. Basically, during and after its standardization, it was clear that national raisons linked to spectrum management and policy, technical idiosyncrasies emerging from the operators' legacy equipment and the different stages of the national switchover processes would have lead to diverging geographical implementations of the same transmission standard.[27] However, in this specific case, it is fair to affirm that most interoperability limitations were rooted in technical and institutional constraints, rather than firms' strategies and IPR incentives. The Italian DTT experience is a case in point, with the R&D centre of RAI (the public broadcaster) having played a qualified role in developing specific adds-on to the basic DVB-T specification; in particular, these adds-on enabled the correction of the high rate of interferences damaging terrestrial transmission, due to the chaotic allocation and usage of the Italian UHF and VHF bands of the TV spectrum (see, for example, the CD3-OFDM equalizer). Later on, a similar beneficial role was also played by the DGTVi association (constituted by the main terrestrial broadcasters), which acted as the main technical body for coordinating and fixing transmission and reception shortcomings affecting the Italian DTT roll-out (for an exam of DGTVi, Matteucci, 2009).

Instead, looking for interoperability limitations directly attributable to IPR incentives, the story of satellite DTV standardization is more illustrative. Since the mid-1990s, digital TV diffusion in Europe was first driven by the expanding pay-TV offers, initially broadcasted via satellite. Basically, the fact that the satellite transmission standard (DVB-S) was the first to be completed responds to favourable technological opportunities. Satellite transmission systems—especially those performing "direct to home" transmission, DTH henceforth—pose inferior technical challenges in terms of interoperability of devices and operations, being the platform/system management fully centralised.[28] Similarly, the switchover process of the satellite platform encountered less coordination problems and switching costs with respect to cable and terrestrial, whose switchovers are technically and normatively more complex and cumbersome. Finally, satellites enjoy huge scope economies, being their fleets regularly updated and incremented to satisfy a large variety of communication purposes (telecom services, military purposes, space-related and broader scientific activities, among others): hence, most R&D and deployment costs were shared. However, despite these favourable pre-conditions

---

[27] For example, two different modulation schemes were developed, differentiating geographical areas starting DVB-T services early (based on 2k OFDM + QAM) and those starting later, and building on single frequency networks (SFN, henceforth). The second required adopting 8k OFDM + QAM modulation, which is backward compatible with 2k/QAM broadcasts. Later on, several EU countries adopted improved algorithms, adds-on and devices to the basic DVB-T standard, in order to cater for their specific transmission and spectrum needs.

[28] In fact, with satellite, a single up-link transmission site (from the earth base station to the satellite transponder) concentrates all system control and signal management functions (Drury, 1999; Reimers, 2006). Here the main problem is transmitting a compatible signal receivable by household devices produced by different vendors, and patronized by different satellite TV operators, while satellite services have transnational coverage and reception.

Table 1 Worldwide shares (%) of total TV receivers diffusion, by platform—2011

	DVB		Other standards	
	2009	2011	2009	2011
Satellite	72	68	28	32
Terrestrial	43	55	57	45
Cable	49	57	51	43

Row/Year sum to 100. All platforms count the first and second DVB generation
Source: DVB and HIS-screen digest

to interoperability, most satellite services have showed a market divergence across Europe, requiring specific equipment and STB; this was mostly due to the 'converging' incentives to incompatibility held by leading pay-TV operators, willing to keep their subscriber bases as walled gardens. Finally, the platform temporal leadership is also a main explanation of why, still in 2011, the DVB worldwide market position remains particularly strong in satellite: in fact, DVB is holding a 68 % market share in the world market of satellite STB, while the corresponding one drops to 57 % for cable STB and to 55 % for terrestrial STB (Table 1).[29]

Has the interoperability and convergence potential progressed in the second generation of DVB transmission standards? On overall, we can respond positively, although the extent of the progresses varies according to the different TV platforms considered—traditional (cable, satellite and terrestrial) and newer or emerging ones (IPTV, mobile TV and hybrid TV).

In detail, on the one side intra-platform compatibility issues were specifically considered while designing the second generation. For example, the DVB-S2 specification (developed in 2003 and ratified by ETSI in March 2005) provides options to make it-self compatible with the installed base of DVB-S STB (thereby supporting backward compatibility),[30] while offering new services to TV viewers agreeing to substitute their STB. Then, DVB-S2 includes new functions connected to the convergence between Internet and TV, such as a generic transport mechanism for IP packet data; additionally, the second generation can accept different input data formats, especially in professional applications.

DVB-S2 market deployment has been fast, and the new standard has basically replaced most of the existing DVB-S services in Europe and US. According to DVB sources, two main drivers pushing this fast process of diffusion and substitution have been the clearly-stated early ITU support and the IPR strategy adopted by the patent pool members. In fact, in 2006 ITU formulated a recommendation of DVB-S2 as the best satellite option for conjugating both audio-video and data

[29] All figures include the DVB second generation of transmission standards. The latter at 2011 is relevant for satellite, still marginal for terrestrial and zero for cable.

[30] Obviously, backward compatibility is a technological possibility offered by the DVB-S2 design, but its actual materialization will always depend on what happens during the commercial implementation, *id est*, on the commercial choices and incentives held, country-by-country, by broadcasters and satellite operators, which interact with CE manufacturers' incentives and market expectations.

transmission needs; at the same time, DVB-S2 IPR holders fixed rapidly and with certainty the licensing terms, which appear to have been considered at least acceptable by the concerned market players.

More generally, a higher interoperability potential seems to mark the second generation of DVB standards, compared with the first. This is certainly rooted in stronger technological opportunities (for an economic analysis of this concept, Klevorick, Levin, Nelson, & Winter, 1995), due to accelerated technical advances pushing the semiconductors industry. In particular, the development of superior CPU and systems on chips (SoC, henceforth) has yielded tremendous gains in computing and memory power (according to the famous Moore 1965's law)—unthinkable 10 years before, at the time of the first generation. Further, the second generation embeds the outcome of important research advances, which pushed the signal transmission performance very close to its theoretical maximum (Siebert, 2011). Finally the second generation, although in varying degrees across its main platforms, was the first to be specifically conceived to enable an efficient transport of IP packets, thereby making the first sizable step in the long journey towards the convergence between the two separated worlds of broadcasting and Internet.

3.3 Standards for Conditional Access and Content Management

The early deployment of the satellite DTV platform was also supported by the strategic choices of pay-TV operators; these, starting from the analogue era, had begun opting for firmly-controlled "vertical" business models,[31] featuring proprietary and *de facto* standards governing the core elements of the transmission and reception devices: primarily, the CAS, API and EPI modules. In the digital era, such a proprietary business model has become the rule, for several reasons. First, operators must control crucial components essential for the viability of the pay business model (for example, the piracy-proof security of the CAS), and have to ensure interoperability with reception equipment produced by different vendors—absent a *de jure* or *de facto* standard. Second, during the analogue-digital switch-over, they need to subsidise prospective viewers, in order to win their reluctance to adopt the new technology, because the retail price of STB and the complementary receiving equipment (antenna, etc.) may represent a significant upfront cost, for the median income household. Last, but not least, a proprietary STB better caters for walled garden strategies against potential entrants, especially when this technical incompatibility strategy is conjugated with exclusivity strategies based on premium content acquisition. However, while at the beginning these proprietary strategies fuelled the take-off of the first digital platforms, later their industry-wide net effect

[31] Nolan (1997) remembers that a strong obstacle for analogue satellite TV diffusion in EU had been the fragmented structure and incompatibilities affecting the reception devices, with the satellite receiver being typically universal and separated from the decoder (signal descrambler), proprietary. Hence, those offering a proprietary unique piece of receiving equipment (STB) enjoyed a strong market boost.

turned negative, because of the dominant positions they created. Matteucci (2004, p. 59), for example, describes the lengthy battles fought by the Italian satellite pay-TV duopolists with the goal of dampening the reciprocal interoperability of their STB bases: they stubbornly resisted the normative provisions and the market pro-compatibility sentiment, even at the cost of being fined by the national regulator.

Indeed, these incompatibility strategies are believed to have caused subscribers' hold up behaviours and prolonged market uncertainty, which might have stalled satellite subscriptions in a favourable period, when DTT offers were still unavailable and satellite was the main multichannel novelty in most EU countries, beside analogue cable. All in all, the satellite platform presents a rather counter-intuitive market outcome of technical standardization, whose precompetitive and pro-convergence trend was reversed by the pro-divergence incentives permeating a small but influential part of the DVB membership. At the end, despite the global reach and convergent potential of the platform, satellite TV services have remained fragmented and relatively underdeveloped across EU, and its market share is now progressively eroded by the other platforms, both traditional (DTT and cable) and new ones (such as IPTV and new hybrid forms of online TV consumption).

Another story featuring lack of user-level interoperability is that of CAS, equally regarding the three traditional TV platforms. Surprisingly, CAS standardization was one of the main remits of the DVB group, and a domain where also the EU Commission put many efforts, revamping a clear interventionist and pro-industry approach, culminating with Directive 98/84/EC on the legal protection of services of conditional access. This story has been widely studied (Cave, 1997; Cawley, 1997; Helberger, 2005; Van Schooneveld, 1996), and will only be tacked here in relation to IPR and its divergence outcomes. Basically, DVB members worked hard to find consensus on a standardised descrambler and CAS, but at the end two alternative systems were submitted to the DVB's Steering Board, Simulcrypt and Multycript, and both were later approved. In fact, while Simulcrypt was patronised by incumbent pay-TV operators willing to retain full market power over their installed base of decoders, on the contrary FTA broadcasters, new pay-TV entrants and CE equipment manufacturers were in favour of the Multycript option (or Common Interface, CI henceforth), enabling direct and fuller CAS interoperability. In fact, CI is a CAS detachable module (hence, CAM) and incorporates both the smart card chip and the descrambler chip, so that it fits a variety of STB from different vendors, rather than requiring lengthy negotiations between entrants and incumbents for obtaining the rights of access to the reciprocal SBT installed bases (as in the Simulcrypt case): hence, CI/CAM qualifies as a proper interface standard, enjoying a higher potential for interoperability. As a result of DVB internal fights and decision impasse, both CAS options were transposed into national laws, following Directive 98/84/EC—that obliging member countries to mandate at least one of the two systems. At that point, market equilibria, unsurprisingly, tipped in favour of Simulcrypt and, despite formal FRAND provisions, new operators willing to enter the pay-TV markets were effectively deterred, so that many consumers across EU were stranded after their initial STB choices, either with Simulcrypt (failing negotiations) or Multycript (unavailability of CAM for certain services). This was also due to the fact that, while the scrambling mechanism was standardised and subject to FRAND, other STB components did not enjoy the same guarantee, as in the case of EPI and API, which however are components equally necessary to reach broad audiences of TV viewers (for an early appraisal, Cave, 1997). In this case, given the proprietary nature of most EPI and API used in pay-TV STB (for the latter, we mention MediaHighway, OpenTV, Liberate, Microsoft TV, NDS, PowerTV), many new operators could not enjoy the same degree of pro-competitive protection and interoperability, with respect to the incumbent' base of STB.

3.4 The First DVB 'Patent' Failure: The MHP Standard[32]

After the noticeable market success obtained by the first generation of transmission standards, DVB activities encountered the first critical steps while moving along the digital TV value chain, at the services/content layer. As a matter of fact, the MHP specification (Multimedia Home Platform) was the first burning failure dampening DVB's reputation. The MHP is an open standard for API, which is a crucial software layer (or middleware) located inside those TV STB which have interactive capabilities (higher quality types; for an example, see Fig. 3). Basically, the API governs the informative fluxes within, from and to the decoder, and provides external applications and services with a compatible interface and a standardized execution environment. In particular, MHP works independently of the underlying STB' system software and hardware, so that third-party content and applications are portable across types of STB produced by different vendors, without need of re-authoring. Moreover, the IPR licensing of MHP, being it an OS, is bound to FRAND terms, differently from proprietary API. Hence, *ex ante*, the leverage effect

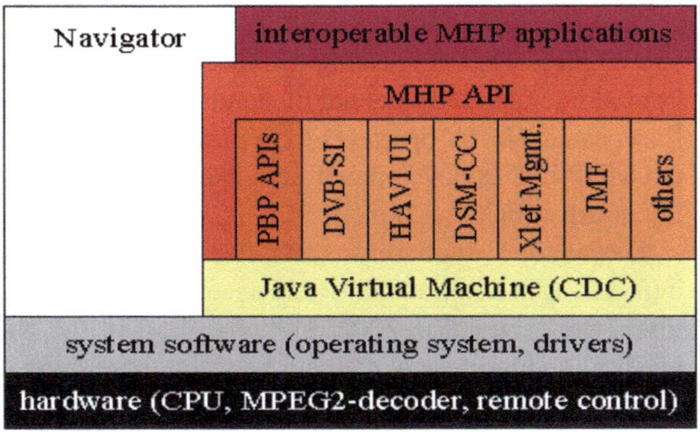

Fig. 3 Software layers and structure of an MHP STB. *Source*: http://en.wikipedia.org/wiki/Multimedia_Home_Platform

[32] This section partly builds on Matteucci (2008, Sect. 3; 2009). We refer to these papers for a comprehensive techno-economic analysis of the EU market experience and policy for Interactive TV, and for a focus on the Italian case.

of a standardised open API such as the MHP should promote the emergence of an horizontal market for STB, originating larger economies of scale and reducing retail prices.

During the second half of the 1990s, several FTA broadcasters and CE manufacturers started to design new enhanced TV services and STB provided with interactivity functions, envisaging interactive TV (ITV, henceforth) as a sort of counter-move against the increasing momentum of the Internet. Further, the contemporary developments of pay-TV were urging a new generation of more intelligent decoders, capable of performing new transactional functions: an adequate API was then needed. DVB embarked in MHP project since 1997; however, at that time, DVB membership did not include major ICT and API producers, so that the original MHP project built around the idea of a simpler low-cost API; later on, the MHP project became more articulated and ambitious, and produced a number of versions and upgrades of increasing complexity.

Meanwhile, European market stakeholders had started to lobby for informing the EU regulator about the need for a EU-wide standardised open API, publicly available for potential implementers. However, at this stage interests and views on MHP were already diverging, within DVB. Strong MHP supporters were found among Scandinavian TV operators and German broadcasters, together with the bulk of major CE manufacturers; later on, also EBU (the association of European broadcasters) would have taken a more explicit position aimed at strong and direct public support for MHP. On the other side, pay-TV operators, having their respective installed bases of proprietary API and STB, did not fancy this public support to an OS; they also claimed that MHP was not so costs-benefits effective, and believed that ITV did not have interesting market potential. Geographically, Scandinavian countries, Benelux and Germany were in general MHP supportive due to their analogue TV legacy, while most of the larger countries (UK, France, Italy, Spain) were opposed, or alternatively were just favouring a long-term migration strategy to MHP, while maintaining compatibility with existing proprietary API and remaining open to alternative future specifications; in particular, in UK and France there was a strong preference for a simpler specification, MHEG-5, better catering for the limited interactivity potential of standard FTA terrestrial services. Equally, several European cable TV operators were preferring a standardised but simpler API, given the MHP higher development and roll-out costs and its technical complexity and immaturity (ECCA, 2003, p. 3).

Despite lack of consensus, efforts towards MHP and the largely utopian idea of introducing at that time ITV were accepted and included in the EU New Regulatory Framework, issued in 2002. The final draft of the NRF incorporates an elaborate political compromise agreement between these diverging views. In particular, Art. 18 of the Framework Directive assigns a special status to open API (and, *de facto*, to MHP)[33] and specifically mentions interoperability as a means to achieve horizontal

[33] Despite the general formulation requested by law, the provisions of NRF were first and foremost inspired—if not specifically crafted—at the benefit of the MHP, which was the first specification to be formally recognized in the list of EU official standards for ITV.

markets, beside other socio-political and cultural goals. Operatively, Art. 18 foresees a system of immediate national market incentives to stimulate spontaneous tipping towards the DVB specification, while does not exclude the "last resort" option for the medium term: old-style direct intervention by the Commission to mandate standardization on MHP.

This complex policy-mix was specified in a series of policy documents. Since the beginning, the Commission had recognised that—due to the diversity of industries, operators and business models called to provide convergent interactive services (deliverable by either digital TV or mobile platforms, such as the UMTS)—achieving effective interoperability of API would have required due time, meant different things and possibly diverging paths. For example, during the mid-2004 public consultation EBU (2004) claimed that, although there were different paths to interoperability (for example, by means of content re-authoring or broadcasting in multiple formats), only the one based on an OS (like MHP but also MHEG-5) would have been rapidly effective at the consumer-level. Further, EBU warned the Commission against a minimalist approach bound to ensure interoperability solely within the DTT platform, because interoperability limitations significantly affected also satellite and cable. On the other side, operators from the pay-TV, telecoms and IT worlds showed a more liberal and market-oriented attitude to interoperability and market convergence of API, claiming that publicly-mandated standardization would have chilled private innovative efforts; moreover, the same subjects observed that interoperability should have been demand-driven, and preferably realised at the content level (for example, with portable content formats, needing re-authoring), rather than at the STB level (with a new API).

Unsurprisingly, face to this plurality of views, in 2004 the Commission (EC, 2004) adopted a cautious waiting approach, to finalise its review in 2005 (EC, 2006), when it decided not to mandate standardization and to maintain a market-led orientation to the promotion of open API, facilitated by public financial incentives. As an example, the Commission pointed to the MHP developments in Italy, where more than two millions of MHP-interactive STB had been sold, after being publicly subsidized.[34] Incidentally, EC (2006) also mentioned the market progress of two other non-DVB standards—MHEG-5 and WTVML[35]; these newer API

[34] At that time, EC (2006) argued that this success was jointly due to a virtuous and synergic public-private mix of factors: the public subsidies to MHP decoders, the voluntary agreement of Italian broadcasters to use MHP, and the definition of common specifications for the national implementation of the MHP standard. As a counterfactual, the Commission noticed that the same degree of stakeholders coordination did not materialize in Germany, nor in the Northern EU markets, despite the early market consensus.

[35] MHEG-5 is a language for presentation of multimedia information, particularly used for ITV services. To be useful for broadcasting, the language has been codified ("profiled") as ETSI standard. WTVML (Worldwide TV Mark-up Language) is an XML-based content format, standardized through ETSI, allowing web designers to create ITV services with minor adaptation requirements.

specifications, despite being still waiting for acceptance as official OS and not being supported by public subsidies, by 2005 managed to total respectively five and seven million units sold. This fact is very curious and interesting, once compared with the institutional fuss devoted to MHP: their formal accreditation as ETSI standards came after spontaneous market acceptance, contrary to the MHP case. Hence, this comparison shows effectively that not all OS are equally market-driven.

Unfortunately, the early Commission (DG Information Society)'s enthusiasms for the Italian MHP experience were too hasty. In fact, the Italian Government's implementation of the EU policy for OS and MHP was highly contentious (Matteucci, 2009, 2010), being centred on public subsidies and advertising support to MHP STB, but limited to terrestrial TV—the incumbent platform. The public measure, opposed by excluded operators, gave rise to a complex and lengthy state aid case: the final decision in 2007 (from DG Competition) was negative, qualifying it as an illegal state aid, being selective and market distortive and tailored to the benefit of one incumbent TV operator, Mediaset (controlled by the family of the then Italian Prime Minister).

Obviously, from a market convergence point of view, the distortionary public aid measure did favoured the widespread diffusion of DTT and MHP in Italy against the satellite competitor and rival API, thanks to the ensuing economies of scale and the falling retail prices of DTT MHP STB. However, from the point of view of the effectiveness of the underlying innovation policy for ITV, the entire pro-MHP public campaign was a waste of tax-payers money. In fact, given the premature stages of the early MHP versions and the small interactivity potential of the subsidized STB (equipped with a narrowband return channel), real two-way interactive services did not take off in Italy. Consequently, Italy has acquired the rather bizarre and embarrassing primacy of having the largest share of the world-wide installed stock of MHP interactive STB (65.8 %, Fig. 4) and one of the lowest rates of actual roll-out and usage of ITV. In fact, the most relevant usage of the

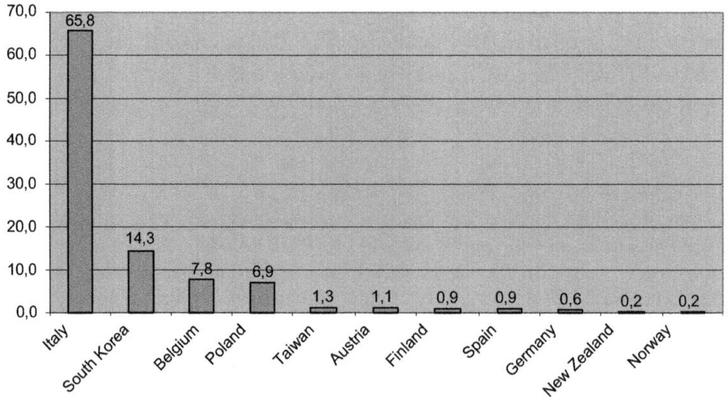

Fig. 4 Worldwide % shares of the installed stock of MHP STB—End 2011. *Note*: % country shares of the total MHP base, equal to 16,725,000 pieces at November 2011. *Source*: Our elaborations on www.mhp.org data

MHP-enabled two-way interactivity is that serving the SBT's pay-TV subscriber management system, for which the public subsidization of MHP acted as free manna from heaven; meanwhile, paradoxically, the country is plagued by insufficient and faulty broadband supply and has one of the lowest rates of Internet usage among OECD countries (for a recent assessment, Matteucci, 2013).

During the last decade, despite (or, perhaps, because of) this intensive institutional support, the IPR and licensing profiles of MHP remained uncertain; the patent pool formation was very late, and the exact terms of licensing were published only in February 2007. This situation also prompted a formal protest of the EU Commission, which wrote a letter to the DVB Chairman to express its disappointment and concerns over the MHP project shortcomings. Moreover, the MHP licensing royalties were considered too high by most broadcasters, whose main association (EBU) announced in March 2008 the discontinuation of its support to MHP. The boycott was later reconsidered, but by that time MHP had already missed most of its opportunities, and did not take-off (apart from Italy).

Face to the MHP failure (resembling in part that of the first DVB standard for mobile TV—DVB-H, not analysed here), the EU Commission started acknowledging how interoperability is difficult to be provided not only by Governments, but also by hybrid settings such as the DVB-ETSI tandem; basically, the latter did not manage to harmonize the consortium members' conflicting visions, incentives and agendas concerning ITV standardization. This story also illustrates how short even OS may fall of guaranteeing interoperability and market convergence, absent a suitable technological landscape and aligned market incentives.

3.5 New Hybrid Forms of TV

So far, we concentrated our analysis on traditional TV platforms; after all, during most of the 2000s, they managed to maintain the bulk of their audience ratings and advertising shares, in aggregate, despite the growing diffusion and usage rates of the Internet. However, starting from the late 2000s, a clear stylised fact of the sector is the mounting audience-stealing effect deriving from the consumption of online TV-related content (videos, TV channels, etc.), as Fig. 5 indirectly demonstrates. It shows that, by the end of 2012, a significant portion of citizens has gained a first experience with online TV content (first histogram); beside this, another significant part (second histogram) exhibits a frequent behaviour (weekly) of on-line TV consumption. UK is the most representative case in EU, since 42 % of the interviewed persons admits to be experienced with online TV, while almost one quarter (23 %) of the population watches it weekly. Other notorious technology-minded countries like China appear to be even heavy on-line TV viewers, with 68 % and 44 % of the interviewed population interested by the same phenomenon.

Indeed, broadband roll-out progresses worldwide, and its increasingly pervasive availability and higher transport capacity are opening revolutionary perspectives for TV convergence, based on the usage of IP networks for content transport and

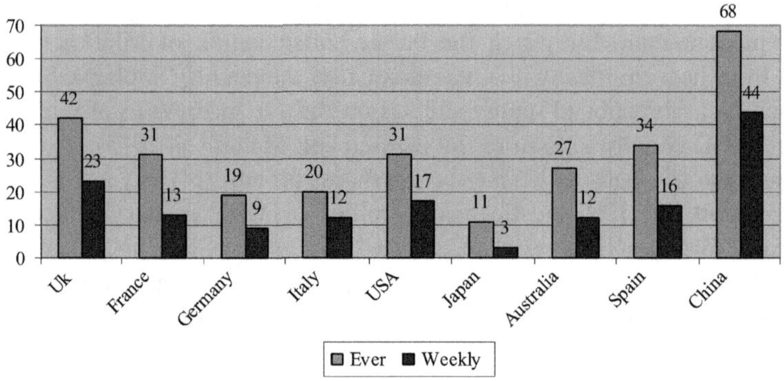

Fig. 5 Usage and frequency of access of TV content over the Internet—% values; October 2012. *Note*: The graph reports the proportion of respondents that accesses online TV in each surveyed country (respectively, usage/weekly frequency). *Source*: OFCOM (2012; Fig. 3.12)

delivery. This trend includes both proprietary networks—like the IPTV—and, increasingly, the open Internet, where OTT operators are active service providers and have been acquiring growing shares of loyal Internet users.

Indirectly, in the previous section, we detected a few signals of this trend commenting on the fact that the second generation of DVB transmission standards, while does not offer a significant quantum leap in inter-interoperability within the TV world (for example, it continues to require specific duplication of electronic components to produce high-end STB capable of multiplatform TV reception), has been specifically designed for enabling a higher potential of IP capacity and interoperability (hence convergence) with the Internet world, with respect to the first generation. This is a clear sign that traditional TV operators and their standardization representatives felt the competitive threat of the Internet, as wisely envisaged by Owen (1999), and have started to harness the opportunities connected to it, responding in several ways to the growing momentum of web-related content consumption.

So far, the main threat for traditional TV players does not seem to come from IPTV operators (with the only exception of France, where the latter has achieved a high rate of diffusion, reaching 28 % of TV homes in 2011, OFCOM, 2012). In fact, the diffusion of IPTV has long been dampened by a few shortcomings affecting its business model: among the most important, we include the shortage of premium content, and the typical usage of strict proprietary and IPR-closed business models, which do not appear conducive to ease the market entry of a new operator (for the considerations spelled out in Sect. 2.2). In the future, however, the latter obstacle might fade away, as soon as IPTV operators continue to develop new hybrid business models and adopt more interoperable types of equipment.

Rather, the main challenge to traditional TV operators seem to come from OTT players. In particular, the main ones (such as Google-YouTube and Apple) since the second half of the 2000s have been introducing a series of initiatives signalling a clear interest to enter the TV sector, either in the form of free (advertising-funded) or pay business models (such as VOD, pay-per-view or subscription based). OTT typically focus on aggregating content and attracting audience, for sale; this, until now, has not included the control of an IP-based transport infrastructure, or a content delivery network (CDN). However, in several cases they have been developing new reception devices, own or in collaboration with major CE manufacturers, to better manage their installed base of viewers/subscribers and provide new converging multimedia services (beside TV content). Main examples are the Internet-connected personal video recorder of TiVo (US), the TV box of Apple or the game-consoles of Sony (Playstation), Microsoft (Xbox) and Nintendo (Wii).

Further, major CE manufacturers have being developing new Internet-connected TV sets, incorporating higher memory and computing power, and featuring original interactive functions and "smart" capabilities (such as voice or eye-based control), teaming up with major content providers or OTT players (like in the case of the software widgets of YouTube, Yahoo and Facebook embedded into the flat panel display (FPD) TV sets of LG, Philips, Samsung and Sony). This has originated a wide and expanding array of accompanying neologisms like "Connected TV", "Smart TV" and so on[36] (for a recent taxonomical and classification effort, Quayle, 2012).

Finally, the latest technological frontier appears to be that of the "Hybrid Broadcast Broadband TV" (or HBB TV), where the two experiences—that of the TV and the Internet, previously offered as separated bundles of services—here are designed and presented to the user as a unique experience and service. Indeed, HBB TV currently represents the most advanced frontier of media convergence between the Internet and the TV. From a technological point of view, the development of multiple competing technical specifications for HBB TV (both in EU and abroad, such as HbbTV, MHP, MHEG, YouView and others) let to prefigure a trend towards accentuated standards fragmentation and industry turmoil, mostly due to the diversity of the existing legacy solutions.

4 Conclusions

This chapter proposed an interdisciplinary perspective on the growing role of technical standards and IPR in the converging TV sector, investigating the main standardization initiatives carried out in the last 20 years, to examine any possible trend toward media convergence or divergence, both within the traditional TV world and in relation to previously separated ICT sectors.

In detail, we first reviewed the economic function of standards to achieve technological and user-level interoperability, to uncover that standards in ICT and media markets open up a series of technical possibilities, as well as strategic incentives, that may enable or foreclose market entry for rival operators, by manoeuvring interoperability and convergence.

[36] Minor variations and logical subsets are also "Web-TV", "Net TV", "OTT TV", "Catch-up TV".

Then, recognising that complex standards may arrive to contain hundreds of IPR, we reviewed the main stylised facts and latest trends in IPR management and policy. In detail, the current hype of open standards (OS) was assessed, and the characteristics of licensing coalitions (patent/IPR pools) were highlighted. This survey uncovered several criticalities affecting OS, FRAND terms and patent pools, associated to contractual loopholes and legal idiosyncrasies. All in all, OS and patent pools cannot completely sterilise the likelihood of anti-competitive and rent-seeking behaviours, and pose new original implementation challenges. In several situations, because of strategic manoeuvring of large IPR portfolios by stakeholders during and after standardization, initial openness and interoperability might fade away, at the expense of smaller competitors, of the market competition and of the consumer welfare. As a result, most of the current technological potential for converging ICT and media markets might be captured for the benefit of a small group of stakeholders, imposing suboptimal 'captive' convergence and stifling wider economic growth and societal progress.

Then, moving to the empirical part of the chapter, we carried out an original long-run empirical analysis of the main standardization initiatives in the TV sector, based on the DVB and EU experience. As a first stylised fact and driver of media convergence, we provisionally accepted the technological explanation initiated by Negroponte (1995). Indeed, the availability of increasingly powerful systems of digital coding and transmission of multimedia content has market the passage from the analogue to digital TV, initiating an era of pervasive dynamism and convergence along its entire value chain. Moreover, thanks to our empirical analysis, we appreciated that, over time, in parallel with the diffusion of the Internet, there has been a second 'reinforced' type of technological confluence of TV standards, with the latter joining a common converging micro-electronic base and adopting the main standardized protocols of the Internet word, such as the TCP/IP.

The in-depth analysis of the work of the DVB consortium supports this view, and uncovers other interesting regularities. The standardization endeavour initially worked well—especially while dealing with the transmission layer (first generation), but has later showed increasing signs of pain and inadequacy when trying to address the more IPR-intensive layers of the value chain—like the stories of the CAS, EPG and API demonstrated. In particular, our analysis of the MHP case highlighted the conflicting visions on interactive TV and the diverging agendas held by the most important DVB stakeholders; this situation, together with the IPR shortcomings of the late patent pool, defeated the big efforts played by EU public bodies toward the deployment of that open API. At the same time, the status of microelectronics and software technologies and the potential for interoperable interactive services were too unripe to enable, during the first half of the 2000s, the successful introduction of truly converging types of interactive or Internet TV, which represents a landmark achievement in media convergence. At the same time, the exam of the case of transmission standards showed us that, on overall, the technological potential for inter-platform interoperability within the traditional TV world has remained under-developed, due to adverse incentives to interoperability and convergence held by traditional TV stakeholders.

Instead, during the second half of the same decade, additional factors ignited new and more powerful convergence trends between the TV and the Internet. Certainly, increasing worldwide Internet penetration and broadband roll-outs set the stage, and put a mounting psychological pressure on TV incumbents, prefiguring increasing audience-stealing effects. This, jointly with the higher technological opportunities unfolding in the same period, contributed to increase the effort towards "external" interoperability of the DVB work, as we detected in the case of the second generation of DVB transmission standards. This trend is also an indirect sign of the fact that traditional TV operators acknowledged early the incoming Internet' threat (see Owen, 1999), and have been working hard, in several domains, to prepare them-selves and accommodate convergence while exploiting their TV market incumbency. Big OTT operators, in turn, seem to be in a strong position to exploit their market incumbency in the online content market, and to transfer part of this to the converging TV one, teaming up with other key players, like CE manufactures producing smart TV sets and other Internet-connected TV "boxes".

Finally, the latest technological frontier of the convergence between TV and Internet appears to be the recent standardization efforts for the hybrid broadcast broadband TV, although here the market reality is still too fluid and unripe to inform a reliable analytical effort aimed at detecting robust convergence or divergence trends.

All in all, so far, despite this high technological turbulence, the established media landscape seems to maintain its main classical characters, with traditional TV and telecoms operators striving to keep market shares through walled garden solutions and lobbying for stronger IPR protection, due to the higher piracy potential of the Internet. Nevertheless, we believe that this apparent market continuity is about to end quickly, as soon as the main and new OTT operators will find new lateral in-roods into established TV markets: audience trends signal already the first traces of the next storm, with on-line TV viewing progressing at high rates at the expenses of traditional TV consumption.

References

Arthur, B. (1989, March). Competing technologies and lock-in by small historical events. *Economic Journal, 99*, 116–131.

Besen, S. M., & Farrell, J. (1994). Choosing how to compete: Strategies and tactics in standardization. *Journal of Economic Perspectives, 8*(2), 117–131.

Blind, K. (2004). *The economics of standards: Theory, evidence, policy*. Cheltenham: Edward Elgar.

Bresnahan, T. F., & Greenstein, S. (1999). Technological competition and the structure of the computer industry. *Journal of Industrial Economics, 47*, 1–40.

Cave, M. (1997). Regulating digital television in a convergent world. *Telecommunications Policy, 21*(7), 575–596.

Cawley, R. A. (1997). European aspects of the regulation of pay television. *Telecommunications Policy, 21*(7), 677–691.

Cerri, D., & Fuggetta, A. (2007). Open standards, open formats, and open source. *The Journal of Systems and Software, 80*(11), 1930–1937.
Choi, J. P. (1996). Do converters facilitate the transition to a new incompatible technology? A dynamic analysis of converters. *International Journal of Industrial Organization, 14*(6), 825–835.
Choi, J. P. (1997). The provision of (two-way) converters in the transition process to a new incompatible technology. *Journal of Industrial Economics, 45*(2), 139–153.
Cowan, R. (1991, July). Tortoises and Hares: Choice among technologies of unknown merit. *Economic Journal, 101*(407), 801–814.
David, P. A. (1985). Clio and the economics of qwerty. *American Economic Review, 75*(2), 332–337.
David, P. A., & Greenstein, S. (1990). The economics of compatibility standards: An introduction to recent research. *Economics of Innovation and New Technology, 1*(1), 3–41.
Drury, G. M. (1999). Interoperability in digital broadcasting systems. In *IEE colloquium on digital television—where is it and where is it going? (Ref. No. 1999/072)* (pp. 6/1–6/7).
EBU. (2004, April 30). *Comments by the EBU on the commission staff working paper on the interoperability of digital interactive television services*. Available at http://europa.eu
ECCA. (2003, February 20). *Comments on the commission staff working document on barriers to widespread access to new services and applications of the information society*. Brussels.
Economides, N. (1996). The economics of networks. *International Journal of Industrial Organization, 14*(6), 673–699.
Eltzroth, C. (2008). IPR policy of the DVB project: Negative disclosure, FR&ND arbitration unless pool rules OK, Part I. *The International Journal of IT Standards and Standardization Research, 6*(2), 21–47.
Eltzroth, C. (2009). IPR policy of the DVB project: Negative disclosure, FR&ND arbitration unless pool rules OK, Part II. *The International Journal of IT Standards and Standardization Research, 7*(2), 1–22.
European Commission. (2004, July 30). *On the interoperability of digital interactive television services*. Communication from the Commission, n. COM 541 final. Available at http://europa.eu
European Commission. (2006, February 2). *Reviewing the interoperability of digital interactive television services pursuant to communication COM[2004] 541 of 30 July 2004*, Communication from the Commission, n. 37 final. Available at http://europa.eu
Farrell, J., & Saloner, G. (1992). Converters, compatibility, and the control of interfaces. *Journal of Industrial Economics, 40*(1), 9–35.
Farrell, J., & Shapiro, C. (1992). Standard setting in high-definition television. *Brookings Papers on Economic Activity: Microeconomics*, 1–93.
Funk, J. L., & Methe, D. T. (2001). Market- and committee-based mechanisms in the creation and diffusion of global industry standards: The case of mobile communication. *Research Policy, 30*, 589–610.
Gandal, N. (2002). Compatibility, standardization and network effects: Some policy implications. *Oxford Review of Economic Policy, 18*(1), 80–91.
Gerbarg, D. W. L. (Ed.). (1999). *The economics, technology and content of digital TV*. London: Kluwer.
Gilbert, R. (2010). The essentiality test for patent pools. In R. C. Dreyfuss, D. L. Zimmerman, & H. First (Eds.), *Working within the boundaries of intellectual property. Innovation policy for the knowledge society*. Oxford: Oxford University Press.
Grimme, K. (2002). *Digital television standardization and strategies*. London: Artech House.
Hart, J. (2004). *Technology, television and competition. The politics of digital TV*. Cambridge: Cambridge University Press.
Hawkins, R. (1999). The rise of consortia in the information and communication technology industries: Emerging implications for policy. *Telecommunications Policy, 23*, 159–173.
Helberger, N. (2005). *Controlling access to content. Regulating conditional access in digital broadcasting*. The Hague: Kluwer Law International.

Jenkins, H. (2006). *Convergence culture*. New York: New York University Press.
Katz, M., & Shapiro, C. (1994). System competition and network effects. *Journal of Economic Perspectives, 8*(2), 93–115.
Klevorick, A. K., Levin, R. C., Nelson, R. R., & Winter, S. G. (1995). On the sources and significance of interindustry differences in technological opportunities. *Research Policy, 24* (2), 185–205.
Krechmer, K. (2006). The meaning of open standards. *The International Journal of IT Standards and Standardization Research, 4*(1), 43–61.
Lemley, M., & Shapiro, C. (2007). Patent holdup and royalty stacking. *Texas Law Review, 85*, 1991–2049.
Lerner, J., & Tirole, J. (2004). Efficient patent pools. *American Economic Review, 94*(3), 691–711.
Matteucci, N. (2004). Diritti Televisivi, oligopolio ed intervento antitrust nella *pay-TV*: il caso Telepiù-Stream. *Quaderni di Dipartimento*, n. 215. Available at http://econ.univpm.it
Matteucci, N. (2008). Open standards and interoperability in EU digital TV: Economics and policy issues. *Review of Economic Research on Copyright Issues, 5*(2), 45–70.
Matteucci, N. (2009). Interoperability provision in NGC: The case of Italian DTV. *Info, 11*(6), 30–50.
Matteucci, N. (2010). Strategies for digital television: DTT in Italy. *International Journal on Media Management, 12*(3), 159–181.
Matteucci, N. (2013). Lo stato della banda larga in Italia: statistiche, modelli diffusivi e implicazioni di policy. *L'industria* (n.s.), Anno *XXXIV*(1), 11–50.
Matutes, C., & Regibeau, P. (1996). A Selective review of the economics of standardization entry deterrence, technological progress and international competition. *European Journal of Political Economy, 12*(2), 183–209.
Moore, G. E. (1965). Cramming more components onto integrated circuits, reprinted from *Electronics, 38*(8), 114ff., *Solid-State Circuits Society Newsletter, IEEE, 11*(5), 33–35, September 2006.
Negroponte, N. (1995). *Being digital*. New York: A.A. Knopf (Random House).
Nolan, D. (1997). Bottlenecks in pay television. Impact on market development in Europe. *Telecommunications Policy, 21*(7), 597–610.
OFCOM. (2012, December 13). *International communications market report 2012*. Available at http://stakeholders.ofcom.org.uk/binaries/research/cmr/cmr12/icmr/ICMR-2012.pdf
Owen, B. (1999). *The internet challenge to television*. Cambridge, MA: Harvard University Press.
Parsons, P. R., & Frieden, R. M. (1998). *The cable and satellite television industries*. Boston: Allyn & Bacon.
PriceWaterhouseCoopers. (2012). *2012 Patent Litigation Study*. Available at http://www.pwc.com/en_US/us/forensic-services/publications/assets/2012-patent-litigation-study.pdf
Quayle, A. (2012). *TV delivery evolution: Hybrid TV, over the top (internet) TV and TV everywhere (Multi-Screen) TV. Status report and projections 2012–2017*. Alan Quayle business and service development consultancy report.
Reimers, U. H. (2006). DVB-The family of international standards for digital video broadcasting. *Proceedings of the IEEE, 94*(1), 173–182.
Rochet, J. C., & Tirole, J. (2003). Platform competition in two-sided markets. *Journal of the European Economic Association, 1*, 990–1029.
Salop, S. C., & Scheffman, D. T. (1983). Raising rivals' costs. *American Economic Review, 73*(2), 267–271.
Shapiro, C. (2001). Navigating the patent thicket: Cross licenses, patent pools, and standard-setting. In A. B. Jaffe, J. Lerner, & S. Stern (Eds.), *Innovation policy and the economy* (Vol. I). Cambridge, MA: MIT Press.
Shapiro, C., & Varian, H. R. (1999). *Information rules*. Cambridge: Harvard University Press.
Siebert, P. (2011, October 24–27). *DVB: Developing global television standards for today and tomorrow* (pp. 171–175). Telecom World (ITU WT), 2011 Technical Symposium at ITU.
Stango, V. (2004). The economics of standards wars. *Review of Network Economics, 3*(1), 1–19.

Swann, P. (2010, May 27). *The economics of standardization: An update*. Report for the UK Department of Business, Innovation and Skills (BIS), Version 2.2. Available at http://www.bis.gov.uk/feeds/~/media/ED32CDA672764D7C8AC2F4E323581010.ashx

Treacy, P., & Lawrance, S. (2008). FRANDly fire: Are industry standards doing more harm than good? *Journal of Intellectual Property Law and Practice, 3*(1), 22–29.

Van Schooneveld, D. (1996). Standardization of conditional access systems for digital pay television. *Philips Journal of Research, 50*(1–2), 217–225.

West, J. (2003). How open is open enough? Melding proprietary and open source platform strategies. *Research Policy, 32*, 1259–1285.

West, J. (2007). The economic realities of open standards: Black, white and many shades of gray. In S. Greenstein & V. Stango (Eds.), *Standards and public policy*. Cambridge: Cambridge University Press.

Converging Multimedia Content Presentation Requirements for Interactive Television

Ioannis Deliyannis

1 Introduction and Problem Discussion

Media convergence may be perceived as phenomenon of the Information Society evolutionary process. Manuel Castells recognizes that "*. . .the growing interaction between horizontal and vertical networks of communication does not mean that the mainstream media are taking over the new, autonomous forms of content generation and distribution. It means that there is a process of convergence that gives birth to new media reality whose contours and effects will ultimately be decided by political and business struggles*" (Castells, 2011). This process is particularly important for interactive television, a domain influenced by technological advances, techno-social interaction and social media that currently demonstrate rapid convergence. One may historically identify the occurrence of these rapid processes precede the establishment of new standards, a time where in most cases opposing technologies try to gain a respectable market share and be adopted by the users. Take for example the battle of formats between the VHS vs. BETACAM and the more recent HD-DVD vs. the BLU-RAY technologies. In the first case the capability of content duplication offered by one of the systems and in the second case the extended availability of a wide number of content titles for one of the standards, forced the public to adopt the technologically less-advanced standard. User choice based on the availability of replication mechanisms in one and content in the other case is indicative of the convergence process that takes place within the information society.

Of course content and technical specifications cannot always be considered as the main deciding factor. One may recall the technological breakthrough that the introduction of interactive-CD and later DVD-interactive media offered to

I. Deliyannis (✉)
Interactive Arts Laboratory, Department of Audio and Visual Arts, Ionian University, Plateia Tsirigoti 7, Corfu, Greece
e-mail: yiannis@ionio.gr

A. Lugmayr, C. Dal Zotto (eds.), *Media Convergence Handbook - Vol. 1*, Media Business and Innovation, DOI 10.1007/978-3-642-54484-2_13

developers. Some identified the positive aspects of these technologies (Taylor, 1999) and predicted their applicability in areas of education (Bush, 2000) and entertainment (Kreuzberger, 2007). Today, although the majority of proprietary DVD-playback devices and computer-based media players support interactive content playback and one may use it to develop interactive movies (Argyriou, 2010) or games (Kreuzberger, 2007), a small number of interactive titles have been produced that take advantage of the content-manipulating capabilities. Clearly this phenomenon is not attributed to the limited user interest or media incompatibility, but to the complex and expensive production process that needs to be actioned in order to produce multimedia titles that increases the cost and production time. Today interactive titles are developed using computer game development environments and are distributed on DVD media or the WWW.

The above examples share common characteristics to interactive television as they describe the introduction of new technologies, which are novel and do not have to deal with already established competitive technologies. The case of Minitel is indicative of the issues arise in the adoption process when new competitive technologies appear. Minitel was a pre-WWW videotext service established at 1982 in France, enabling users to perform actions similar to those offered by the WWW services that include: messaging, bookings, reservations and purchases over their phone line. The availability of this service delayed the absorption of the WWW in France and in 2009 Minitel still totalled ten million users. The service was officially terminated in 2012.

Whether interactive broadcasting systems will eventually catch-up with society is a non-use non-adoption issue (Deliyannis, Antoniou, & Pandis, 2009) affected by the availability of other technologies, marketing, technological infrastructure cultural factors and consumer behavior (de Mooij & Hofstede, 2002). In order to research the factors of adoption or non-adoption of technological developments, the evolution of television is in this work addressed from the technological (Lugmayr, Niiranen, & Kalli, 2004; Zheng, Fu, & Lugmayr, 2012), social and ethnographic perspectives (Tay & Turner, 2009; Tsekleves, Whitham, Kondo, & Hill, 2009). Under this standpoint it is easier to examine how users often combine communication media and engage them constructively within their social context (Paradise, 2012). For example, in the case of television, evolution is influenced by a number of interrelated technological, content and user factors. Today the majority of networked mobile devices allow standard stream playback facilities, covering the content access user demands. At the same time, users capable to utilize the advance features of their mobile devices are able create, process, enhance with metadata and stream/broadcast/share audiovisual content on the WWW. This clearly transforms the user from content receiver to content provider, triggering an interactive content publishing cycle.

Political forces and media-providers do try to affect the users, but in the end the ability to retrieve and enhance content with metadata such as comments and links to other comments enables a new reality to evolve. The news reading service available from Google at http://news.google.com is a typical example that may be used to demonstrate the shift towards a more organized way to access and evaluate news feeds. This service triggered by content published on websites providing news feeds, organizes content thematically allowing direct access to the sources. The user is able to scan the article titles and access the text across different sources, a process that often reveals that in some articles specific information may not be disclosed in its entity, offering a biased interpretation of facts. User comments and thoughts can then be posted on the source website, or other social media services. Clearly, social media may be considered superior to traditional broadcasting media as their interactive capabilities cover the need for immediate access to information while users are allowed to contrast, interpret and criticize published content on the same subject from various sources.

Interactive television functions in a similar way as it allows users to share their feedback and influence the live broadcasting process. The main aim of this work is to identify the key factors that affect the converging process in the field of interactive television, and provide solutions that may be applied in the media industry. Section 2 researches the literature, identifies and analyses the governing factors of the convergence process. As new technologies introduce advanced interaction capabilities, we do not limit this research to existing content types. Instead, advanced content case studies are developed in an attempt to reveal the converging multimedia content presentation requirements for interactive television content. Section 3 presents and evaluates the findings of these case studies designed to test the limitations of the interactive broadcasting architecture and Sect. 4 summarizes the results, proposes further research and development directions that if followed by content providers can aid convergence by advancing their products and services while directing users to voluntarily adopt this new medium.

2 Literature State of the Art/Literature Review

In the introduction we identified a number of standpoints under which one needs to examine convergence in the interactive television sector: the technological, content and user perspectives. Close interaction between these three interrelated entities is observed that influence each other and trigger evolutionary changes. A diagram (Fig. 1) is used to describe interrelationship using area-inclusive representation between the entities involved and represent the opposing interests introduced in the process of convergence. We choose to examine convergence from different standpoints: society, technology and content. It is informative to present the evolutionary perspective for each standpoint and contrast how it affects the other domains, in an attempt to view the changes from this collective prism.

We classify the convergence problem as a typical "*chicken or the egg*" causality dilemma. The transition to interactive television is a process that requires constant adaptation and resolution of technological, content and social issues. Within this context, one would expect various divergence issues to surface. From the technological perspective this is clearly not the case for a number of reasons: interactive television is an extended service designed to offer backward compatibility, enabling encapsulation of existing content, alongside the new interaction features.

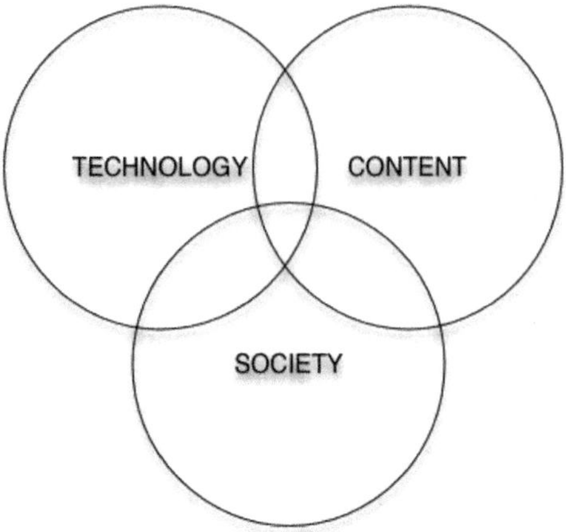

Fig. 1 Interrelationship diagram between society, technology and content

However, divergence is evident on the content and social perspectives, as content developers and users have to evolve and re-design their content in order to be benefited by the advanced content-projection capabilities offered by the proposed system. These issues are reviewed in the present section and specific application issues are discussed further within the case studies (Sect. 3). In order to describe the mechanics of the process and predict future developments, we start by examining the technological advances that spark the changes. Then we review the social implications introduced and investigate how users use this new technology to view, enrich, transform and redistribute content. This is accomplished by examining the content-related issues that arise and reveal the converging requirements, within the prism of related theories.

2.1 Technological Advances

Various theories and processes may be used to analyse the mechanics of the changes that occur during convergence. *Activity Theory* introduced by Lev Vygosky, Alexander Luria and Alexei Leont'ev in the 1920s describes convergence from the socio-technical standpoint. This is achieved by identifying and analyzing the influence between humans and interactive systems through a temporal perspective (Bertelsen & Bødker, 2003). "*A person is influenced and in turn influences the environment and others as human mind and human activities are linked under the model. As a result, the changes that a person introduces to the environment influence humans that are born within this environment*" (Deliyannis, 2012). It is informative to contrast the technological capabilities during the years where television was the only way for the general consumer to receive audiovisual information and today, where there are many different information sources. In the past, the most important issues that a content provider had to constantly balance in order to *stay on the air* included high technological cost, the need for an independent broadcasting infrastructure, costly content production and licensing.

Today, the shift to digital production and broadcasting has greatly reduced the technological cost, the quality of reception has improved and there is no need to use costly broadcasting infrastructures as the Internet may be used to relay and distribute content to the public. Modern television stations are purely digital and various case studies have evolved offering high-quality and low-cost in-house broadcasting solutions (Deliyannis et al., 2009). This is identified as a stage of technological convergence where broadcasters had to invest into new technologies and provide content for other platforms. It is interesting to contrast how the media industry reacted to technological changes that provide additional communication channels to the public.

By reviewing the websites of major news-based broadcasting networks including CNN, NBC and BBC, one can easily observe that the WWW is used as a media-extension platform employed to stream and reference content. In addition to live or offline streaming, a small number of interactive options are available to the user. Commenting on pre-recorded video streams is sometimes permitted, while stream sharing via URI-linking for varying platforms and social media applications allows content to be shared further. In special interactive implementations, interactive content broadcasting is supported enabling users to select the camera viewpoint during live events. In other cases live videoconferencing or text-based user-feedback is linked directly to the studio (Long, Chei Sian, & Goh, 2012). The cycle continues as content that originates from broadcasting networks is combined with user-generated content and redistributed through the use of social media sites including, blogs, content-communities, collaborative projects, Internet forums, wikis, podcasts, video/photo sharing websites, review sites, virtual social and game worlds. In these environments, users capture content, enhance it by adding metadata and share it to a closed social circle or the whole networked world, triggering interactive dialogues, responses and reactions.

Interestingly, novel content types have already been produced and delivered using interactive experimental broadcasting systems based on multicast technologies that take advantage of the wider interaction spectrum offered by bidirectional communication (Chorianopoulos & Spinellis, 2006; Munson & Pimentel, 2012; Prata, 2009; Song, Moustafa, & Afifi, 2012; Ursu et al., 2008). Interactive television may safely be considered today as the leading candidate technology that may be used as a platform to cover extended user and content requirements, beyond the support for interactive broadcasting and selection/evaluation of content. We have already developed a candidate system termed iMediaTV (Fig. 2), which consists of a hardware and software architecture (Deliyannis et al., 2009; Deliyannis & Kanellopoulos, 2008; Deliyannis & Karydis, 2011) that may be adjusted to cover the requirements for user interaction within multiple content types, including interactive art, live performances and experimental interactive multimedia content (Deliyannis, 2012).

In our earlier research we investigated the implications including content-access and copyright issues that arise when users interact with the studio participants and

Fig. 2 The iMediaTV control panel featuring the audio mixer, the video mixer, the server and cameras that connects to the video mixer and the direct video server

alter the course of the show (Deliyannis, Karydis, & Karydi, 2011). The results indicate that two shifts are implemented simultaneously: the content is being gradually transferred through channels that do not offer exclusivity and that the users are able to develop low-cost broadcasting infrastructures offering new interaction capabilities that may in the future compete with existing broadcasting networks. Having equaled the technological production and broadcasting issues, it is a matter of time before content availability will shape the domain.

2.2 Social Implications

"*Diffusion of innovations*" is a process introduced in 1962 by Everett M. Rogers, used by many as an essential tool for media convergence analysis (Srivastava & Moreland, 2012). It describes the process with which an innovation (an idea, practice, or object perceived as new by an individual or other unit of adoption) is communicated through certain channels over time among the members of a social system (Rogers, Singhal, & Quinland, 2009). The Internet is clearly the way that new systems, ideas and most importantly content is transported and re-used today. Blogs, wikis, forums, codebases and repositories set a solid foundation enabling users to develop, search, communicate content (Brennan & Resnick, 2013). Users perceive the WWW as a content-sharing platform, where information is published, distributed, enhanced with metadata and recycled. It offers the ability for content to be shared via URI-linking. This is a powerful feature as it provides a universal way to link content, enable communication and trigger innovation, provided that the correct applications are available at the user end.

Content transformation and technological advances create new products, user trends, demands and markets, which are mutually dependent (Elfving & Urquhart, 2013). Take for example the evolution of the music industry where technology combined with the portability of content transformed the whole industry: talent scouting, production, promotion and distribution (Tschmuck, 2012). Today users purchase music tracks electronically at reduced cost, as they do not pay the costs of storage media, distribution and retailer profit. The same applies for movies, television shows, podcasts, books and applications that are accessible through electronic stores like iTunes, enabling synchronization and playback throughout the user's media devices.

The social implications should be traced much deeper than the changing commerce and economy layers (Dutton, 2013), as networked users have already moved to "*platformed sociality*" (van Dijck, 2013) where content sharing is an essential interactive process. According to Jenkins, user-participation through the Internet is defined as convergence culture: "*where old media and new media collide, where grassroots and corporate media intersect, where the power of the media producer and the power of the media consumer interact in unpredictable ways*" (Jenkins, 2006). Clearly the social implications are different for each user group as new communication dynamics are introduced. Those who appreciate the new opportunities offered by technology and invest time to use media to their advantage, often increase their influence, for example when politicians use interactive and personalised online communication methods (Kruikemeier, van Noort, Vliegenthart, & de Vreese, 2013).

In interactive television, various technologies have been introduced offering various interesting social characteristics. Today, YouTube Live offers a basic infrastructure that may be employed to support such functionality without the need to setup a private multicast-broadcasting network (http://www.youtube.com/live/). A developer may deploy a web-based platform that links different services under a unique system similarly to iMediaTV. The most important feature at the user level is the integral system capability that allows them to broadcast back to the interactive TV studio their live streams. This phenomenally two-way communication between them and the studio is in fact a many-to-many communication instance, as it is also made available and can be re-broadcasted to the other viewers. Social software services utilize a similar information broadcasting process, which triggers other user's interest indicated by various digital responses. Similarly with social software, interactive TV users they may link via URI to the generated responses, search them using appropriate metadata and share them other users.

2.3 Content Transformation and Transportation

Our research in media policies shows that these are often outdated by technological developments that open new content processing and distribution routes. This becomes clearly evident when contrasting current research on the content aspects introduced by interactive television (Deliyannis et al., 2011) to the Australian reports on future media policies published that include the Convergence Review (Review, 2012) based on an independent media inquiry (Finkelstein, 2012) and the *Review of the National Classification Scheme* (ALRC, 2012). In these reports is recognized the technological infrastructure shift from broadcasting-centers that utilize telecommunications and radio technologies to new technological infrastructure, networks, devices, applications and content, a converging state where content providers need to adapt. However, the issue of content ownership remains and the reports do not address the main question: how can the industry distribute content to society using alternative technologies and methods and keep the strategic advantage of distribution control?

The main content issue observed today is that it can be transformed and transported using multiple ways. Digitization, streaming and file sharing enable direct or indirect content exchange, while application-layer protocols like the RTSP are openly available for network-based multimedia-content broadcasting. Take for example the case where geographical-based licensing is required in order to view a specific broadcast. Clearly, for those who know better how computer-broadcasted content is distributed today, it is apparent that the policy is already outdated. For a broadcast licensed to a network in order to be broadcasted only on a specific country on television and the WWW, all that is required for the stream to be relayed to another country is access on a proxy server hiding the user's true address, enabling content transfer to any other location worldwide. Beyond content access, the above policies ignore the user ability to capture, decode, possibly enhance, encode and re-broadcast content creating a new type of broadcasting that cannot easily be controller by content owners. Sometimes the situation is reversed. Take for example various traditional television programs that collect, edit, comment on and re-broadcast WWW-based videos. Content re-use is a new development that from our experience will raise serious content ownership issues, particularly as today cloud-based video editing software such as *JW Player Online Edition*, *YouTube Editor*, *Video Toolbox* any many more such applications are already freely available.

3 Methodology and Approach

iMediaTV offers an interactive broadcast infrastructure (Deliyannis et al., 2009, 2011) developed using low cost equipment, that provides a platform for new content development (Deliyannis & Karydis, 2011). The same system can also be used to house the requirements of experimental multimedia systems (Deliyannis, 2012) that present novel content and interactive features. The platform consists of

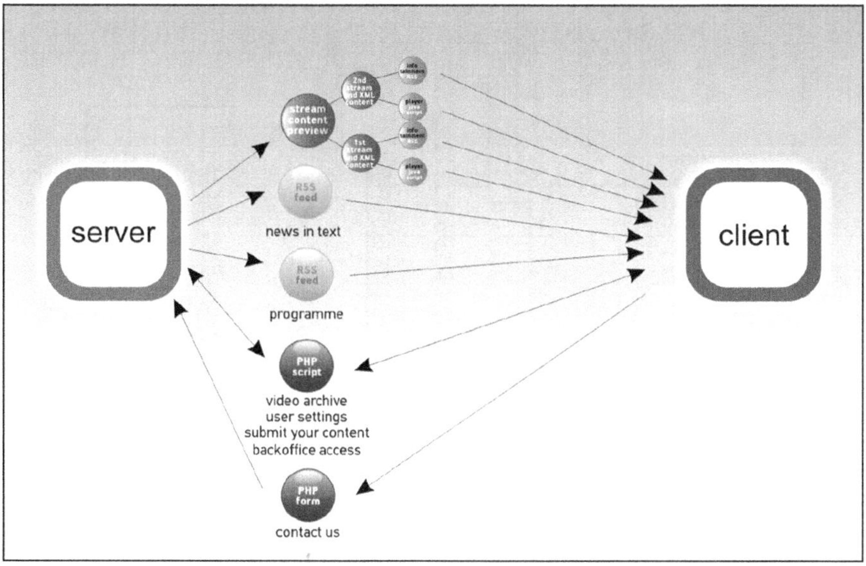

Fig. 3 The iMediaTV server providing various linking options to the client

hardware, software and content. Its hardware implementation depicted in Fig. 2 consists of a video and an audio mixer connecting two standard-definition cameras, a DVD-player and a computer used as a video source, linked through an appropriate standard-definition video/audio card. For each camera the video signal is directed both to the console and an individual streaming server using appropriate splitter connections, offering to the remote user direct camera access. On the software forefront, system-user linking is implemented using a WWW client-server-based user-interface structure that connects autonomous software communication components, as shown in Fig. 3.

Here RSS, PHP-MySQL technologies and multicast broadcast streams are combined under the user-interface enabling client-server linking. The user interface menu options shown in Fig. 4 (top) links the software and hardware components (bottom). Video-streaming flow is described in Fig. 5. In order to speed-up streaming, separate multicast RTSP streaming servers using the H264 codec (Richardson, 2006). The interactive nature of the station is supported on various content distribution levels. Offline content is categorized according to its nature to single or multiple streams, enabling the appropriate number of servers for its broadcast. Multi-stream broadcasting in parallel over the network enables multiple interactive modes to be implemented. Users can access offline content directly from the multimedia-database, while online content supports a number of interactive options. Originally the system was implemented using a single server. This computationally proved to be a quite demanding process introducing high latency rates of over 15 seconds.

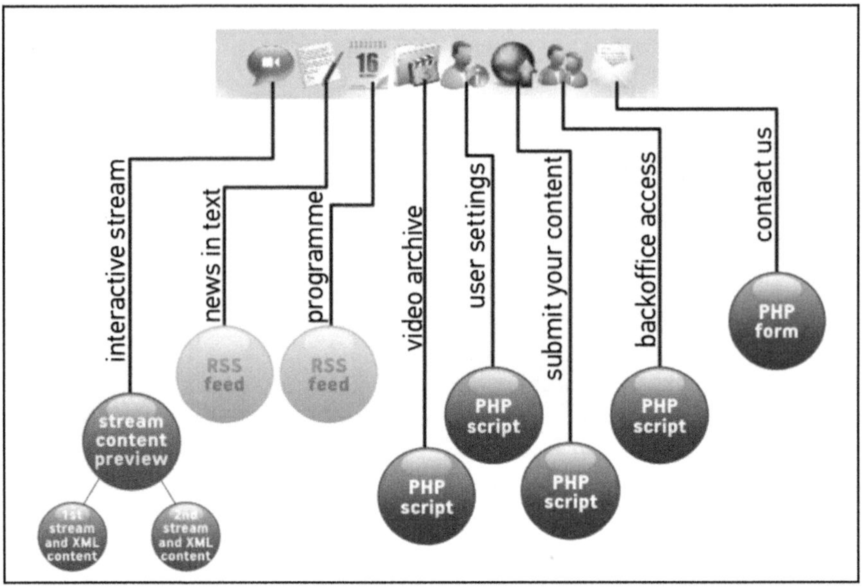

Fig. 4 iMediaTV user-interface options to data stream linking

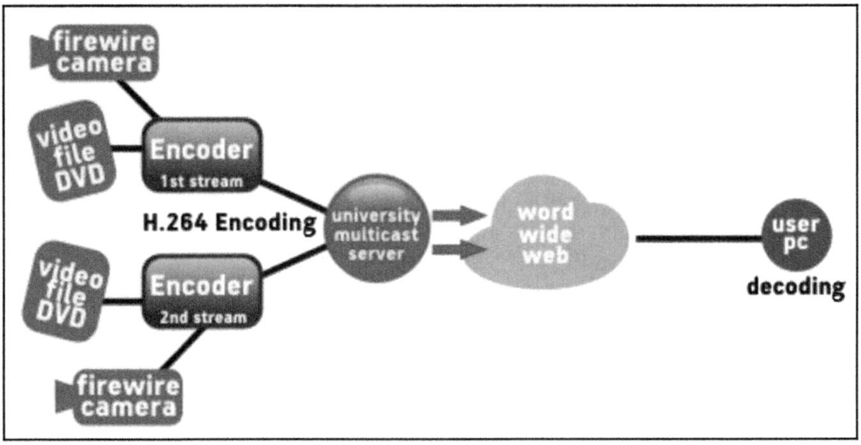

Fig. 5 iMediaTV video encoding and streaming diagram

As a result separate servers under the same network were employed for different streams: two servers support streaming content delivery via the use of separate but synchronized encoders (Fig. 5) and one server is dedicated for the creation of the RSS, PHP-MySQL pages and provides streaming access to offline content. This distributed access based on TCP/IP linking enables certain services to be relocated dynamically, easing the workload. Currently, all offline streaming content and feeds are being relocated in order to increase the response time and save on bandwidth.

This platform enables the implementation of user interaction scenarios whose complexity may vary from linear stream switching to user-generated content. The following case studies present a wide range of cases that were implemented as an extension of existing streaming media, under the prism of convergence, while backward compatibility issues are discussed, in an attempt to cover the divergence cross-technology issues that often arise. This process is essential as it is important to provide a smooth transition to the new medium in order the convergence to be successful.

3.1 Broadcasting Interactive Content

The film REC. (Argyriou, 2010) is a multi-camera DVD-based title which as presented at the *Short Film Corner* of *Festival de Cannes* in 2011. It film consists of three simultaneous video streams and the user is able to alter the view interactively at any point, an action that changes the film's context as each view displays a different reality. Figure 6 displays the set during filming. The movie was originally produced for DVD media, enabling camera switching via the camera-view selector of the remote control. Our research indicated that content that is destined to be edited by the user interactively should contain as little as possible pre-defined

Fig. 6 Filming of REC. using three camera-view scenarios enabling a different reality to be viewed according to the camera selection

transitions, as they can actively be used to affect the meaning. This can be clearly demonstrated through an example: A person arguing loudly with his boss at work that also mumbles while returning home followed by a quiet scene where he makes a cup of tea conveys a totally different meaning when contrasted to a sequence of scenes where a person arguing loudly at work is followed by a scene where he is quietly brewing the same cup of tea at home. This is a problem that introduces backward compatibility problems, particularly if the interactive content has to be broadcasted through traditional television. One has to either view all three camera-views, or a director's cut version that highlights via editing the most important scenes for each view, a process that introduces the editor's perspective. This is clearly a compromise that reduces the value of the interactive audiovisual title, and it has to be downgraded before being shown in this medium. Transmitting it in interactive mode via the interactive broadcasting infrastructure was straightforward. The process involved copying and linking of the three MPEG-2 streams to three streaming servers, enabling the users to access each individual stream over the iMediaTV interface via selection of the appropriate stream URL.

Similar problems are introduced when interactive New Media Art is broadcasted through traditional television. Similarly to the REC. case study, editing is required in order to capture various perspectives of the work and transmit that to the viewer. Often, and for complex artwork instances this is not possible (Strapatsakis, 2008), as there is interaction between different viewpoints while capturing and broadcasting these requires long time periods. The use of multiple camera views offer within iMediaTV provides a flexible tool that allows viewers to experiment and explore with pre-set cameras allowing them to direct their attention to specific aspects of the work, while they can also observe the director's cut complete with commentary if they desire. As a result one may argue that the use of interactive television broadcasting provides the backward compatibility functionality that avoids divergence as it allows users to view standard non-interactive content, while at the same time it offers the freedom to both content developers and users to experiment and explore further the technological capabilities offered by this new medium.

3.2 Interaction and User Involvement

Within traditional television, user involvement is a powerful aspect that has in the past been implemented through a wide variety of technologies: mail, phone, voting systems, polls, e-mail, SMS messaging, blogs, Internet chat, video chat are all methods employed to support user to studio interaction. These methods when combined with a medium that offers only one feedback channel back the user often create communication bottlenecks, rendering them unusable or raising questions about the level of feedback. On the other hand, the personalised viewing experience offered by interactive television can be aided by the availability of user to studio feedback. To demonstrate how interaction can increase the level of user involvement a number of case studies featuring different content types and interaction have been developed.

3.2.1 Live User Participation

Students at the studio implemented a number of live programs in order to study user involvement, investigate how the pre-programmed sequence of content is altered via user-studio interaction and study the content implications. The most representative case study featuring a free interactive mode included a concert case study featuring a solo musician who performed at the studio while viewers watched the performance over the network. A number of interactive feedback channels were activated including the forum, text-based chat, video conferencing and SMS/MMS, and all screens were projected in front of the musician who engaged into a dialogue with the audience. Although there were difficulties in the communication process as they were not all immediate but ranged between a few seconds or even minutes for the forum feedback and interaction depended on the level of feedback attention the performer was prepared to receive, the users managed to alter the original concert program with their requests. This is a new development on the copyright forefront, as the broadcasted content sequence cannot be predicted. In our previous work we have proposed the development of a dynamic licensing system that enables studio staff to receive content licensing on-the-fly in order to cover interactive presentation requirements (Deliyannis et al., 2011). Content licensing is also an issue in other areas including music information research (Karydi, Karydis, & Deliyannis, 2012).

The next case study combines artistic creation and cross-studio interaction. Here, four preschool children participated in a remote "Art Class" program where an artist at the studio instructed them to each draw a flower that would be used to create a larger painting. For this purpose the network-drawing application COLLABORADOODLE was used shown in Fig. 7. The artist interacted with the children from the studio in a two-way communication mode via videoconferencing, while children used shared their creations interactively and accepted feedback from the artist while drawing. This example demonstrated how a traditional show could be converted to an interactive collaborative experience that enables multiple viewing and participation modes to the viewer. They may choose to view what other viewers are doing, interact by offering suggestions, sharing experience and personal preferences or collaborate in an active manner and share their creativity with others. This scalability of the medium in terms of interaction is important as watching television in lazy mode may be what the user requires at a certain time period. This introduces both technical and practical questions and issues, often observed in real-life shared-space creative classes. How does one deal with the refusal of users to participate, or the case where there are too many participants? The answer in the first part lies with the creativity of the presenter as within a traditional class, while for the latter case a maximum number of participants may be set for a show. Interestingly, these are questions that occur in real-life art classes and this indicates the communication and creative aspects of interactive television. In conclusion, previous case studies featured specific user-content-studio communication and interaction scenarios. "Experimental Multimedia Systems" (Deliyannis, 2012) that feature novelty in both interaction and content may also use the interactive television infrastructure to cover their presentation requirements, as the adaptability

Fig. 7 The COLLABORADOODLE application used to implement within iMediaTV the creative interactive drawing mode across the network

of the medium is not limited to content sharing but with the appropriate control software and hardware it can support adaptive sensory experiences, under the MPEG-V standard (Waltl, Rainer, Timmerer, & Hellwagner, 2012).

4 Viewpoint on Convergence and Conclusion

The first television sets were advertised as *Radios with a Screen*, and their advertising campaign stated that viewers could both hear and view the orchestra performing. Advertising a new medium as an extension of an already established technology enables users to clearly view the technological advantage, although in the case of television, it did not predict its future uses and capabilities. Similarly, interactive television is today perceived by the general public as a mechanism that offers dynamic audiovisual content streaming, enabling users to select content or access live television over the network. The idea of dynamic content, user interaction and the communication aspects offered by interactive television are not well known.

Speeding up the convergence process and reveal the true potential of interactive television, requires further research on the user and content aspects, coupled with the development of new deployment strategies. Infiltration and wider acceptance of interactive television standards requires careful strategic planning aiming at users, targeting their familiarization with this new technology. Convergence is clearly aided by the fact that interactive television does not require specialized hardware beyond a multimedia-enabled personal computer and Internet access. It is therefore critical for content providers and developers to reduce the non-use non-adoption factor. This work proposes the establishment a common standard designed to promote the advantages of interactive television, while it preserves traditional television features such as lazy playback.

A strategy designed to speed up the process should involve the deployment of a dynamically expandable service (offered as an open standard) designed to allow users to tune-in to specific broadcasts and automatically configure their devices to support interaction for each content instance. This development must be based on modular software components and codecs. Beyond the provision of direct access to content, it should allow its software components and interaction structures to be upgraded at any time, enabling new interactive features to be supported at the clients end. In addition to the proposed service, an open content indexing/filtering/suggestion mechanism (interface) should be developed to complete the user experience, enabling users to access broadcasts through an interactive TV content-access gateway that promotes the content and simplifies the "tuning" process. Appropriate strategies can be built upon this infrastructure, permitting the much-needed infiltration to take place, as this will allow the new medium to be perceived as a superior and cost-effective solution.

Successive to the infrastructure for content publishing and access described above, a number of content-based approaches should also be realized. The development of new content targeting to demonstrate the novel technological features and the social aspects of interaction is an essential system component that can speed-up the convergence process. The proposed content is categorized under three basic types:

"Hot" content that includes live broadcasting of sport events, concerts, performances, exclusive shows and unedited footage, offering when possible interactive features that include user-selected camera views, additional information and access to extra information.

"Early preview" content offering web-based viewing for series, documentaries and movie premieres before they are broadcasted on television.

"Purely-interactive content" production featuring interactive access that permits the user to take part in interactive talk shows, studio-based games where users are allowed to interact directly or with the players by providing their feedback, live performances where the audience is digitally present and user-generated news allowing commenting attachments.

New scientific challenges are introduced (Deliyannis & Karydis, 2011) when users and content creators are presented with novel expressive capabilities (Deliyannis et al., 2011). This work presented interactive television convergence issues and proposes numerous strategies that may be implemented to speed-up the process, based on system and content adaptation. It was suggested that in order interactive television convergence requires three major stages to be completed: a dynamically-adjusting standard combining all the required technologies under a single specification; a user interface that will organise and provide centralised access to users, offering metadata-search capabilities and finally, in addition to traditional content, novel and innovative interactive content that will engage users in this new communication process. The case studies presented in this work and the experience shared publically clearly demonstrates the new capabilities offered by interactive television.

References

ALRC. (2012). *Classification—Content regulation and convergent media*. Australian Law Reform Commission, Australian Government, Sydney, Australia.

Argyriou, D. (2010). *Rec. Interactive Movie on DVD*. Department of Audio and Visual Arts, Ionian University, Corfu, Greece.

Bertelsen, O. W., & Bødker, S. (2003). Activity theory. In J. M. Carrol (Ed.), *HCI models, theories, and frameworks: Toward a multidisciplinary science* (pp. 291–324). Amsterdam: Morgan Kaufmann.

Brennan, K., & Resnick, M. (2013). Imagining, creating, playing, sharing, reflecting: How online community supports young people as designers of interactive media. In C. Mouza & N. Lavigne (Eds.), *Emerging technologies for the classroom. Explorations in the learning sciences, instructional systems and performance technologies* (pp. 253–268). New York: Springer. doi:10.1007/978-1-4614-4696-5_17.

Bush, M. (2000). Digital versatile disc (DVD): The new medium for interactive video. *Calico Journal, 17*(3), 453–474.

Castells, M. (2011). *The rise of the network society: The information age: Economy, society, and culture* (Vols. 1 and 12). Oxford: Wiley-Blackwell

Chorianopoulos, K., & Spinellis, D. (2006). User interface evaluation of interactive TV: A media studies perspective. *Universal Access in the Information Society, 5*(2), 209–218.

de Mooij, M., & Hofstede, G. (2002). Convergence and divergence in consumer behavior: Implications for international retailing. *Journal of Retailing, 78*(1), 61–69. doi:10.1016/S0022-4359(01)00067-7.

Deliyannis, I. (2012). From interactive to experimental multimedia. In I. Deliyannis (Ed.), *Interactive multimedia* (pp. 3–12). Rijeka, Croatia: Intech.

Deliyannis, I., Antoniou, A., & Pandis, P. (2009). *Design and development of an experimental low-cost internet-based interactive TV station*. Paper presented at the 4th Mediterranean Conference on Information Systems (MCIS), Athens.

Deliyannis, I., & Kanellopoulos, N. (2008). *Interactive communication methodology for social software and multimedia applications*. Paper presented at the IADIS International Conference IADIS e-Society, Algarve, Portugal.

Deliyannis, I., & Karydis, I. (2011, June). *Producing and broadcasting non-linear art-based content through open source interactive Internet-TV*. Paper presented at the ACM EuroITV, Lisbon, Portugal.

Deliyannis, I., Karydis, I., & Karydi, D. (2011, May). *iMediaTV: Open and interactive access for live performances and installation art*. Paper presented at the 4th International Conference on Information Law (ICIL2011), Thessaloniki, Greece.

Dutton, W. H. (2013). *The Oxford handbook of internet studies*. Oxford: OUP.

Elfving, S. W., & Urquhart, N. (2013). Product service system challenges within telecommunication: Reaching the era of mutual dependency. In Y. Shimomura & K. Kimita (Eds.), *The philosopher's stone for sustainability* (pp. 269–274). Berlin: Springer. doi:10.1007/978-3-642-32847-3_45.

Finkelstein, R. (2012). *Independent media inquiry report*. Australian Government, Australia.

Jenkins, H. (2006). *Convergence culture: Where old and new media collide*. New York: New York University Press.

Karydi, D., Karydis, I., & Deliyannis, I. (2012). *Legal issues in using musical content from iTunes and YouTube for music information retrieval*. Paper presented at the International Conference on Information Law, Corfu, Greece.

Kreuzberger, G. (2007). *Game state vs. Play state: From DVDi games to a language of gaming experience*. Retrieved from http://www.aaai.org/Papers/FLAIRS/2007/Flairs07-113.pdf

Kruikemeier, S., van Noort, G., Vliegenthart, R., & de Vreese, C. H. (2013). Getting closer: The effects of personalized and interactive online political communication. *European Journal of Communication, 28*(1), 53–66. doi:10.1177/0267323112464837.

Long, M., Chei Sian, L., & Goh, D. H. L. (2012, April 16–18). Sharing in social news websites: Examining the influence of news attributes and news sharers. In *2012 Ninth International Conference on Information Technology: New Generations (ITNG)* (pp. 726–731). doi:10.1109/itng.2012.143

Lugmayr, A., Niiranen, S., & Kalli, S. (2004). *Digital interactive TV and metadata: Future broadcast multimedia*. New York: Springer.

Munson, E. V., & Pimentel, M. G. C. (2012). Issues and contributions in interactive multimedia: Photos, mobile multimedia, and interactive TV. *Multimedia Tools and Applications*, 61(3), 519–522.

Paradise, A. (2012). Picture perfect? College students' experiences and attitudes regarding their photo-related behaviors on Facebook. In L. A. Wankel & C. Wankel (Eds.), *Misbehavior online in higher education (Cutting-edge technologies in higher education)* (Vol. 5). Bingley: Emerald Group Publishing. doi:10.1108/S2044-9968(2012)0000005015.

Prata, A. (2009). Interactive television evolution. In M. Pagani (Ed.), *Encyclopedia of multimedia technology and networking* (2nd ed., pp. 757–762). Hershey, PA: IGI Global. doi:10.4018/978-1-60566-014-1.ch102.

Review C. (2012). *Convergence review*. Australian Government, Department of Broadband, Communications and the Digital Economy, Sydney, Australia.

Richardson, I. E. G. (2006). *H.264 and MPEG-4 video compression: Video coding for next-generation multimedia*. Chichester: Wiley.

Rogers, E. M., Singhal, A., & Quinland, M. M. (2009). Diffusion of innovations. In D. W. Stacks & M. B. Salwen (Eds.), *An integrated approach to communication theory and research* (2nd ed.). Mahwah, NJ: Lawrence Erlbaum Associates.

Song, S., Moustafa, H., & Afifi, H. (2012). A survey on personalized TV and NGN services through context-awareness. *ACM Computing Surveys, 44*(1), 1–18. doi:10.1145/2071389.2071393.

Srivastava, J., & Moreland, J. J. (2012). Diffusion of innovations: Communication evolution and influences. *The Communication Review, 15*(4), 294–312. doi:10.1080/10714421.2012.728420.

Strapatsakis, M. (2008, November). *Invisible places—Immense white*. Interactive Video Installation, Apollonia Art Exchanges, Strasbourg.

Tay, J., & Turner, G. (2009). Not the apocalypse: Television futures in the digital age. *International Journal of Digital Television, 1*(1), 31–50.

Taylor, J. (1999). DVD-Video: Multimedia for the masses. *Multimedia, IEEE, 6*(3), 86–92.

Tschmuck, P. (2012). Creativity and innovation in the music industry's value-added chain. In P. Tschmuck (Ed.), *Creativity and innovation in the music industry* (pp. 253–271). Berlin: Springer. doi:10.1007/978-3-642-28430-4_13.

Tsekleves, E., Whitham, R., Kondo, K., & Hill, A. (2009). Bringing the television experience to other media in the home: An ethnographic study. In *Proceedings of the seventh European conference on European interactive television conference* (pp. 201–210). New York: ACM.

Ursu, M. F., Thomas, M., Kegel, I., Williams, D., Tuomola, M., Lindstedt, I., et al. (2008). Interactive TV narratives: Opportunities, progress, and challenges. *ACM Transactions on Multimedia Computing Communications and Applications, 4*(4), 1–39. doi:10.1145/1412196.1412198.

van Dijck, J. (2013). *The culture of connectivity: A critical history of social media*. New York: OUP.

Waltl, M., Rainer, B., Timmerer, C., & Hellwagner, H. (2012). An end-to-end tool chain for sensory experience based on MPEG-V. *Signal Processing: Image Communication, 28*(2). doi:10.1016/j.image.2012.10.009

Zheng, H., Fu, Y., & Lugmayr, A. (2012). *The development of ambient TV in 2016*. Paper presented at the Proceeding of the 16th International Academic MindTrek Conference, Tampere, Finland.

The Twenty-First Century Television: Interactive, Participatory and Social

Pauliina Tuomi

1 Introduction

Media synergy is on the increase, with media companies releasing their media content across different delivery channels. Nowadays, popular TV programs are multi- and cross-media projects that are not restricted to the television (Simmons, 2009; Tuomi, 2010; Ytreberg, 2009). Television viewing has fragmented to several platforms and the use of web elicits new uses of TV and audience behaviors. Audiences are now combining different mediums and content in order to gather a coherent media experience. This means a lot changes also in the fields of TV and audience research. The focus of the research is no longer on the mass consumption; on the contrary, it has shifted to study audience as individuals. It raises the question whether it is a matter of convergence after all or do these new consumption models rather lead to divergence? (Fig. 1).

This article will cover approximately 10 years of recent Finnish TV history, 2000–2012. It sets up to answer *how does media convergence and the use of second screens affect the TV watching experience (what extra features does it offer) and what is the standpoint on audience convergence/divergence?*

Interactivity, participation and sociality are features that characterize today's TV. At first the TV watching experience is analyzed through interactive elements and the focus is on the **SMS**-based iTV-entertainment. The second analysis elaborates on the features of **Web 2.0** and the dimensions it brought to the TV watching experience concentrating on the participatory online features (basically on different websites, blogs, discussion forums and such) that became more common around the years 2008–2010. The third concentrates purely on **social media** (e.g., Facebook and Twitter), it's features and what this has brought to the TV watching.

P. Tuomi (✉)
Tampere University of Technology, Pohjoisranta 11 A, PO Box 300, 28101 Pori, Finland
e-mail: sptuom@utu.fi

A. Lugmayr, C. Dal Zotto (eds.), *Media Convergence Handbook - Vol. 1*, Media Business and Innovation, DOI 10.1007/978-3-642-54484-2_14

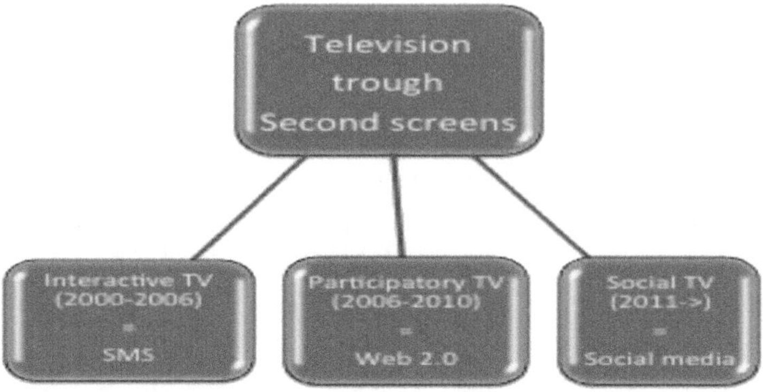

Fig. 1 The main eras of twenty-first century television

2 Literature State of the Art

Television has enlightened people for decades. That warm heart of the living room has both brought families together and given possibilities to consume entertainment individually. It has been seen as a complex medium, at the same time collective and disjunctive (Heinonen, 2008). Television is overall a very versatile broadcast medium (Näränen, 2006, 29). TV means different things to different people and there are several ways how to consume it, especially nowadays. Information- and communication technology differs from other equipment at home by their ability to connect the people and the outside world—either interactively or passively (Silverstone & Hirsch, 1994, 15).

Nowadays television is constantly being developed with new technical and social features that are widely acknowledged and the idea of being interactive comes through a convergence between additional devices and television. Jenkins (2006) has defined convergence by following: *the flow of content across multiple media platforms, the cooperation between multiple media industries, and the migratory behaviour of media audiences.*

Today's interaction, or rather participation, comes from multiplatform formats that combine TV and Web platforms in order to create a wider TV experience (Tuomi, 2010). Since 1990s interactivity has been one of the most significant features of the future media culture. The growth of media culture's playability is one of the results that is triggered by interactivity; cultural products are no longer just products that are watched, read or listened—nowadays products enable active participation as well (Parikka, 2004, 84). "TV meets the internet" is a global expression that characterizes digital and interactive TV (e.g., Jensen & Toscan, 1999). A user can be active in many ways, on various platforms, without breaking the link to a TV program format. Audience involvement has widened to include "passive" ways of using a broadcast platform, "active" uses of Web platforms, and in-between phenomena such as voting via mobiles and participating in net meetings

(Tuomi & Bachmayer, 2011, 55–56). In fact, multiplatform formats seem to stretch the whole somewhat ancient passive/active dichotomy well beyond the breaking point (Ytreberg, 2009, 483).

Previous studies in the field of today's television concentrate heavily on interactive features from the technological point of view (e.g., Bachmayer, Lugmayr, & Kotsis, 2010; Cesar, Bulterman, & Soares, 2008). Television has been the model example of a 'push' media and one-way mass communication. Nevertheless, TV has been interacting with viewers in many different ways for years. Very often, the interactive features of television are seen as a novel phenomenon or something that will happen sometime in the future. In reality, interactive television has a long history—just as long as television itself. Due to the convergence of TV and the internet, several research projects have appeared in the last few years aimed at finding ways of combining TV and web content, with informational or communicational purposes, solely using iTV or being cross media (Prata, Chambel, & Guimarez, 2012, 335; Rodriguez-Alsina & Carrabina, 2012). Participatory TV content has also been studied especially after the triumph of reality TV (e.g., Hill, 2005; Murray & Ouellette, 2008; Von Feilitzen, 2004; Ytreberg, 2009). The different hooks to entice the audience to take part in TV formats have been explored to some extent (e.g., Hautakangas, 2008), as well as the roles of different platforms as part of participatory TV (e.g., Sundet & Ytreberg, 2009; Ytreberg, 2009). The linkage between television and the internet has been widely addressed over the years (e.g., Antonini et al., 2013; Cortez, Shamma, & Cai, 2012; Rautiainen et al., 2013; Tuomi, 2013). Social TV as a theme of its own has been the centre of TV research, especially from the technological aspect, for years (e.g., Cesar, Geerts, & Chorianopoulos, 2009; Harboe et al., 2008). Social networking sites like Facebook and Twitter have been associated with social television since they enable remote viewers to interact socially with each other through television or mobile devices in general, in situations where viewers often are separated in time and/or in space. Since becoming a topical subject, social media and its role in TV production has been widely studied during the past few years as the numbers of users of, for instance, Twitter and Facebook have greatly increased (e.g., Gruzd, Wellman, & Takhteyev, 2011; Highfield, Harrington, & Bruns, 2013; Harrington et al., 2012; Marwick & Boyd, 2010; Weller, 2011; Weller, Bruuns, Puschmann, Burgess, & Mahrt, 2013).

All of these features based on different convergences, have a tremendous impact on audience research as well. The more technologies there are available through which television texts can be consumed, the more different contexts of television consumption there are, and the more difficult audience research in this area becomes (Silverstone & Hirsch, 1994; Simmons, 2011).

3 Methodology and Approach

Television viewing has fragmented to several platforms and the possibilities to connect with TV have expanded greatly during the last 10 years. In this chapter the three most dominate and recent phases of television will be introduced. The three phases of twenty-first century television—interactive, participatory and social—will all be analyzed through three criteria of convergence, which are: *technology and interaction*, *additional value* and *social aspect*. By answering these questions the chapter will elaborate on the actual level of convergence on audience level and making a stand whether these converging steps actually support convergence or divergence instead. The study bases on media convergence, participatory audience theories and TV watching experience. The research material covers for example taped iTV-formats, media observations around Web 2.0 and Tweet-analyses of TV based Twitter conversations. iTV data is taped during the years 2000–2011, the web 2.0 is studied through Internet Archive's *Wayback Machine*, which is a service that enables users to see archived versions of web pages across time[1] and the Twitter tweets are from the Yle's website (the tweets that were published on the TV screen) concerning the events "Linnan Juhlat" (2011) and Eurovision Song Contest (2011). The three phases will be represented more or less through different case studies in order to combine both theoretical and actual steps of convergence together.

4 Interactive Television

During the interactive phase of television in the beginning of the year 2000 the technology that provided the interaction were mobile phones and to be more precise SMS messages. The screen of the mobile phone acted as second screen for television. It gave to the TV watching experience more playfulness and playful elements in general. It also brought a huge change into the communication between TV stars and audiences since it offered a real time based communication through the SMS function.

The iTV-experience to take place: the viewer needs in addition to television, a mobile device that supports SMS-function. It is surprisingly difficult to try to define TV mobile games, particularly because of the term *mobile*. We are used to associating different kinds of feelings with this term. Usually *mobile* means something that one can carry and use whenever, and more importantly, wherever one wants to. In these TV mobile games this was not an option because one cannot play games at any time. On the contrary, these games could only be played when they are on the air on television. On the other hand, we really cannot use the term *TV game* either because it suggests that these games are similar to the console-based TV games. Probably the best term is still *TV mobile game*s since these games are played using mobile phones and watched from the TV screen (Tuomi, 2008, 68).

[1] Internet Wayback Machine: http://archive.org/web/web.php

Interaction is best seen in cross media formats. Cross media entertainment means connection between for example mobile devices, Internet or/and television. Cross media enabled gaming experiences between TV and mobile phones. At first, one could participate in different TV-chats—one could send greetings with a text message (SMS) and almost immediately see his/her text on the TV-screen. This new form of entertainment soon became adopted by SMS game show producers. It could be said that Finland is a forerunner when it comes to interactive, TV-linked, mobile technology. Of course these formats have also spread to Europe and Asia (Tuomi, 2008, 67). From the year 2002 there have been different kinds of TV-mobile games on television. At first they were games one could participate in with a text message—just by choosing the right coordinates to hit a certain target. Games were often based on problem solving and the interaction between the player and the game was limited to text messaging. Later on (2004), games developed further and a live human host stepped onto the playing field. Hosts were now playing against people on their sofas. For example, games were based on getting a football past the host or trying to hit her with a snowball. Since the participation took place via SMS or IVR,[2] literally all viewers were able to participate (Tuomi, 2008, 68).

The level of interaction grew enormously, especially after the chat-function was added to the games. Now it was possible to play against a live host and talk to him/her—and most importantly: to get a response to one's action! TV was also used daily as marketing place for different mobile accessories—ring tones, backgrounds etc. It was also possible to donate to charity via SMS. This option was advertised on TV after different disasters all over the world. One was also able to get a quick loan with a text-message, but the interest rates were very high. Different kinds of mobile services were advertised on TV. One can order weather information or information on an unknown phone number via SMS. These examples are just the tip of the iceberg. There were multiple choices of how to interact with TV and its contents (Tuomi, 2008, 69).

Interactive TV mobile games were games one could participate in by text messaging on a mobile phone. The mobile phone brought the viewer and the TV screen together. The games were usually based on coordinates that one must choose in order to throw, for example, a snowball towards the host or kick a football past her. It was exciting to participate from one's sofa, with one's own phone, in a live TV show. It is as though the TV had became a game console and the mobile phone a game pad (Tuomi, 2008, 68). For example, in a TV-mobile game called Horse Derby, the player was supposed to press phone buttons one and three as fast as possible in order to make his horse (trotter) run on the screen. There was a clear resemblance to old Commodore 64 sport game Track and Field, in which players were obligated to hit joystick buttons constantly (Tuomi, 2008, 69). In these games, one message/game move cost approximately 1 euro. There were no physical prices

[2] Interactive voice response is a phone technology that allows a computer to detect voice and touch tones using a normal phone call.

and therefore it is a spiritual battle between the player and the host/other players. It is motivating to get on the TOP 10-list and pursuing to get better results (Hanski & Kankainen, 2004, 75). Every sent message activates the host and the feedback is instant (Tuomi, 2008, 69).

iTV brought a playful experience to TV watching. Playing on the TV screen is however nothing novel. Interactive game shows have appeared on the Finnish TV before such as Hugo the Troll (1993) and Vito—Game Over (1994). Hugo the troll was an international game where viewers were able to make a phone call to the TV-show and play the game via number buttons on a phone (Tuomi, 2009, 16). However, the interaction enabled by the Hugo troll was something spectacular since the viewers were now able to control the happenings on the real-time TV. The viewer was able to guide the animated game world from one's own coach.

In the end, consumers seemed to want true 'power' to have an effect on TV formats and contents.[3] In the era of iTV, consumers did not have any effect on significant matters in the society, only, for example, on reality TV formats by voting. iTV entertainment was a huge field that could have been used much more effectively to facilitate people's influence on the important matters of society (Tuomi, 2008, 70).

5 Participatory Television

After the interactive phase (based especially on iTV content and SMS), Finnish television evolved into participatory TV by offering plenty of TV-related material on dedicated websites as well as through online magazines as additional news material. The phase of participatory TV is mainly the era of the coming of Web 2.0. Participatory television exploited the convergence between TV broadcasts and the internet, especially Web 2.0 features. Through this convergence, the audience was able to communicate with each other, create content, and enjoy the material provided by the broadcasters. This, however, led to the situation where audiences could no longer be seen as the masses. This era emphasised the individual by offering the capability of watching TV whenever (online/net TV etc.) and wherever you wish. This era also saw more social TV-related activity amongst the online participants. The time when internet changed to be somewhat more social than before and it started to accept and be based of more and more user generated content. This had an impact on television viewing experience as well. Web 2.0 acted as second screen for television. Internet became an archive for all the TV format's other related material and viewers were able to browse through asynchronous discussions, blog texts and dedicated web pages offered by the TV channel. There were a lot of features and activities that served individuals particularly, but also lots of user-generated content that enabled asynchronous communication

[3] This was seen from the internet inquiry (https://www.webropol.com/P.aspx?id=210451&cid=20911803) answers.

between the viewers. UGC is the term used to describe any form of content such as video, blogs, discussion form posts, digital images, audio files, and other forms of media that was created by consumers or end-users of an online system or service and is publically available to others consumers and end-users (e.g., Kaplan, 2010, 61). Viewer needed in addition to television—not necessary though—internet connection. This could be established of course through mobile devices as well.

What are the ways participatory media, especially different TV and media spectacles, use when activating people to take part? Usually the repetitive features are similar in almost every participatory TV format. Hautakangas (2008) states that there are clear hooks in the interactive TV formats that also live in other mediums as well. These hooks are used to involve people into the narrative of the participatory media spectacle (Hautakangas, 2008). One of the prime features is naturally the referendums aka votings that enable almost every citizen to participate. Another, not so apparent, feature is intermedialism. The whole event is built up in lots of different mediums simultaneously for example in the press, on the net and on TV.

Because of the media convergence there are lots of different intermedial products for sale nowadays, which means that the contents are available in many mediums (Hautakangas, 2008, 161). One essential part of the breakage of the digitalization has been the growth of the internet and mobile media becoming a significant part of the everyday media consumption and—markets. On the other hand, because of the seamless co-operation between the web press and the dedicated web page it is possible to get news feeds and stories in real time. A committed TV format follower can easily dive into the world of different news, extra-material and gallups. Quite often the tabloid magazines actually offer different shadow votes where one can participate in online (Tuomi, 2013, 140).

The participatory nature of convergence between TV and internet before the coming of social media was based purely on combining TV and web-content asynchronously together. First, before the web 2.0, there were simple web pages that offered some background information about the TV format, but these were platforms that were updated by the dedicated people, audiences were not able to affect these sites. After that the webpages turned into web platforms where the users control their own data (O'Reilly, 2005). Usual core features of a Web platform are services, participative architecture (e.g., uploading videos, comment functionality, votes) and availability of social features (e.g., blogs, discussion forums, like/dislike functionality, polls) (Tuomi & Bachmayer, 2011, 58). There are then different types of linkage between TV and internet content, which can be defined on a thematic or technical level. Especially the purpose of the linkage between Web and TV content tells us what is the additional value to the audiences (Tuomi, 2013, 140).

Also there were a possibility given to the audience to influence the Content on TV via webplatforms, but this was really minimum compared to the next step of convergence between television and social media, which will be dealt with in the next chapter. Webplatforms give the viewers the possibility to influence the course and/or characteristics of the TV content (e.g., by a voting mechanism) and to offer viewers the possibility to participate in the show (e.g., by concurrent forum and live TV discussion; the outcome of the forum discussion is introduced to the discussion on TV). Influence can happen active or passive. However, in contrary to interaction, a participating viewer might not get an immediate feedback about his/her participation. It even might not be clear whether the participation had any effect or not—for example Big Brother and Idols. The webplatforms also supported web transmission of content that is broadcast on TV. For instance, a newscast is offered as a video-on-demand stream after the (live) broadcast on TV (Tuomi & Bachmayer, 2011, 60). Also the webplatforms were used to shift content. The content is broadcast alternate on different platforms (it is shifted from TV to the Web and back again). There are multiple ways how webplatforms were engaging the audiences also outside the television broadcasts. Next the nature of participatory era will be presented through an example.

5.1 Emmerdale Online in 1999: An Example of Participatory Television

As an example we can use Finnish TV channel MTV3's drama/soap opera Emmerdale and analyze how the content on TV programme related websites invited audiences to take part in before the time of social media. The participatory online features, resulting from media convergence between TV and web, can be addressed and analysed also through the use and purposes of websites. As an example, the Finnish TV channel MTV3's website for the British TV programme Emmerdale was retrieved from the Internet Wayback Machine from the year 1999. This is a good example of a website that has been established after the wave of Tim O'Reilly's idea of Web 2.0. The term Web 2.0 became known in 1999 to describe web sites that use technology beyond the static pages of earlier web sites. It is closely associated with Tim O'Reilly because of the O'Reilly Media Web 2.0 conference, which was held in late 2004 (O'Reilly, 2005).

The website material was analysed and features relevant to audience participation were categorised based on the criteria of content and purposes. Overall, the purposes were dealt into (1) Up-to-date info, additional information and extra material (photos, background images, episode descriptions, spoilers, trivia and character presentations etc.), (2) Advertisement (use of site to advertise programme- and channel-related merchandise), (3) Interaction with the audience/ Participation in content (chats, discussions forums, blogs, polls and votes, feedback etc.), and (4) Transmission of content (video clips, online TV application etc.).

The example webplatform of Emmerdale in 1999[4] is constructed in a way where the upper banner is holding the links to additional information and extra-material. The right side of the site is offering for example polls and shadow votes. From the up left the offering is the following: (1) front page, (2) episodes, (3) characters, (4) trivia and background, (5) the village, (6) background images, (7) Kaarina's

[4] See page at: The Emmerdale weplatform from the year 2009: http://web.archive.org/web/20090817111252/http://www.mtv3.fi/emmerdale/

(maintainer) blog, (8) discussion, (9) Katsomo—net TV application and (10) feedback.

This emphasizes lots of the criteria listed before. The webplatform, acting as a second screen, offers additional information such as episode and character information and trivia and background. It also offers a place to viewers create content in discussions and blog-sections + there is a possibility to participate in weekly poll questions and give feedback. Katsomo offers the transmission of content since viewers are able to watch episodes outside the actual broadcast times. Also the aspect of advertisement can be found.

6 Social Media Television

There are different ways audience converge social media and television together. At first, it can be used in a same way webplatforms acted in the era of participatory television. To give viewers the possibility to gain and gather extra-material and synchronous news in particular. If the participatory TV was more asynchronous communication, social media brings the real-time based interaction to the scene. Basically the same way iTV-entertainment did. Secondly, social media can be used in order to enhance the viewing experience by allowing viewers to communicate in real-time during the TV broadcast on TV. In this chapter the focus is in the latter function of combining TV and social media, Twitter in this case.

As known, television is nowadays a part of the big picture of interconnected devices, operating on several platforms. Espen Ytreberg (2009) states that multiplatform formats[5] are a type of media output that does not seem to fit well with thinking in terms of basic differences between broadcast and digital media (Ytreberg, 2009, 16). This means that the broadcasters are eager to provide new ways to drive viewer engagement. According to the company Twitter itself, Twitter provides a forum for real-time context and commentary that turns watchers into participants. For example, the royal wedding (for example already in the year 1981 when Charles and Diana got married the festivities were televised around the world and again, with Twitter possibility, in 2011 when William and Kate were married) is just one example of how real-time Twitter integration can enhance TV coverage and help drive viewership.

The aim of this section is to elaborate on the connection and linkage between television and social media in 2011. This is done through a case study that focuses on the experiments done by the Finnish national broadcaster YLE. In Finland there is a more innovating way TV content and watching experience can be supported with the assistance of the Web platform on the TV screen itself than just having the Twitter feed on different second screens for example smart phones. In Finland the

[5] Formats that operate on several platforms e.g., internet, mobile phones, TV being the central cluster.

national broadcaster company YLE has done experiments in creating social TV by combining teletext and social media for a couple of years now.

From the year 2010, YLE (The Finnish Broadcasting Company) has enabled people to watch Eurovision Song Contest and Independence Day celebration (Linnan juhlat) and same time follow the Twitter conversation around it on the teletext (or teletext)—in real time. Teletext as a phenomenon is mostly European and nowadays it is popular for example in Britain (BBC) and Sweden in addition to Finland (Turtiainen, 2010, 33). Teletext is a television information retrieval service and the very first tryouts took place in Britain in the early 1970s. Teletext is a means of sending text and diagrams to a properly equipped television screen by use of one of the vertical blanking interval lines that together form the dark band dividing pictures horizontally on the television screen. It offers different text-based information, typically including news, weather and TV schedules. YLE started regular teletext broadcasts in Finland 7th of October in 1981 and as a curiosity at the end of 1990s, 70 % were using teletext regularly (Turtiainen, 2010, 34). For some reason teletext has not really been studied academically at all (Turtiainen, 2010). Despite the fact that teletext has survived as an existing feature all these years.

There are hundreds of tweets published yearly. YLE's try out of using Twitter tweets on teletext application during the TV show is experimental, it broadens the TV experience and at the same time brings the Web and social media closer to the TV content. What is noticeable is that it is actually done through a quite old medium, Teletext, which emphasizes a level of remediation and the fact that new bridging connections between TV and social media can also be established with the help of the older devices as well (Tuomi & Bachmayer, 2011, 63).

In these experiments Twitter is introduced as a participatory feature to traditional TV broadcast in real-time. The tweets are delivered to audience through an older medium teletext. Tweets are then shown on the TV screen similarly as subtitles do.

7 Example: Teletext Meets Twitter and Social Networking

For this service to take place, viewer needs a television set that supports teletext. Television broadcast and secondary screen with the internet connection. The second screen is acted by social media and by through different mobile and non-mobile devices. Social media sites like Facebook and Twitter have changed the idea of secondary screens totally. Now the TV watching experience is gathered from scattered media field. Lots of extra-material and real-time based/synchronous activity takes place in social media. Facebook- and Twitter threads are also used for viewers to communicate with others—known and unknown people.

Teletext and Twitter-experiment is technologically executed with the appropriate hash tag # for example #euroviisut by which the published tweets end up through the RSS-feed into moderated publishing system. By choosing the teletext page 398, the tweeting of the audience absorbs onto the TV screen as a part of Eurovision TV broadcast viewing the same way as texting does.

Remediation is the incorporation or representation of one medium in another medium. According to Bolter and Grusin, the newer media exploits the older one and justifies its existence (Bolter & Grusin, 2000). Twitter exploits the older medium teletext, but the teletext itself also benefits from this symbiosis; it gives it yet another purpose. The use of teletext could easily carry the un-media sexual label being an old medium but it is actually well accepted medium despite the fact that it is somewhat ancient compared to social media and Twitter. Like mentioned before, teletext has a solid ground in Finnish TV usage, which probably explains the reason why it is so easily adopted and not questioned (even) by the (younger) audience (Tuomi, 2012, 252).

"Linnan Juhlat" is a Finnish yearly event that is held to celebrate and honour the independence of Finland. The president of Finland invites Finnish people that have been noteworthy and have forwarded equality and good values during the present year. This event is one of the most followed TV broadcast in Finland. The Twitter meets Teletext-experiment is also been done during the media spectacle Eurovision Song Contest that is a popular yearly song contest that all the active European Broadcasting Union (EBU) member countries are allowed to participate in (Tuomi, 2012, 249).

After analyzing the 3005 tweets of both of the media spectacles, seven different categories were found in order to emphasize for what the audiences are using the Twitter opportunity for.

The categories are (1) *Eurovision Song Contest artists and performances/Independence Day quests and appearances,* which addresses all the tweets concerning the appearance of the people on TV—the general habitus and clothing. (2) *Nationality* gathers the tweets emphasizing the collective feelings as being a Finn during the media spectacle. (3) *Teletext and Twitter—experience* handles all the content that somehow deals with the experiment—content and/or technical issues. (4) *Interaction and dialogue* analyses the tweets and re-tweets that contain clear dialogue and tweets that appear to sent in order to arise discussion. (5) *Other notions and irrelevant comments* elaborate every tweet inappropriate to other categories. (6) *Hosts and general arrangements of the event* organizes the tweets that contain opinions and statements concerning the overall arrangements and hosts. (7) *Media spectacle and atmosphere* presents the tweets that deal with the traditions and experiences that are clearly connected to media spectacle (Tuomi, 2012, 250).

These results emphasize the usage of this second screen service to social purposes among the audience. In the iTV era, the SMS was limited into 160 characters, which obviously districted what, how and how much audience actually would have wanted to say. Naturally tweets are also short, but they are un-chargeable and provide however more space to express oneself. Also the use of social media like Facebook give more opportunities to act on matters with more space, in real-time and whenever and wherever.

7.1 Viewpoint on Convergence/Divergence Through the Eyes of the Audience

The trend towards converging media environments across devices as TV sets, PCs, smart mobile phones, intelligent gadgets, and tablets is unstoppable. Internet, broadcasting networks, and service ecosystems become more and more a single service space, rather than having strict boundaries between each medium (Lugmayr & Zheng, 2012, 49). The audience does not perceive the TV stream as single medium—the consumed media is a service eco-system actively reacting on implicit or explicit consumer interactions (Ollikainen et al., 2011). The fact that the role of television as a physical object is changing, also influences other aspects of the context of the whole television viewing experience. This can be noticed in the fact that television viewing is becoming more and more individualized (Simmons, 2009). Not only is television viewing evolving into a more individual activity, also the social aspect afterwards is being influenced. A very important part of watching television is the possibility to talk about it the next day. According to Simmons (2009) time-shifting technologies interrupt the temporal flow of television and therefore the feeling of a shared experience might disappear. On the other hand, audience groups might also be reinforced: instead of talking about television programs the next day at work or at school, viewers of internationally successful programs can go online and talk about the episodes with viewers from around the world (Simmons, 2009, 220).

iTV both enabled and disabled social activity on TV screen. At first different TV chats actually offered social activity between the TV viewers through discussions that took place on the TV screen. It also acted as place for seeking company and dating as well as for getting advice from other viewers on different problems viewers had come up with. On the other hand, the connection between the TV stars mentioned before influenced on this social activity between viewers since after this communication with TV persons was possible, the interaction mainly took place between the viewers and the TV hosts. The social activity did not stop, it just changed.

During the participatory phase users/viewers were offered a possibility to socialize and share. From this point of view convergence of TV and internet did empower the social activity among audience. However, like mentioned this also emphasized more individualistic use of television physically, but also at the same time the activity might have been really social between the other online users.

During the social media era, the real-time based communication has boosted the social activity around TV formats. Now it even includes, thanks to Facebook etc., communication between the broadcaster and the audience. Social media has opened the lines to contact broadcasters quickly and the broadcasters are obligated due to the reputational issues to respond to these requests.

The idea of converged audience is not actually occurring through a traditional TV viewing but more taking place in audiences' online activity. Especially important for this gathering TV experience is the convergence between social media and television. It gives the opportunity to viewers to connect TV related in real-time and also the possibility to contact the TV producers as well. The analyzed tweets reveal that social communication that take place during the actual live broadcast tweeting via teletext is mainly used to express opinions, something one would like to say to the person sitting next to one on sofa. The most valued features tweets on teletext did brought to the audience were the idea of presence, a possibility to 'take part' in a new way and the possibility to tweet was seen overall as a positive extra value to more traditional ways of watching TV.

Use of social media combines the interactive and participatory elements wisely. However, use of the second screens in cooperation with the television does not lead to convergence, not technically at least, since there are several different devices to be used and the television cannot offer these features without the help of mobile devices. The idea of smart TV[6] is however trying to bridge this gap and with the smart TV and internet connection viewer is actually able to go online with the same physical device, television. Still there are features that really do not support functional social communication on smart TV's either. There is a possibility for example on Samsung to actually use one's mobile phone as a remote control but basically the use of independent keyboard is required when using online features of the television. This emphasizes that the technology convergence is still defining its limits and possibilities while content and the audiences are converging more painlessly. Also the different temporal changes in consuming TV content affect to that audiences are consuming TV more independently and individually. Online video stores such as Netflix and Viaplay allow people to watch the content whenever, also the recording choices of broadcasted content have developed a lot. Offering audiences the possibility to watch TV broadcasts whenever most suitable.

8 Conclusions

Interactive SMS-based entertainment was based on the convergence between mobile phones and television. This era was mainly entertaining and offered no actual and factual influence to the audiences. The era mainly offered material to play games, take part in quizzes and to the order SMS-material to mobile phones for example. iTV entertainment did not take place in vain, however. On the contrary it was an important phase in the continuity of building interactive television. Nowadays this entertainment has mainly vanished mainly due to the problem areas that occurred: high participation prizes, low quality and especially problematic target groups (children and youth). Participatory television exploited the convergence between TV broadcasts and internet, especially web 2.0 features. By this convergence, audience was able to communicate with each other, create content and enjoy

[6] A smart TV describes a trend of integration of the Internet and Web 2.0 features into television sets and set-top boxes, as well as the technological convergence between computers and these television sets/set-top boxes.

material provided by the broadcasters. This however led to the situation where audiences no longer could be seen as masses. This era emphasized the individual by offering watching TV whenever (online/net TV's etc.) one wishes to and wherever. Although this era also brought more social TV related activity amongst the online participators.

Social media and television convergence offers both social activity between the viewers and the broadcasters, even TV stars. TV broadcasters and—channels are communicating more and more with audience via Facebook and Twitter. The audience is being heard and at the same time they are able to interact in real-time with each other TV content related. Twitter and TV—combination is seen as a very functional way of creating the long wanted form of social TV. One reason is the low costs for tweeting (Zhao & Rosson, 2009, 252). Other one is the fact that Tweets do not need much time to write or read and one has an easy access via PC, notebook or smart phone (Jungnickel & Schweiger, 2011). The coming of the social TV is as anticipated as other forms of activating television. The current reality proves that people cherish the opportunity to communicate with others, expose aspects of their lives in public and even participate in virtual communities (Obrist, 2007). Will it happen through convergence or divergence is yet to be seen. Obviously, convergence is also one of the terms that carry notions of utopia and hype, which is something that should be acknowledged when the term convergence is addressed.

In conclusion, TV-screen has often been defined in relation to other technologies such as mobile devices or game consoles. Often add on devices are created precisely to give extra-value to TV's purposes of use (Hellman, 1988, 165). Nowadays, it can be said that television is actually giving extra-value since watching TV has changed so dramatically after the becoming of web 2.0 and social media. To sum up, TV's role in creating social experiences is not diminishing. It can be said that people feel TV as some kind of a mystical machine, yet in twenty-first century. We are used to idea that TV is a broadcast medium and when being able to step on a flow—whether it is through SMS, real-TV vote, online action or commenting on social media; it feels as we are breaking a magic and public boundary (Tuomi, 2009).

References

Antonini, A., Pensa, R. G., Sapino, M. L., Schifanella, C., Teraoni Prioletti, R., & Vignaroli, L. (2013). Tracking and analyzing TV content on the web through social and ontological knowledge. In *Proceedings of the 11th European conference on interactive TV and video (EuroITV'13)* (pp. 13–22). New York: ACM. doi:10.1145/2465958.2465978.

Bachmayer, S., Lugmayr, A., & Kotsis, G. (2010). Convergence of collaborative web approaches and interactive TV program formats. *International Journal of Web Information Systems, 6*(1), 74–94.

Bolter, J., & Grusin, R. (2000). *Remediation: Understanding new media.* Cambridge ja Lontoo: MIT Press.

Cesar, P., Bulterman, D. C. A., & Soares, L. F. G. (Eds.). (2008). Human-centered television. *ACM Transactions on Multimedia Computing, Communications and Applications (TOMCCAP), 4*(4): 1–7.

Cesar, P., Geerts, D., & Chorianopoulos, K. (2009). *Social interactive television: Immersive shared experiences and perspectives* (pp. 1–362). Hershey, PA: IGI Global. doi:10.4018/978-1-60566-656-3.

Cortez, J., Shamma, D. A., & Cai, L. (2012). Device communication: A multi-modal communication platform for internet connected televisions. In *Proceedings of the 10th European conference on interactive TV and video, EuroiTV'12* (pp. 19–26). New York: ACM.

Gruzd, A., Wellman, B., & Takhteyev, Y. (2011). Imagining twitter as an imagined community. *American Behavioral Scientist, 55*(10), 1294–1318.

Hanski, M.-P., & Kankainen, A. (2004). Pelien laadun kehittäminen käyttäjien näkökulmasta. Teoksessa Marja Kankaanranta, Pekka Neittaanmäki & Päivi Häkkinen (toim.): *Digitaalisten pelien maailmoja*. Jyväskylä: Jyväskylän yliopisto, 67–76.

Harboe, G., Metcalf, C., Bentley, F., Tullio, J., Massey, N., & Romano, G. (2008). Ambient social TV: Drawing people into a shared experience. In: *Proceeding of CHI 2008*, 1–10.

Harrington, S., Highfield, T., & Bruns, A. (2012). More than a backchannel: Twitter and television. In J. M. Noguera (Ed.), *Audience interactivity and participation* (pp. 13–17). Brussels, Belgium: Transforming Audiences, Transforming Societies.

Hautakangas, M. (2008). Yleisöä kaikki, tuottajia kaikki: toimijuuden neuvotteluja Suomen Big Brotherissa. In K. Nikunen (toim.), *Fanikirja: tutkimuksia nykykulttuurin fani-ilmiöistä*. Jyväskylä University, Jyväskylä.

Heinonen, U. (2008). *Sähköinen yhteisöllisyys. Kokemuksia vapaa- ajan, työn ja koulutuksen yhteisöistä verkossa*. Kulttuurituotannon ja maisemantutkimuksen laitoksen julkaisuja 14. Turun yliopisto, Pori.

Hellman, H. (1988). *Uustelevision aika?* Helsinki: Hanki ja jää.

Highfield, T., Harrington, S., & Bruns, A. (2013). Twitter as a technology for audiencing and fandom: The #Eurovision phenomenon. *Information, Communication & Society, 16*(3), 315–339.

Hill, A. (2005). *Reality TV – audiences and popular factual television*. Abingdon: Routledge. 231 pp.

Jenkins, H. (2006). *Convergence culture: Where old and new media collide*. New York: New York University Press. 308 pp.

Jensen, J., & Toscan, C. (1999). *Interactive television: TV of the future or the future of tv?* Aarhus: Aarhus University Press.

Jungnickel, K., & Schweiger, W. (2011, March 14–16). *Twitter as a television research method.* General Online Research Conference GOR 11, Heinrich Heine University, Düsseldorf.

Kaplan, A. M. (2010). Users of the world, unite! The challenges and opportunities of Social Media. *Business Horizons, 53*, 59–68. Elsevier, Indiana University.

Lugmayr, A., & Zheng, H. (2012). Forecasting the usage of ambient media in TV broadcasting in the year 2016. In *Proceedings of the 5th International Workshop on Semantic Ambient Media Experience (SAME)*.

Marwick, A., & Boyd, D. (2010). I tweet honestly, I tweet passionately: Twitter users, context collapse, and the imagined audience. *New Media and Society, 13*(1), 114–133.

Murray, S., & Ouellette, L. (2008). *Reality TV: Remaking television culture*. New York: New York University Press.

Näränen, P. (2006). *Digital television – analyses on history, politics and TV-system*. Doctoral dissertation, University of Tampere, Tampere.

O'Reilly, T. (2005). *What is Web 2.0? – Design patterns and business models for the next generation software*. http://oreilly.com/Web2/archive/what-is-Web-20.html

Obrist, M. (2007). Finding individuality in the technological complexity: Why people do it themselves? *The International Journal of Interdisciplinary Social Sciences, 2*(1), 203–212.

Ollikainen, V., Aalto, E., Norros, L., Seisto, A., Lugmayr, A., & Lindqvist, U. (2011). *NELME—Next generation electronic media multiproduction and multicasting processes, services, solutions and tools*. Espoo: VTT Technology.

Parikka, J. (2004). Interaktiivisuuden kolme kritiikkiä. *Lähikuva, 2004*(2–3), 83–97.

Prata, A., Chambel, T., & Guimarez, N. (2012). Personalized content access in interactive TV-based cross media environments. In Y. Kompatsiaris, B. Merialdo, & S. Lian (Eds.), *TV content analysis: Techniques and applications series: Multimedia computing, communication and intelligence*. Boca Raton, FL: CRC Press/Taylor & Francis group. ISBN 9781439855607.

Rautiainen, M., Heikkinen, A., Sarvanko, J., Chorianopoulos, K., Kostakos, V., & Ylianttila, M. (2013). Time shifting patterns in browsing and search behavior for catch-up TV on the web. *EuroITV, 2013*, 117–120.

Rodriguez-Alsina, A., & Carrabina, J. (2012). Analysis of the TV interactive content convergence and cross-platform adaptation. In Y. Kompatsiaris, B. Merialdo, & S. Lian (Eds.), *TV content analysis: Techniques and applications series: Multimedia computing, communication and intelligence*. Boca Raton, FL: CRC Press/Taylor & Francis group. ISBN 9781439855607.

Silverstone, R., & Hirsch, E. (Eds.). (1994). *Consuming technologies. Media information in domestic spaces*. New York: Routledge.

Simmons, N. (2009, June 3–5). *Me TV: Towards changing TV viewing practices?* Euroitv09, Leuven. ACM 978-1-60558-340-2/09(06).

Simmons, N. (2011, June 29–July 1). Television audience research in the age of convergence: Challenges and difficulties. In *Proceedings of the 9th International Interactive Conference on Interactive Television* (pp. 101–104). Lisbon, Portugal.

Sundet, V., & Ytreberg, E. (2009). Working notions of active audiences: Further research on the active participant in convergent media industries. *Convergence, 15*(4), 383–390.

Tuomi, P. (2008). SMS-based human-hosted interactive TV in Finland. In *ACM international conference proceeding series: Vol. 291 archive* (pp. 67–70). UxTV, Silicon Valley, CA.

Tuomi, P. (2009). Television interaktiivinen pelihetki – Television pelillisyys ja merkitys pelikokemusten tuottamisessa. Teoksessa: J. Suominen, R. Koskimaa, F. Mäyrä, & O. Sotamaa (toim.), *Pelitutkimuksen vuosikirja 2009* (s. 34–48). Tampere: Tampereen yliopisto.

Tuomi, P. (2010, June 9–11). The role of the traditional TV in the age of intermedial media spectacles. In *Proceedings of the 8th international interactive conference on interactive TV & video 2010* (pp. 5–14). Tampere, Finland.

Tuomi, P. (2012). *Teletext + Twitter =a new form of social TV? MindTrek'12 Proceeding of the 16th international academic MindTrek conference* (pp. 249–254). New York: ACM. ISBN: 978-1-4503-1637-8. doi:10.1145/2393132.2393188.

Tuomi, P. (2013). TV-related content online: A brief history of the use of webplatforms. *Euroitv '13 Proceedings of the 11th European conference on interactive TV and video* (pp. 139–142). New York: ACM. ISBN: 978-1-4503-1951-5. http://dl.acm.org/citation.cfm?doid=2465958.2465974

Tuomi, P., & Bachmayer, S. (2011, June 29–July 1). The convergence of TV and Web (2.0) in Austria and Finland. In *Proceedings of the 9th international interactive conference on interactive television* (pp. 55–64). New York: ACM.

Turtiainen, R. (2010). Tulos ei päässyt edes teksti-TV:lle. Miksi vanhanaikainen teknologia on säilyttänyt asemansa digitalisoituneessa mediaurheiluympäristössä? Tekniikan Waiheita 4/2010. 32–48.

von Feilitzen, C. (Ed.). (2004). *Young people, soap operas and reality TV*. International Clearinghouse on Children, Youth & the Media, Gothenburg University, Nordicom, Yearbook 2004.

Weller, K. (2011, June 28). *Approaches to analyzing scientific communication on twitter*. The World According to Twitter Workshop, Brisbane, Australia. http://www.slideshare.net/katrinweller

Weller, K., Bruuns, A., Puschmann, C., Burgess, J., & Mahrt, M. (2013). *Twitter and society*. New York: Peter Lang.

Ytreberg, E. (2009). Extended liveness and eventfulness in multiplatform reality formats. *New Media and Society, 11*(4), 467–485.

Zhao, D., & Rosson, M. B. (2009). *How and why people twitter: The role that micro-blogging plays in informal communication at work*. Proceedings of the ACM 2009 International Conference on Supporting Group Work. Retrieved March 4, 2012, from http://www.personal.psu.edu/duz108/blogs/publications/group09%20-%20twitter%20study.pdf

Why Did TV Bits and Radio Bits Not Fit Together? Digitalization and Divergence of Broadcast Media

Marko Ala-Fossi

1 Introduction

Back in 1995, Nicholas Negroponte assumed that because all spectrum used in a digital world would be bits, "they may be radio bits, TV bits or marine communications bits, but they are bits nonetheless, subject to the same commingling and multi-use that define multimedia" (Negroponte, 1996: 54). Now, almost 20 years later it is easy to see that while he recognized the potential of digitalization, he also misidentified its true essence and neglected the political, economic, and cultural aspects of system standardization. Digitalization as such is not a natural force producing convergence; rather, it is a tool, which can be used in blurring market boundaries for implementing expansive business strategies and more market-oriented policies. For example, in terrestrial broadcasting, digitalization has not resulted in any significant convergence, but rather increased divergence. In the analog days, there were only three major systems for television and two for radio broadcasting. Now there are altogether five types of digital TV broadcasting systems and four types of digital systems for terrestrial radio broadcasting worldwide. And even if both radio and TV are sometimes broadcast in digital, radio bits and TV bits are in most cases still strictly separated.

However, it would be unfair to argue that the development of digital broadcasting systems has failed to increase convergence, because that has not been its original goal. A historical analysis reveals how the policies and politics for developing new digital broadcasting systems have been driven by strong economic and political motives since the beginning, but usually not by intentions for digital media convergence. Both broadcasters and other organizations involved in the development have usually been more interested in protecting and maintaining the existing structures than striving for convergence or merging mediums together.

M. Ala-Fossi (✉)
University of Tampere, Tampere, Finland
e-mail: marko.ala-fossi@uta.fi

A. Lugmayr, C. Dal Zotto (eds.), *Media Convergence Handbook - Vol. 1*, Media Business and Innovation, DOI 10.1007/978-3-642-54484-2_15

Neither convergence nor divergence even requires digitalization. European countries were unable to agree on a common standard for monochrome TV in the 1950s and for color TV in the 1960s. However, in the early 1980s, the representatives of the leading TV broadcasting organizations from the US, Japan, and Europe made preparations for a joint proposal for new and improved global standards for analog TV production and transmission. But just before the proposal was about to be accepted in 1986, the European countries and the European Commission turned against it. It was felt that a global standard for a global market was too risky, while a separate standard for the common European market would help to protect the European electronics manufacturing and broadcasting industries against foreign competition. That decision to strive for divergence has then largely determined the later course of events also in digital broadcasting systems development.

Now, China is the largest selling and production market for all consumer electronics (Frost & Sullivan, 2012) and South Korea dominates the global TV receiver markets. Although electronics production still exists in Europe, there are no more state-owned television factories in France—and even the Dutch giant Philips has sold its TV manufacturing (2012) to China and the rest of its consumer electronics capabilities (2013) to Japan. TV standards have also lost most of their former cultural and political importance as means to control the flow of information, not least because the Internet provides an alternative access to all sorts of audiovisual content across national borders. As the habits of media consumers are changing, broadcasting is also losing ground as a socially privileged cultural form and technology on both sides of the Atlantic. The FCC already favors broadband over broadcasting in the US while, after 2015, the European TV broadcasters are about to lose also the 700-MHz band for mobile broadband besides the 800-MHz band, which was originally considered as "the digital dividend."

Now, the leading TV broadcasting organizations of the world have finally joined forces again and established a new project for the Future of Broadcast Television (FOBTV). After almost 70 years of fierce competition with separate standards, their goal is now to develop together a unified and global next-generation digital system for all broadcast television that could replace the current selection of incompatible digital television systems (Mouyal, 2012). However, even this project striving for increased convergence still concentrates only on broadcast television and not on digital broadcasting in general—so it seems that radio bits and TV bits might not fit together even in the next-generation global system for digital broadcasting.

2 Defining and Protecting Markets with Analog Standards

For most of us today, RCA is familiar as the name of a simple audio connector, but once, the Radio Corporation of America was the mightiest broadcasting organization on the planet. It was not only a major American electronics manufacturer and the owner of the most important patents on radio technologies, but also the main owner of National Broadcasting Company (NBC), a commercial radio network based on AM (amplitude modulation) technology. As an electronics company, RCA was of course investing in new technologies. Originally, it had also initiated the development of the static-free FM (frequency modulation) radio system, but this project was stopped in 1936. At that point RCA decided to concentrate on developing electronic television and creating new lucrative businesses in both TV manufacturing and broadcasting instead of just improving radio. This strategic choice caused RCA later to use its power and influence to oppose the introduction of FM radio broadcasting in the US. While the FM system was seen as a threat to the domestic radio business, it was however perfect to carry the sound of TV, while pictures were broadcast using the quadrature amplification modulation (QAM) system. RCA had a major impact on the US standard for monochrome TV (525 lines with 30-Hz frame frequency with FM sound), adopted by the National Television System Committee (NTSC) just before the war in 1941. And largely because of the urgent commercial business interests of RCA, a backward compatible version for the NTSC color TV standard was approved as early as in 1953. And this is why the US ended up with the lowest resolution analog standard for color TV in the world (Ala-Fossi, 2005: 100; Berkman, 2000: 49–56; Chandler, 2005: 16–17).

Not only business strategies but also politics and policies as well as the outcome of WWII had an impact on the adoption of new broadcasting systems and standards originally developed and patented in the US. At the European Broadcasting Conference of 1948 in Copenhagen, the Allies left Germany with very few AM frequencies, while in Finland the AM reception was ruined after 1950 by the powerful propaganda stations of the Cold War era (Ala-Fossi, 2005: 149). This is why both the Federal Republic of Germany and Finland adopted FM broadcasting on a large scale as early as the 1950s. On the Pacific front, Japan was occupied by the US forces until 1952—and the foundation for the rise of the Japanese electronics industry and rapid development of broadcast services was laid during the US-led reconstruction (Cringely, 2000). Interestingly enough, although Japan soon became the leading manufacturer of new transistor radio sets, FM radio broadcasting did not become commercially significant in Japan or in the US until the late 1960s.

In Europe, Germany, Great Britain, and France had introduced their own standards for TV broadcasting as well as the first regular TV services already before the war, but the national public service broadcasters were not able to re-launch their regular television operations until the late 1940s. The UK started in 1946 with their own pre-war system (405 lines with 25 Hz frame frequency and AM sound) but their French neighbors had been experimenting already during the war with higher picture resolution. So in 1948, after a 4-year-long debate about a standard for monochrome TV, a new "high-definition" system (819 lines with 25 Hz frame frequency and FM sound) was chosen for France by the Minister of Information, Francois Mitterrand, and broadcast services were started the following year. The French politicians did not even try to hide the fact that the deviant TV standard was not only providing better image quality but also about protecting national interests and domestic markets (Fickers, 2006: 18–20; Fickers & O'Dwyer, 2012: 65–66).

However, after the Consultative Committee of International Radio (CCIR) conference in Stockholm 1952 failed to agree on a common European TV standard, most of the European countries including the Federal Republic of Germany ended up using a variant of NTSC monochrome standard (625 lines with 25 Hz frame frequency and FM sound), originally developed by Soviet engineers right after the war. The frame frequency—or picture repetition rate of NTSC—was based on the 60 Hz power frequency used in the US, and when the standard was adjusted for the European 50 Hz power frequency, the number of lines was also increased, resulting in better picture resolution. The large and influential Dutch electronics manufacturer Philips had also created its own 50 Hz variant of NTSC, but gave its full support for the 625 line system, which was taken into use in the Netherlands in 1951. In bi-lingual Belgium, both the French and the "continental" system were used side by side via complex and expensive multi-standard receivers (De Bruin & Smits, 1999: 5–7; Fickers & O'Dwyer, 2012: 67–68).

In Japan, the nation was not divided by the language, but by the frequency of the electric power system. Since the 1890s all eastern Japan (including Tokyo) has been using European-type 50 Hz power systems, while Osaka and all western Japan has been using US-type 60 Hz power systems (Owen, 1997: 12), so the Japanese decision about the TV standard was obviously not based on any considerations over the power frequency issues and the problem must have been solved some other way. Regular TV broadcasting in Japan using NTSC was started in 1953—and in 1960 it was the second country after the US to start TV broadcasting in color. However, the engineers at the technical research laboratory of the national public service broadcaster NHK were not particularly satisfied with the picture quality of the American color TV during the 1964 Tokyo Olympics, which were televised worldwide via satellite (Hart, 2004: 84; NHK 1977: 251).

The NTSC color TV had problems with color stability—that is why it was sometimes referred as the system with Never Twice Same Color. The European countries had already opted out of adopting the NTSC color standard, and developed two competing 625-line standards for improved color television instead. The French SECAM (Système Électronique Couleur Avec Mémoire) was originally designed in 1956 to fix the phase shifting problems of NTSC, but it also became a highly politicized expression of national pride and an instrument of French foreign policy. The German PAL (Phase Alternating Line) system introduced in 1961 was basically a NTSC version inspired by SECAM, although it used a different method for stabilizing the colors. Both systems included new inventions protected with patents and some SECAM patents were used in PAL also. At the CCIR Oslo conference in 1966 it became clear that there would no common European standard for color TV, either. France, Luxemburg, Greece, and the Soviet Union with its satellites including East Germany chose SECAM, while the rest of the European countries decided to implement PAL (De Bruin & Smits, 1999: 8–9; Fickers, 2010: 97–111).

Accordingly, NHK decided to aim even higher and instead of just improving NTSC, to develop a new analog system for High Definition Television (HDTV). The original name of the system—Hi-Vision—suggests that it was understood as a visual equivalent to hi-fi sound. By the end of the 1960s, Japan was already a major player in the worldwide consumer electronics market, but it was losing its competitive edge in television set manufacturing against even cheaper production from Eastern Asia. The NHK proposal for a new TV system was exactly what the major Japanese electronics manufacturers had been looking for, so in 1970 they formed a national coalition with NHK for developing HDTV (Curwen, 1994: 17; Hart, 2004: 84–87).

3 Why Did the PSBs also Invest in Developing New Broadcast Technologies?

From a politico-economic perspective it is obvious why large electronics manufacturers invest in the research and development of new technologies. New or improved systems can be turned into new interesting and competitive products for consumer markets and result in increasing sales and profit for the shareholders. But why did a national public service broadcaster like NHK—as well as its counterparts in Europe—become heavily involved in the development of new broadcasting technologies with the national consumer electronics industries after WWII? There are at least three reasons for that.

First of all, most public service broadcasters (PSBs) used to have a national monopoly on broadcasting as well as their own, well-resourced technical research laboratory. It was part of their remit to strive for better quality and availability of the national broadcast services. Second, most governments used PSB monopolies as tools to support their national industrial policy projects by requiring the PSB to cooperate with the national champions (Galperin, 2004: 132–135). And finally, license-fee funded PSBs were interested in developing new broadcasting systems, because they were financially dependent on people's willingness to buy and use new broadcast receivers (Ala-Fossi, 2012a).

During the radio era, the price of a license fee remained rather low, but as the broadcasters also started to organize TV production, they needed more money for their operations. For example, in the UK, an additional and higher license fee for the use of TV receiver was introduced in 1947. Based on the BBC statistics on the number of valid receiver licenses and their prices,[1] the amount of BBC revenue from TV licenses surpassed radio license income within 7 years—and an exactly similar pattern was repeated in Japan five and in Finland 10 years later: TV licenses became the main source of revenue of NHK in 1959 and of YLE in 1964. Since 1968 in the UK and Japan and 1969 in Finland, the highest possible license fee was charged for the use of a color TV receiver. A color TV set became a true hit product in the consumer market—and in less than 10 years, color TV license fee revenue was the main source of income for all of these PSBs (Ala-Fossi, 2012a: 35–37; Hart, 2004: 54–59; NHK 1977, 246–247; Salokangas, 1996: 144). So the PSB

[1] http://en.wikipedia.org/wiki/Television_licensing_in_the_United_Kingdom_%28historical%29

organizations also had a strong internal motive for the development of new and better systems for broadcast reception—especially in countries like the UK, Japan, and Finland, where they could not have any direct advertising income.

However, in Germany, public service broadcasting had a more diverse source of revenues. Advertising was allowed up to a certain limit and, moreover, there was still a separate license fee for radio—a practice that had already been abolished in Japan in 1968, in the UK in 1971, and in Finland in 1976. So it might not be a pure coincidence that it was the technical research department of German public service broadcasters, Institut für Rundfunktechnik (IRT) that first became interested in developing a digital system for radio in 1981. The project was inspired by the recent introduction of the Compact Disc (CD), a new digital format for audio recordings, but digital audio broadcasting was actually not a completely new idea (Ala-Fossi, 2010: 47). The basic principles of digital audio communication like Pulse Code Modulation (PCM) had been developed during WWII and already implemented in digital telephony during the 1960s. By the late 1970s, some European broadcasters including the BBC had developed their own systems for Near Instantaneous Companded Audio Multiplex (NICAM), which were used for transferring digital audio between fixed links within broadcasting networks (Gilchrist, 1978: 7). The BBC R&D department was also developing NICAM stereo sound for TV broadcasting to be used in addition to a FM sound carrier, but because the system was suited only for fixed reception, it could not become a system for digital radio—and the digital sound broadcasting systems for satellite delivery had similar problems (Hoeg & Lauterbach, 2003: 5; Lax, 2010: 77). A new approach was needed.

This was why the alliance of national PSBs in Europe, the European Broadcasting Union (EBU) had commissioned the Centre Commun d'Études de Télévision et Télécommunications (CCETT) in France—another R&D laboratory of a national PSB—to do research on satellite delivery of digital sound broadcasting to mobile receivers. By the mid-1980s they had found a solution: a completely new digital multicarrier transmission system (OFDM), which was soon introduced also to the IRT. So in 1986, there was an idea or a draft of digital audio broadcasting, but an operational digital radio system did not yet exist (Gandy, 2003: 3; Hoeg & Lauterbach, 2003: 5; O'Neill, 2010: 32; Schulze & Lüders, 2005: 262).

4 No Global TV Standard for the Common Market: Europe Opts to Diverge

The political climate in the industrialized countries of the Western world had started to change with the economic downturn in the early 1970s, when the governments were forced to seek new ways to fight the recession. After the conservative governments of Margaret Thatcher in the UK (1979) and Ronald Reagan in the US (1980) had come to power, the neo-liberalist approach of reducing both public spending and government control over the markets was already becoming the new dominant economic policy paradigm in most advanced capitalist countries—but not yet in France (Harvey, 2005: 18–34; Hesmondhalgh, 2007: 82–86).

The economic crisis and rising unemployment had hit also France, and although the fifth non-socialist president of the so-called Fifth Republic, Valéry Giscard d'Estaing, and his administration had tried to improve the national competitiveness through modernization projects and by following the Japanese example in creating a computerized information society, the majority of voters were not happy or convinced. The candidate of the French Socialist Party, Francois Mitterrand, won the presidential elections of 1981 by promising massive social and economic reforms, including the nationalization of nine large industrial groups. One of the companies taken over by the state after the elections was the large electronics manufacturer Thomson, which soon became an instrument of French protectionist trade policy. President Mitterrand himself instructed Thomson to find a "European solution" for video recorder production instead of an agreement with the Japanese JVC (Holton, 1986: 71–74; Nevalainen, 1999: 7–8).

While the French government was still trying to protect its domestic electronic industry even with desperate tricks, like slowing down the imports of Japanese video recorders by sending them all to a single customs office in Poitiers (Fridenson, 2012: 2), in the US the government had no interest or will anymore to intervene in the markets, which had been taken over by imported consumer electronics. The American television manufacturing industry had declined to a shadow of itself and even the saga of RCA was coming to its end. The company had spent millions of dollars on unsuccessful projects, and huge losses from an analog Selectavision videodisc system (1981–1986) were part of the reason why General Electric (GE) took over RCA in 1986 and divested the company in small pieces. In the following year, only one larger US TV manufacturer remained under American ownership (Abramson, 2003, 212–216; Chandler, 2005: 61, 78; Hart, 2004: 74; 78–82; USC 1990: 34;).

So the US Department of State did not have too much to lose when it decided to follow the formal recommendation of the Advanced Television Systems Committee (ATSC) and support the NHK proposal for analog HDTV as the world standard in April 1985. After its first presentations in 1981, NHK had even modified its original Hi-Vision system to match with the American requirements, and the US broadcasters welcomed the wide-band HDTV as a good excuse to keep their vacant broadcast channels. In addition, a unified global TV standard would have been a benefit for the US movie industry (Alvarez, Chen, Lecumberri, & Yang, 1999; Brinkley, 1998: 16–19; Hart, 2004: 196, 1994: 215; USC 1990: 36).

Originally, also the EBU and the European countries had been working together with ATSC and NHK since 1983 to produce world standards for TV program production and transmission within an interim working group of the CCIR. However, in January 1986 when the preparations were almost ready, the French government announced that the NHK proposal was "inequitable, premature, unfair and unsuitable for Europe." Soon also the major European electronics manufacturers, Thomson, Philips, and Bosch group from Germany found it harmful for their interests and the European Commission asked the member states to oppose it. The formal reason for opposition was that the proposal was based on 60 Hz and not the European 50 Hz (Cianci, 2012: 48; Dai, Cawson, & Holmes, 1996: 153–154; De Bruin & Smits, 1999: 10–11, 104).

But the reasons were much more complex. Besides being incompatible with the existing equipment, the NHK system would have given a competitive advantage to the Japanese TV manufacturing industry over European production and opened gates for even easier distribution of American TV programming. And this unholy alliance of Japanese and US businesses and governments supporting the NHK system was too much, especially for France, but also for its partners in the new efforts to increase European integration and create a common market. The German patent rights for the PAL system had expired in 1982 and Japanese manufacturers had already been able to increase their market share in Europe (Gaillard, 2007: 17). So the European countries, led by France, refused to accept the NHK initiative and suggested instead a different approach at the CCIR Dubrovnik meeting in May 1986. France was also the initiator for EUREKA, a new pan-European organization for research and development coordination established by 17 European states and the European Commission, which proved to be a useful tool for defending European electronics industries against the Japanese Hi-Vision. The development project (Eureka 95) for the European version of HDTV to replace SECAM and PAL was launched in July 1986. Major European electronics manufacturers, such as the French Thomson, Bosch and the main coordinator Philips, were centrally involved in the project along with the technical research organizations of several European public service broadcasters and the EBU (Chandler, 2005: 77–78; Evain, 1995: 39; Hart, 2004: 68, 119, 126–127; Sandholtz, 1992; USC 1990: 32).

In December 1986, the same three major European manufacturers were joined by the BBC, IRT, and CCETT as well as 13 other European organizations to launch another Eureka project (Eureka 147). The formal initiative for the project came from the German Federal Ministry for Research and Technology, and it was coordinated by the German Space Agency (DLR) (Evain, 1995: 39; Hoeg & Lauterbach, 2003: 6). The project for developing a new Digital Audio Broadcasting (DAB) system was obviously a part of the protectionist European counterattack (Ala-Fossi, 2010: 46). The original Eureka project form states that "the drawing up of a new digital audio broadcasting standard will therefore provide a long term counterbalance to the increasing dominance of the countries of the Far East in the consumer electronics sector." The text refers also to the concept of "integrated services digital broadcasting" (ISDB), an idea for a multipurpose digital broadcasting system drafted at NHK in the early 1980s, but the original main focus in DAB development was in improving radio and not in extending it to multimedia (Eureka, 1986; O'Neill, 2009; Yoshino, 2000).

Although Japan and the US were striving together for a single global standard for TV broadcasting, the NHK proposal in 1986 was much more about market convergence than media convergence. A single analog standard would have effectively blurred the strict national market boundaries in TV manufacturing and content distribution, but its idea of broadcasting was strictly media-specific. The European governments, electronics manufacturers, and PSBs preferred not to fully open their domestic markets to Japanese and American imports, but they shared a similar idea of broadcasting: TV and radio were considered as independent entities. Although in Europe radio was supposed to be the first mass media to go digital, the future of television was thought to be not only separate, but also analog.

The HDTV systems of the 1980s are sometimes described as direct precursors to digital television, but European HD-MAC (Multiplexed Analogue Components) as well as Hi-Vision were nothing more than advanced analog television systems although they complemented analog video with digital sound. It should be noted also that in the late 1980s, a completely digital television system was still seen primarily as a theoretical option for the future (Brinkley, 1998: 94). This is why such a system could not have been even seriously considered as a useful short-term tool for defense against the Japanese initiative.

5 The Future of TV in the US Becomes Digital: Europe Has to Start Over

The defeat of the NHK system as the worldwide HDTV standard in 1986 forced the US government agencies, electronic industry, and broadcasters to reconsider their strategies. It was noted that the use of Japanese or European HDTV systems for satellite broadcast delivery could have been a threat to the stability of the US TV market structure, based on local broadcast licensees (De Bruin & Smits, 1999: 12; Reimers, 2004: 4). This is part of the reason why new initiatives were set up for domestic HDTV technology research and new policies were drafted for a regional terrestrial HDTV broadcasting standard. In 1987, the FCC started to look for industry proposals for a NTSC-compatible analogue HDTV; and finally, in 1989, the US reversed its official support for the NHK system. If it was not going to be a world standard, there was no good reason to hand over the competitive advantage to the Japanese manufacturers (USC 1990: 35–36).

By that time, a completely digital television system was considered as an impossible goal within the US broadcast industry, although there was already a group of professors (including William Schreiber & Nicholas Negroponte) from the Massachusetts Institute of Technology (MIT) urging for digital solutions and complete integration of television sets and computers (Brinkley, 1998: 93–95; Neil, 2010; Neuman, 1988). VideoCipher, the San Diego division of General Instruments (GI), a small producer of satellite signal scrambling devices, was not part of the broadcast industry, so a group of engineers led by Woo Paik took a digital approach and made a breakthrough in developing the first all-digital HDTV system in early 1990. Both the FCC and the US industry rapidly swapped their approach from analog to digital—while GI allied with MIT to develop a fully computer-compatible digital HDTV system with progressive image scanning. Suddenly, the US was leading in the race for the future of television and by May 1993, the developers of four competing American digital systems joined forces in the Digital HDTV Grand Alliance (Alvarez et al., 1999: 6–10; Brinkley, 1998: 120–134; Galperin, 2004: 74–78; Negroponte, 1996: 37–40). Two members of the alliance were actually US subsidiaries of the largest European TV manufacturers: Philips and Thomson, which had bought the remains of RCA television production in 1987.

It seems that despite all the new possibilities a more convergent digital approach could have offered for the future of broadcasting, the US developers were still striving primarily for HDTV but now in an all-digital format and secondary for compatibility with the computer world. It was not necessary to think about mobile reception—or sound broadcasting—especially since other projects were working on those issues already (Yamada, 2006: 28).

Meanwhile, the first common European TV standard, the HD-MAC system, was in trouble despite intergovernmental support and a special directive (92/38/EEG) of the European Commission. The analog technology was not only expensive but also inclined to have interference problems: in addition, it did not appeal to the increasing number of new private broadcasters and required a broad spectrum. So it did not take long after the news from the US had reached Europe before the first unofficial studies for digital television systems in Europe were started. HD-DIVINE, a joint Scandinavian project launched in early 1991, was able to demonstrate an operational but incomplete digital terrestrial HDTV system already by the following year, but there were at least five other small European projects working with other aspects of digital TV. Most of them were based on ODFM modulation, but in France, CCETT was able to combine digital image coding with an adaptation of the COFDM system already developed for DAB. Based on the experience with audio broadcasting, three different modes of reception were taken into account: mobile-like DAB, portable or "plug-free," and fixed reception with roof-top antenna (Abramson, 2003: 248; Reimers, 2004: 7–8). However, HD-MAC was officially expected to become a success in the consumer markets during the Olympic year of 1992. After a bitter failure in all fronts, the European Commission was still ready to pour in even more money and issue a new directive in order to save the project. Largely because of the UK opposition and Martin Bangemann taking responsibility for the European technology policies in the new Commission, the EU finally abandoned analog HDTV. A new industry-led and Europe-wide project for Digital Video Broadcasting (DVB) was launched in September 1993. The project was based on an earlier German initiative and it consisted of 84 European broadcasters, manufacturers, telecom organizations, and regulatory authorities (De Bruin & Smits, 1999: 14; Galperin, 2004: 132–134; Kemppainen, 2008: 32–35; Näränen 2006: 42–43).

The failure of the HD-MAC system set a new course both for European broadcasting policies and technology initiatives in general. On the European level, the traditional neo-mercantilist high-tech rivalry of the analog era was now replaced by more market-driven, neo-liberal policies drafted in the Bangemann Report (EC 1994). There were now less intergovernmental arrangements with the national champions over the increasingly complicated and privatized broadcasting markets and more reliance on private standards development instead (Galperin, 2004: 133–135; Michalis, 2007: 149–151). This experience with HDTV made the Commission also reluctant to openly support any particular technology, and led to the adoption of the principle of technological neutrality a few years later (Lembke, 2002: 227, 240). Although the European Commission also financially supported digital video broadcasting (DVB) development and in 1995 created a special directive (95/47/EC) on television standards, this time the directive was aimed at structuring the market and not setting the standard itself. On the industry level, the failed HD-MAC was seen as the result of a technology-led project, indicating a need for a more consumer-driven approach. This is why the new digital DVB standard was designed to meet the European market demand and to provide more channels on a reduced bandwidth rather than to improve the picture quality as in the HDTV approach (De Bruin & Smits, 1999: 137; Fagan, 1994).

The first system specifications for DVB use in satellite broadcasting (DVB-S) were drafted and adopted by the project group already by the end 1993 and for cable distribution (DVB-C) in early 1994. The standard for terrestrial broadcasting (DVB-T) was approved in late 1995 and the first regular broadcasting services on DVB-T were introduced in the UK 3 years later, in November 1998 (Reimers, 2004: 8, 256).

6 First But Not Global Standard for Digital Radio: The US Opts to Diverge

The first public demonstration of the DAB system was given in Geneva during the World Administrative Radio Conference of 1988, less than 2 years after the Eureka 147 project had been launched. During the following years, several similar demonstrations and exhibitions were arranged not only in Europe, but also in Singapore and Tunisia as well as in Canada and the US (Hoeg & Lauterbach, 2003: 8–9). Despite the simultaneous dramatic transition of broadcasting policies, regulatory practices and even industry structures of European radio broadcasting in the late 1980s and early 1990s, the DAB development remained firmly in the hands of PSBs and electronics manufacturers. This is partly because the alliance of European private and commercial broadcasters, the Association of European Radios (AER), was not established before 1992—and it was not politically active until the mid-1990s (Lembke, 2002: 219–220). So the numbers and market share of private broadcasters were increasing, but no private radio broadcasters were involved in either of the two phases of Eureka 147, the project developing DAB digital radio in 1987–1991 and 1992–1994 (Rissanen, 1993).

The new DAB digital radio system was supposed to be superior to analog radio systems; it was not only to do everything better, but also more than before—such as to deliver several technically high-quality program signals with additional services on a single transmitter on a single frequency. However, channel multiplexing and Single Frequency Network (SFN) capabilities, which made perfect sense for cost-effective nationwide and regional broadcasting by PSBs, did not meet at all the needs of the increasing numbers of local and small-scale broadcasters. The serious mismatch between DAB and local radio was pointed out in a report for the Council of Europe (Gronow, Lannegren, & Maren, 1992) already 2 years before DAB was formally recognized as an official standard for digital radio in 1994. But even the European spectrum allocations for digital radio in 1995 were tailored for national and regional broadcasting.

However, the first warning signal came from the US, when the American radio industry decided to develop its own digital radio system based on the existing frequency allocations. Interestingly enough, the course of events had a rather similar pattern as in the case of the NHK proposal: at first, in 1991, the National Association of Broadcasters (NAB) gave an initial endorsement for DAB, but within a few months the US radio industry decided to reject the whole system. The system design reflected the power structures of the European broadcasting and it would have required changes that were considered to be too risky for the economic stability of American radio markets. The US radio industry started to develop its own digital radio system based on an IBOC (In-Band, On Channel) approach and in July 2000 IBOC developers merged into a single development company, iBiquity Digital Corporation. Perhaps the most important feature of the system—now marketed as HD radio (!)—is that it uses existing broadcasting bands and present FM channel allocations. An operational, but not a complete version of the system was introduced in April 2002, and a couple of months later it was approved as the technical standard for terrestrial digital radio broadcasting in the US (Ala-Fossi & Stavitsky, 2003: 62–67, 74).

The development of digital computer technologies, which later became the basis for the Internet, had started in the US in the mid-1960s, but the Internet did not have any serious effects on traditional media before the introduction of the World Wide Web in 1992 and the first web browser in 1993 (Henten & Tadayoni, 2008: 49–51). All this happened at the same time the US National Information Infrastructure (NII) initiative was launched by the newly elected Clinton administration in September 1993. Its vision about the "information superhighway" was largely based on the ideas of the development of computer technology and the Internet. However, it also was expected that a new generation of interactive, computer-compatible digital TV sets could provide the most natural gateway to the information society in the nation, where more households already had television sets than telephones or personal computers. Once again, the future was seen as an extension of the present situation (Aufderheide, 1999: 43; Galperin, 2004: 37–39; Negroponte, 1996: 42–43, 54). A group of MIT academics and the US computer industry, especially Apple Computers, had been promoting the compatibility of the new US digital television standard with computer technology (Hart, 2004: 155, 165; Neil, 2010). However, US broadcasters were reluctant to abandon their interlaced display formats. This is why the preconditions for the digital convergence of TVs and computers did not exist until the Advanced Television Systems Committee (ATSC) standard for digital television was approved by the FCC in December 1996 (Brinkley, 1998: 393).

The European Union reacted rapidly to the NII initiative in order to protect the competitiveness of its own member states. As of December 1993, the Delors White Paper suggested a large development program for telecommunications, computer networks, and other information infrastructure in Europe. The following implementation plan, the so-called Bangemann Report (1994) largely adapted the American NII program for Europe and gave a number of recommendations for the member states on how to pave their way for the information society. Also the development of digital television (interactive video) was included in the European action plan (Galperin, 2004: 39–40, 134–135; Michalis, 2007: 164–165). It is worth noting that although digital radio was an already existing European invention and the first of its kind in the world, it was not mentioned at all in the Bangemann Report. In addition, there was only one reference to digital radio in the European Green Paper on Convergence (EC 1994, 1997: 5).

7 Supplementary Versus Convergent Approach to Digital Broadcasting

As described earlier, the developers of European digital TV did not have to start from scratch in 1993, when the standard development was officially launched. They had been working on digital broadcasting issues already for several years and had developed many working solutions and one fully functioning system, DAB, which was already about to be approved as the first official standard for digital radio in the world. So it was not only handy but reasonable to utilize the already-existing European research on digital broadcasting for digital television development, like CCETT was doing in its STERNE project. This is why DVB terrestrial digital television has so many technical similarities to terrestrial DAB digital radio, including MPEG Layer 2 audio coding, modulation system COFDM, multiplexing, and SFN. Actually, under certain conditions "DVB-T is effectively a wide bandwidth version of the DAB system" (Laven, 1998: 5–6). But there were also reasons why these two closely related digital systems were made fully incompatible: they were supposed to supplement each other.

In Europe, the national PSBs still had organizational structures based on the division between radio and television. Both divisions usually had their own distinctive development strategies, and sometimes they were also competing for power, money, and resources although the size of radio could no longer match with that of TV. These organizations had no particular internal desire to challenge the existing market boundaries and merge the new digital broadcast systems despite the increasing popularity of the idea of convergence and possible awareness of the Japanese ISDB concept. The members of the EBU saw the two digital systems mostly as independent digital replacements: DAB was designed for radio and the new consumer-driven approach was supposed to lead the DVB developers to aim exactly at what the majority of consumers were expecting from television.

So, instead of HD-quality pictures, DVB was designed to offer more channels to provide more choice (De Bruin & Smits, 1999: 14; Fagan, 1994). And because most European consumers at the time were not watching television on the move, DVB-T was designed for fixed roof-top antenna reception of video content on wide channels, and the options for mobile or portable reception capability, which has already been studied in 1992, were completely ignored on purpose (Abramson,

2003: 248; Yamada, 2006: 28). In other words, the incompatibility with DAB and severe unsuitability of the DVB-T standard for mobile use and for any radio-type services were not any accidental deficiencies of the system design, but its characteristics, which were considered and chosen according to market research and best possible understanding concerning the future needs (Baumgartner, 2005: 1–2; Laven, 1998: 5–6; Wood, 1995, 2001: 2).

In Japan, the switch from analog to digital approach was a very difficult process—primarily because it meant that most of the time, money, and effort invested in Hi-Vision during the past 25 years had been wasted. NHK set up a new department for digital broadcasting research in 1991 (Yamada, 2006: 32), but most Japanese industry leaders and politicians still officially supported analog HDTV in 1994 (Fagan, 1994; Hart, 2004, 199–203; Negroponte, 1996, 37–40). However, the possibilities for a fully digital television system had been studied unofficially also in Japan—especially because several European subsidiaries of Japanese electronics manufacturers were now allowed to become members of the DVB project (Reimers, 2004: 8). The head of the new NHK digital broadcasting research department, Osamu Yamada, had become interested in OFDM and DAB as early as the late 1980s, but at that time OFDM studies at CCETT were strictly classified and Yamada was turned away. In 1992, he found out that mobile reception was going to be been left outside of both the DVB and ATSC standard specifications for digital TV. Based on this piece of information, he knew that mobile reception using OFDM could offer a unique competitive edge for the forthcoming Japanese digital broadcasting system (Yamada, 2006: 28).

Another new competitive advantage was found in Japan from the earlier concept of ISDB: there would be only one system for broadcasting instead of many. By utilizing already-existing research on digital broadcasting, the Japanese were able to introduce the technical specifications of their new digital broadcasting system in 1997. ISDB-T has been described as "very closely related to DVB-T" and it has several technical commonalities with both DAB and DVB, most notably the modulation method COFDM (Reimers, 2004: 9). But thanks to its convergent technological approach on digital broadcasting, it was able to offer mobile reception of both radio and television in standard quality as well as in HD quality, unlike its European relatives. The digital signal carries all these forms, and the output is mainly dependent on the characteristics of the receiver set (Miyazawa, 2004: 6, 26; Nakahara, 2003: 5, 27, 30–33; Yokohata, 2007: 1). In this way, it was able to challenge the conventional conceptions of both broadcast radio and television.

The ISDB approach was possible mainly because radio and television were not treated in the design process as completely separate entities with unique identities, but rather as two forms of broadcast media based on different degrees of technical complexity. From the Japanese perspective, radio was not only culturally but also in every other way a minor and subordinate media when compared to television (Kato, 1998: 177). In addition, there was no need to reinvent any DAB-type system for digital radio. However, this integrated digital platform blurring the traditional boundaries between radio and television seems to be more like a common ground for all types of broadcast media than a basis for a new converged medium in itself.

In the European context, perhaps it is not surprising that the strategy of creating two complementary digital broadcasting systems led to an internal race of standards by the end of the 1990s. Instead of just supplementing and supporting each other, DAB was in fact losing a competition with DVB over political and industrial support in Europe. Some broadcasters were so impressed by the capability of new and shiny DVB that they wanted to use it for all digital broadcasting. The EBU had to come out and defend DAB by reminding that the systems were designed to be complementary, and there was no way they could replace each other (Laven, 1998; Wood, 2001). David Wood from the EBU predicted also that "ultimately, radio is likely to survive and prosper better in the digital age, if it is the master of its own environment" (Wood, 2001: 3).

At same time the increasing importance of the Internet as the first converged digital media platform and the ISDB mobile TV experiments in Japan made European developers of digital television understand that they had actually designed a future standard using contemporary market specifications. Digital broadcasting could perhaps offer more than just traditional analog media concepts—and they had made a strategic mistake by excluding mobile reception from the DVB standard. The very first prototype of a European mobile TV receiver was built by Nokia's Finnish DVB unit in Turku. Nokia also set up a new EU-supported project (AC318 Motivate) to compete against ISDB and to develop a European system for mobile television. Nokia's investment in DVB-H must have been significant, because hundreds of people worked in the project for almost 10 years (Kemppainen, 2008: 23; Motivate, 1998; Talmola, 2005: 2; Torikka, 2007).

As noted earlier, the DVB system was made for fixed reception only, so the power consumption of receivers was quite high and mobile reception was unreliable. Nokia solved the problems by dividing the signal into short bursts and by reducing the picture quality a bit. But after these modifications, it was not possible to receive any regular DVB broadcasts with a handheld DVB-H receiver or vice versa (Ala-Fossi, 2010: 52–56; Högmander 2003; Torikka, 2007; Vihma, 2007). The new modified and backwards-incompatible version of DVB was considered to be a brilliant piece of European engineering, and Nokia's DVB-H project team even received the Finnish Engineering Award in 2007. Later in the same year, the European Commission gave its support for DVB-H and a year later it became one of the Official Standards of the EU (EC 2007; EU 2008; Torikka, 2007).

However, despite high expectations and political support, DVB-H never became a success in the marketplace. In Finland, program providers had problems with dual copyright payments because incompatible signals for mobile TV were considered simulcasts, and in Germany telecom operators left without DVB-H licenses started to sell Korean mobile phones with regular free-to-air DVB-T receivers. In addition, European consumers were really not so interested in mobile TV especially after they were hit by a severe economic recession (Ala-Fossi, 2012b; Pearce, 2008; Yoshida, 2008). Nokia decided to sell its mobile TV unit (Mobile Broadcast Solutions) to the Indian company Wipro in April 2009. Gradually, it became clear that DVB-H mobile TV would not make a breakthrough in Europe and even the last remaining commercial DVB-H services in Italy and Finland were shut down in 2012 (Mannila, 2009; Winslow, 2012).

The latest new versions of DVB-T have expanded the family of European digital TV standards even more toward the convergent functionality of ISDB-T design. While DVB-T was primarily about increasing the number of channels with less spectrum space, the DVB-T2 standard published in 2008 can finally offer both HD- and SD-quality pictures as well as further increased channel capacity thanks to a more effective coding system. In addition, DVB-T2 is designed both for fixed and portable receivers, but for mobile and handheld reception there is yet another new system version called DVB-T2-Lite, which was introduced in November 2011 (EBU 2012: 107; Wells, 2011: 4–6). In addition, also the first mobile TV version of the American ATSC system has been introduced for the US markets in January 2010.

8 Discussion and Conclusions

As Andreas Fickers (2006) has pointed out, early analog radio and television were developed in Europe in very different ways. When radio broadcasting was first introduced in the early 1920s, there were no competing alternatives to AM broadcasting and the technology adopted for radio was practically similar throughout the world. Moreover, the broadcast signals did not stop at national borders. Accordingly, an increasing number of listeners were soon enjoying programs from far away, while the national authorities had to cooperate on an international level to prevent chaos and interference. But when TV was introduced after WWII, it was in practice a "medium without a public," the implementation of which was possible to design in advance according to the cultural, political, and economic goals of each nation (Fickers, 2006: 16–18). This was also the case when digital radio and TV were first introduced: there was no need to compromise with the past.

The technical pretext for not adopting any American (or Japanese) TV standards as such anywhere in Europe was path dependency caused by the different power frequencies of electricity, but especially France—and to some extent also the other leading European TV nations—created and used divergent, incompatible standards to promote their national interests as well as to protect their domestic electronics industries. On the other hand, Japan adopted US industrial practices and US broadcasting standards for its domestic use—which soon gave it a competitive advantage in taking over the US TV receiver markets. By the mid-1980s, the electronics industry in the US had already lost the game, while in Europe there was much more left to protect, partly thanks to earlier protective measures.

In the process of European integration toward a common market, the traditional French approach of protecting the home markets and industries with a divergent TV standard became also the European approach. France took the lead in opposing the global standard proposal and others followed. A new joint organization was created for international research and development cooperation between European countries and, for the first time in history, a common European technological standard for analog HDTV was created. Unfortunately this great political success turned out to be a big cultural, economic, and technological failure after the US breakthrough with digital TV in 1990.

Although the European project for DVB development was in many ways different from earlier efforts in Europe, there was still no intention of digital broadcast media convergence. DVB was drafted not only to protect the market but also to meet the European market demand and supplement the already-existing DAB radio. This approach has later resulted in a large number of complementary and partly competing new digital systems. Also, the US developers were originally not seeking any convergence between TV and radio, but between digital TV and computer, and that is what they were able to achieve. Only the Japanese have been determinedly seeking to utilize the convergent potential of digital technology in bringing the different forms of broadcasting together—and they also produced a fully functional digital system to make it happen. Convergence or divergence does not just happen by itself; they must be first pursued by people in power.

References

Abramson, A. (2003). *The history of television, 1942–2000*. Jefferson, NC and London: McFarland.

Ala-Fossi, M. (2005). *Saleable compromises. Quality cultures in Finnish and US commercial radio* (Academic dissertation, Department of Journalism and Mass Communication, University of Tampere, Tampere, Tampere University Press).

Ala-Fossi, M. (2010). The technological landscape of radio. In B. O'Neill, M. Ala-Fossi, P. Jauert, S. Lax, L. Nyre, & H. Shaw (Eds.), *Digital radio in Europe. Technologies, industries and cultures* (pp. 43–66). Intellect: Bristol.

Ala-Fossi, M. (2012a). Social obsolescence of the TV fee and the financial crisis of Finnish Public Service Media. *Journal of Media Business Studies, 9*(1), 33–54.

Ala-Fossi, M. (2012b). For better pictures on radio: How Nokia's efforts for multimedia radio have shaped the radio landscape in Finland. In J. A. Hendricks (Ed.), *The Palgrave handbook of global radio* (pp. 109–128). Basingstoke & New York: Palgrave Macmillan.

Ala-Fossi, M., & Stavitsky, A. G. (2003). Understanding IBOC: Digital technology for analog economics. *Journal of Radio Studies, 10*(1), 63–79.

Alvarez, S., Chen, J., Lecumberri, D., & Yang, C. (1999). *HDTV: The engineering history. Project paper for course. The structure of engineering revolutions*. Massachusetts Institute of Technology. Accessed February 23, 2013, from http://web.mit.edu/6.933/www/HDTV.pdf

Aufderheide, P. (1999). *Communications policy and the public interest: The Telecommunications Act of 1996*. New York: Guilford Press.

Baumgartner, B. (2005). *The implementation of HDTV in the European Digital TV environment*. Paper presented at NAB 2005. Accessed July 14, 2005, from http://www.broadcast.harris.com/nab2005/technical_papers/pdf/Baumgartner.pdf

Berkman, D. (2000). The battle over TV resolution, history repeats. *Television Quarterly, 30*(3), 49–56.

Brinkley, J. (1998). *Defining vision: The battle for the future of television*. San Diego: Harcourt Brace.

Chandler, A. D. (2005). *Inventing the electronic century. The epic story of the consumer electronics and computer industries. Harvard studies in business history* (Vol. 47). Cambridge, MA: Harvard University Press

Cianci, P. J. (2012). *High definition television. The creation, development and implementation of HDTV technology*. Jefferson, NC: McFarland & Co.

Cringely, R. X. (2000, May 25). *Stranger in a strange land: How Homer Sarasohn brought industrial quality to Japan and why it took Japan so long to learn*. The Pulpit. Accessed February 23, 2013, from http://www.pbs.org/cringely/pulpit/2000/pulpit_20000525_000408.html

Curwen, P. (1994). High-definition television. A case study of industrial policy versus the market. *European Business Review, 94*(1), 17–23.

Dai, X., Cawson, A., & Holmes, P. (1996). The rise and fall of high definition television: The impact of European technology policy. *Journal of Common Market Studies, 34*(2), 149–166.

De Bruin, R., & Smits, J. (1999). *Digital video broadcasting: Technology, standards, and regulations*. London and Norwood: Artech House.

EBU. (2012, May 2012). *Frequency and network planning aspects of DVB-T2* (EBU TECH 3348). Geneva: EBU. Accessed February 23, 2013, from http://tech.ebu.ch/docs/tech/tech3348.pdf

EC. (1994). *Europe and the global information society. Recommendations to the European Council*. Report from the High Level Group on the Information Society (Bangemann Report).

EC. (1997). *Green paper on the convergence of the telecommunications, media and information*. EC

EC. (2007, July 18). *Commission Opens Europe's Single Market for Mobile TV Services*. A press release, European Commission. Accessed February 23, 2013, from http://europa.eu/rapid/pressReleasesAction.do?reference = IP/07/1118&format = HTML&

EU. (2008, March 17). *Mobile TV across Europe: Commission endorses addition of DVB-H to EU List of Official Standards*. A press release. Accessed February 23, 2013, from http://europa.eu/rapid/pressReleasesAction.do?reference = IP/08/451

Eureka. (1986). *Eureka 147 Project Form*. Accessed October 10, 2008, from http://www.eureka.be/inaction/AcShowProjectOutline.do

Evain, J. (1995). Ten years of EBU participation in European collaborative projects. *EBU Technical Review, Summer, 1995*, 32–42.

Fagan, M. (1994, February 24). Japan will continue to back high-definition TV plans. *The Independent*. Accessed October 10, 2008, from http://www.independent.co.uk/news/business/japan-will-continue-to-back-highdefinition-tv-plans-1396151.html

Fickers, A. (2006). National barriers for an imag(e)ined European Community: The techno-political frames of postwar television development in Europe. In L. Hojbjerg & H. Sondergaard (Eds.), *European film and media culture, northern lights, film and media studies yearbook 2005* (pp. 15–36). Copenhagen: Museum Tusculanum Press.

Fickers, A. (2010). The techno-politics of colour: Britain and the European struggle for a colour television standard. *Journal of British Cinema and Television, 7*(1), 95–114.

Fickers, A., & O'Dwyer, A. (2012). Reading between the lines: A transnational history of the Franco-British 'Entente Cordiale' in post-war television. *VIEW Journal of European Television History and Culture, North America, 1*(2). Accessed February 23, 2013, from http://rdbg.tuxic.nl/euscreen-ojs/index.php/view/article/view/jethc019/41

Fridenson, P. (2012). *Japanese-French business encounters: Two opposite relations*. Paper presented at EBHA conference in Paris 2012: Business enterprises and the tensions between local and global. Accessed February 23, 2013, from http://ebha-bhsj-paris.sciencesconf.org/6450/document

Frost & Sullivan. (2012, August 30). Measuring China's dominance in Electronics Manufacturing Market. *Market Insight*. Accessed February 2013, from, http://www.frost.com/sublib/display-market-insight-top.do?id=265692709

Gaillard, I. (2007). *The CCIR, the Standards and the TV Sets' Market in France (1945–1980)*. Paper presented at the EBHA 11th annual conference, Geneva, 13–15 September 2007. Accessed 23 February, 2013, from http://www.ebha.org/ebha2007/pdf/Gaillard.pdf

Galperin, H. (2004). *New Television, Old Politics. The Transition to Digital TV in the United States and Britain*. Cambridge: Cambridge University Press.

Gandy, C. (2003). *DAB: An introduction to the Eureka DAB system and a guide to how it works* (BBC R&D White Paper, WHP 061).

Gilchrist, N. H. C. (1978). *Digital sound signals: Tests to compare the performance of five companding systems for high-quality sound signals* (BBC Research Department Report 1978/26). Accessed February 23, 2013, from http://downloads.bbc.co.uk/rd/pubs/reports/1978-26.pdf

Gronow, P., Lannegren, G., & Maren, L. (1992, September 22). *New technical developments in the sound broadcasting sector and their impact on mass media policy*. CDMM, Strasbourg: Council of Europe.

Hart, J. A. (1994). The politics of HDTV in the United States. *Policy Studies Journal, 22*(2), 213–228.

Hart, J. A. (2004). *Technology, television, and competition: The politics of digital TV*. Cambridge: Cambridge University Press.

Harvey, D. (2005). *A brief history of neoliberalism*. Oxford: Oxford University Press.

Henten, A., & Tadayoni, R. (2008). The impact of Internet on media technology, platforms and innovation. In L. Küng, R. G. Picard, & R. Towse (Eds.), *The internet and the mass media* (pp. 45–64). London: Sage.

Hesmondhalgh, D. (2007). *The cultural industries* (2nd ed.). London, LA and New Delhi: Sage.

Hoeg, W., & Lauterbach, T. (Eds.). (2003). *Digital audio broadcasting. Principles and applications of digital radio* (2nd ed.). Chichester: Wiley.

Högmander, T. (2003, November 1). *Monitievastaanotin sietää vauhtia*. [Multipath receiver can stand the speed] (pp. 24–25). Prosessori.

Holton, R. (1986). Industrial politics in France: Nationalisation under Mitterrand. *West European Politics, 9*(1), 67–80.

Kato, H. (1998). [Television in] Japan. In A. Smith (Ed.), *Television. An international history* (2nd ed., pp. 169–181). Oxford: Oxford University Press.

Kemppainen, P. (2008). *RTT: ensimmäisenä monessa. Radio-ja televisiotutkimuksen nousu Suomessa 1988–2008 [RTT: first in many issues. The rise of radio and television (technology) research in Finland 1988–2008]*. Helsinki: RTT and Viva Media Ay.

Laven, P. (1998, Winter). DAB—Is it already out of date? *EBU Technical Review*. Accessed February 23, 2013, from https://tech.ebu.ch/docs/techreview/trev_278-laven.pdf

Lax, S. (2010). 'A vision for radio': Engineering solutions for changing audiences—From FM to DAB. In B. O'Neill, M. Ala-Fossi, P. Jauert, S. Lax, L. Nyre, & H. Shaw (Eds.), *Digital radio in Europe. Technologies, industries and cultures* (pp. 67–84). Intellect: Bristol.

Lembke, J. (2002). *Competition for technological leadership. EU policy for high technology*. Cheltenham, UK and Northampton, MA: Edward Elgar.

Mannila, M. (2009, April 21). Nokia myy tv-teknologiansa Wiprolle [Nokia sells tv-technology unit to Wipro]. *Digitoday*. Accessed February 23, 2013, from http://www.digitoday.fi/bisnes/2009/04/21/nokia-myy-tv-teknologiansa-wiprolle/200910100/66

Michalis, M. (2007). *Governing European communications: From unification to coordination*. Lanham: Lexington Books.

Miyazawa, S. (2004). *The trend of digital broadcasting in Japan*. Accessed July 15, 2005, from http://www.dibeg.org/PressR/seminar_in_indonesia2004/presentation-1.pdf

Mosco, V. (1996). *The political economy of communication: Rethinking and renewal*. London: Sage Publications.

Motivate. (1998). *AC318 motivate. Overall description and project website*. Accessed February 23, 2013, from http://cordis.europa.eu/infowin/acts/rus/projects/ac318.htm

Mouyal, N. (2012, April). Developing a single global DTT standard. *DigiTAG Web Letter*. Accessed February 23, 2013, from http://www.digitag.org/WebLetters/2012/External-Apr2012.html

Nakahara, S. (2003). *Overview. ISDB-T for sound broadcasting. Terrestrial digital radio in Japan*. Accessed July 15, 2005, from http://www.rthk.org.hk/about/digitalbroadcasting/DSBS/ABU_AIR_2003/ses2.pdf

Näränen, P. (2006). *Digitaalinen televisio. Analyysejä alkuhistoriasta, viestintäpoliittisista haasteista ja tv-järjestelmän muuttumisesta [Digital television. Analyses on early history, challenges to media policy and transformation of television]* (Academic dissertation, Tampere University Press, Tampere).
Negroponte, N. (1996)[1995]. *Being digital*. New York: Vintage Books.
Neil, S. C. (2010). *The development of high definition television: An ecology of games*. Thesis (PhD D). Massachusetts Institute of Technology, Dept of Political Science. Accessed July 6, 2011, from http://dspace.mit.edu/handle/1721.1/62476
Neuman, W. R. (1988). *Testimony before the Subcommittee on science, research and technology, committee on science, space and technology, US House of Representatives*. Hearings on New Directions in Video Technology. Accessed July 6, 2011, from http://www.wrneuman.com/nav_pub_79_2956783307.pdf
Nevalainen, R. (1999). *Suomi Tietoyhteiskunnaksi—Eespäin tiedon poluilla ja val tateillä. Tietoyhteiskuntatoiminnan lyhyt historia [Making Finland an Information Society—Forwards on the Information pathways and highways]*. Helsinki: SITRA.
NHK. (1977). *50 Years of Japanese broadcasting*. Tokyo: Nippon Hoso Kyokai, Radio & TV Culture Research Institute.
O'Neill, B. (2009). DAB Eureka-147: A European vision for digital radio. *New Media and Society, 11*(1&2), 261–278.
O'Neill, B. (2010). 'Sounding the future': Digital radio and CD-quality audio. In B. O'Neill, M. Ala-Fossi, P. Jauert, S. Lax, L. Nyre, & H. Shaw (Eds.), *Digital radio in Europe. Technologies, industries and cultures* (pp. 85–98). Intellect: Bristol.
Owen, E. L. (1997, November/December). The origins of 60-Hz as a power frequency. *IIEEE Industry Applications Magazine,* 8–12. Accessed February 23, 2013, from http://www.sos.siena.edu/~aweatherwax/electronics/60-Hz.pdf
Pearce, J. Q. (2008, July 31). Operators Kill DVB-H in Germany. *The Washington Post*. Accessed July 6, 2011, from http://www.washingtonpost.com/wp-dyn/content/article/2008/07/31/AR2008073101709.html
Reimers, U. (2004). *DVB—The family of international standards for digital video broadcasting*. New York: Springer.
Rissanen, K. (1993, May 1993). *Digital audio broadcasting (DAB) TUIKE*
Salokangas, R. (1996). The Finnish broadcasting company and the changing Finnish society, 1949–66. In R. Endén (Ed.), *A history of broadcasting in Finland*. Yleisradio: Helsinki.
Sandholtz, W. (1992). *High-Tech Europe: The politics of international cooperation*. Berkeley: University of California Press.
Schulze, H., & Lüders, C. (2005). *Theory and applications of OFDM and CDMA*. Chichester: Wiley.
Talmola, P. (2005). DVB-H standard overview. A powerpoint presentation. Technology sectors, and the implications for regulation. *COM* (97) 623. Accessed July 13, 2005, from http://www.abo.fi/~jbjorkqv/digitv/DVB_H_Standard_2.pdf
Torikka, M. (2007, May 24). Insinöörityöpalkinto meni mobiilitelevisiolle [Engineering Award was given to Mobile TV]. *Tekniikka & Talous*. Accessed February 23, 2013, from http://www.tekniikkatalous.fi/ict/insinoorityopalkinto+meni+mobiilitelevisiolle/a38703
USC. (1990). *The big picture: HDTV and High-Resolution Systems, OTA-BP-CIT-64*. Washington, DC: US Congress, Office of Technology Assessment, US Government Printing Office. Accessed February 23, 2013, from http://digital.library.unt.edu/ark:/67531/metadc39964/m2/1/high_res_d/9007.pdf
Vihma, P. (2007, May 24). *Kännykkätelkkari sai alkunsa junassa. [Mobile TV was conceived on a train] Talouselämä. [in Finnish]*. Accessed February 23, 2013, from http://www.talouselama.fi/uutiset/article169256.ece
Wells, N. (2011). *DVB-T2 in relation to the DVB-x2 Family of Standards*. Presentation at ATSC Symposium on Next Generation Broadcast Television, February 15, 2011, Washington DC,

USA. Accessed February 23 2013, from http://www.atsc.org/cms/pdf/pt2/04-Wells_ATSC_T2_v2.pdf

Winslow, G. (2012, April 23). Mobilizing for Mobile DTV. Broadcasters, wireless carriers eye new technologies that will break limits. *Broadcasting & Cable*. Accessed February 23, 2013, from http://www.broadcastingcable.com/article/483472-Mobilizing_for_Mobile_DTV.php

Wood, D. (1995, Winter). Satellites, science and success. The DVB story. *EBU Technical Review*. Accessed February 23, 2013, from http://www.ebu.ch/trev_266-wood.pdf

Wood, D. (2001, March). *Is DVB-T A Substitute for DAB? Memorandum of the Technical Department of EBU*. Accessed July 20, 2005, from http://forum.europa.eu.int/irc/DownLoad/kjeYAKJSmtGuG3jEtqpqaZLkGuM0hH6vV_4DRRUS7cuqGQKvSlot-Bq2VFhqJRRFZhMN/ONP-DBEG%2001-16 %20.PDF

Yamada, O. (2006). The reminiscence of the development of terrestrial digital broadcasting. *Journal of the Institute of Image Information and Television Engineers, 60*(9):1373–1378. Accessed February 23, 2013, from http://www.jstage.jst.go.jp/article/itej/60/9/1373/_pdf/ English version: http://www.mtc.gob.pe/portal/tdt/Documentos/reminiscence_development_digital_broadcasting.pdf

Yokohata, K. (2007). *ISDB-T transmission technology*. Paper presented at Summit on Information and Communication Technologies, September 27–28, 2007, Tashkent, Uzbekistan. Accessed October 17, 2008, from http://ru.ictp.uz/summit2007/yokohata.pdf

Yoshida, J. (2008, September 9). Dying is easy, mobile TV is hard. *EE Times*. Accessed September 13, 2010, from http://www.eetimes.com/design/microwave-rf-design/4005807/Dying-is-easy-mobile-TV-is-hard

Yoshino, T. (2000). *Looking back at the Development of ISDB. A web page*. Accessed October 13, 2008, from http://www.nhk.or.jp/strl/publica/bt/en/tn0005-2.html

Part IV

Social Media Aspects of Convergence

Recommend, Tweet, Share: User-Distributed Content (UDC) and the Convergence of News Media and Social Networks

Mikko Villi, Janne Matikainen, and Irina Khaldarova

1 Introduction

The media industry is challenged by a complex environment. This is mainly due to a multitude of technological developments (digitalization, internetization, the emergence of new communication devices) and the growing significance of the participatory audience assisted by various web services and platforms, mostly referred to as the social media. In social media, the mass media and the horizontal communication networks of the users[1] are converging in increasingly multifarious ways. They interact on the individual, group, organisational and societal level (van Dijk, 2006). Therefore, it is not surprising that most of the so-called legacy media companies are keen on adapting their offerings to the networked communication environment and playing an active role as part of the social networks. A key approach used by news media to incorporate social media features into their online practices is enabling the audience to share and distribute news content.

Mass media consumption does not remain on the individual level alone. It flows from personal media use toward more collective experiences and social sharing (Heikkilä, Ahva, Siljamäki, & Valtonen, 2011). Accordingly, Jensen (2010, p. 14) proposes a shift of focus from media to communication, an agenda emphasising the recombination and reconfiguration of one-to-one, one-to-many and many-to-many communication. This poses challenges to media studies, especially that in the social media environment, mass communication is increasingly subject to interpersonal communication; that is, the mass media rely increasingly on the interpersonal

[1] 'Users' are generally referred to as Internet actors; in the Internet environment, members of the audience and media consumers are increasingly also users.

M. Villi (✉) • J. Matikainen • I. Khaldarova
Department of Social Research, Media and Communication Studies, University of Helsinki, Helsinki, Finland
e-mail: mikko.villi@helsinki.fi; janne.matikainen@helsinki.fi; irina.khaldarova@helsinki.fi

A. Lugmayr, C. Dal Zotto (eds.), *Media Convergence Handbook - Vol. 1*, Media Business and Innovation, DOI 10.1007/978-3-642-54484-2_16

networks of audience communities in distributing their content. As Livingstone (2004, p. 75) notes, 'Mediated communication is no longer simply or even mainly mass communication (from one to many), but rather, the media now facilitate communication among peers (both one to one and many to many).'

When studying the role of the audience in the online practices of media companies, emphasis has often been placed on user-generated content (UGC) (Napoli, 2010; Singer et al., 2011; Thurman, 2008; Wardle & Williams, 2010). To offer an innovative take on mass communication and the participatory audience, the paper examines the significance of *communication in audience communities* for news media, instead of focusing on the role of the audience in *producing content* with or for the media. This emphasis enables the mapping of an emerging conceptual terrain by highlighting the growing importance of user-distributed content (UDC). The concept of UDC (Napoli, 2009; Villi, 2012) refers to the process by which the mass media converge with the online social networks through the intentional use of social media features and platforms in an effort to expand the number of content delivery channels and maximize visibility online. Practices related to UDC (although not using the concept of UDC) have been discussed in works by Glynn, Huge, and Hoffman (2012), Lee and Ma (2012), Bechmann (2012), Hermida, Fletcher, Korell, and Logan (2012), Himelboim and McCreery (2012), and Weeks and Holbert (2012).

By utilizing UDC as a conceptual framework, this paper examines how news media can tap into the communicative dimensions of participatory audience communities. Many legacy media have already launched online sites in the latter half of the 1990s, but for a long time, their online practices were rather a matter of performing one-way mass communication and did not engage in participatory processes with the audience. It is only during the last several years that the development of social media tools and platforms has cultivated new forms of relationships between the producers and consumers of professional media content. The legacy media have integrated social media tools, such as social plugins[2] like the *Facebook Recommend* and *Twitter* buttons, into their sites, making their content more readily available for horizontal distribution among the audience. In this study, such integration of social media features into the news media's own websites is defined as the *internal* use of social media. This includes the possibilities offered on the websites for the audience to comment on articles, provide photographs and, in the context of this paper, the possibility to share content using social plugins. The news media also take advantage of the social media platforms outside of their own domains. This is done by using social media networks, such as Facebook pages or Twitter accounts. In this paper, this is referred to as the *external* use of social media.

This paper contributes to the study of the media industry by empirically examining UDC. An important goal in the study is to develop a metric for the collection

[2] The social plugins (Kontaxis, Keromytis, & Markatos, 2012) have also been referred to as 'social bookmarking tools' (Messner, Linke, & Eford, 2011) or 'social buttons' (Gerlitz & Helmond, 2011).

of data on the internal and external use of social media in the rapidly changing online environment. The study draws on the examination of 15 news media, which are based in seven different countries—Finland, Sweden, Germany, Russia, UK, US and Qatar. By analysing the sizeable amount of data, the study sheds light on how these news media have implemented social plugins on their own websites, how the audience uses the social plugins and how the media build their presence on the social media platforms. In relation to the external use of social media, the study provides information on how the selected news media use Facebook pages and Twitter accounts and how active audience participation is in this context.

Based on the analysis of the empirical data, three main findings are presented. First, the audience shares news content actively by using social plugins. Second, by being active in social media (especially on Facebook and Twitter), the news media can promote the distribution of their content by the audience. Third, news media are more active on Twitter than on Facebook despite the fact that the audience is more active on Facebook. The main conclusion in the paper is that the audience has an important role in the distribution of news content and that this role can be substantially supported by the news media themselves.

2 Literature Review

In the paper, the plural *audience communities* is used when stressing that instead of a mass audience, the social consumption of media content takes place in smaller, networked audience communities that maintain contact through several channels and communication tools (Marshall, 2004, p. 103). One's Facebook friends are an example of such a community (for a more detailed discussion of audience community see Malmelin & Villi 2015a, 2015b; Villi & Jung 2015). A close point of reference to audience communities is the concept of 'networked publics', especially when it is used to refer to a networked collection of peers (boyd 2011). In social media, the dynamics of publication and distribution of news are being reshaped by networked publics (Hermida et al., 2012).

Social media have become the top outlet for people to access news stories shared by other Internet users (Ma, Lee, & Goh, 2012). In a world of links and feeds, it is often easier to find the next thing to read, watch or listen to from friends than it is to stick with any given publication (Anderson, Bell, & Shirky, 2013, p. 8). Online media and the particular applications and services designed for sharing media content enable and encourage the users to share their media consumption experiences. As a result, the importance of audience communities in distributing and marketing professional media content is growing (Newman & Dutton, 2011).

An important context for UDC is 'socialized media' (Jenkins & Deuze, 2008, p. 5). Socialized media represents the many shifts in contemporary (online) communication, such as the participatory culture, connected communities of interest and fluidity across platforms (boyd 2008). This may equally refer to both the development of the social media as such and, in the context of our paper, to the process when news media gradually incorporate the features and tools of social media into their practices and include the audience as a participatory partner in content distribution. According to Castells (2006), socialized communication beyond the mass media system is one of the main characteristics of the network society. The networking technologies allow the emergence of so-called 'mass self-communication', which is both mass communication and self-communication at the same time (Castells, 2009).

The participatory audience is a substantial phenomenon in the context of the contemporary media industry. The audience cannot be regarded as a group of passive consumers, but as active participants in the processes through which value is created in the market. The transformations in the relationship between media producers and consumers suggest a shift in which the frameworks of analysis and categories that worked well in the context of an industrial media economy are no longer helpful (Banks & Humphreys, 2008) as the consumption of media content is not restricted anymore to the reception of ready-made mass products (Chan-Olmstead, 2006). Importantly, this change is not only about the production of media content (UGC), but also the participatory processes that rely on the communication structures in audience communities (UDC).

In connection to this shift, Jenkins (2006, p. 20) has introduced the concept of affective economies, which refers to how 'the ideal consumer is active, emotionally engaged, and socially networked'. To motivate such an ideal consumer, Jenkins and Deuze (2008) encourage media companies to accelerate the flow of media content across different delivery channels for them to be able to expand revenue opportunities, broaden markets and reinforce consumer loyalties and commitments. Here, the role of the audience is to participate in the processes of the media industry by distributing media content and enclosing it with social relations (Villi, 2012, p. 619).

In this paper, we take a step forward by empirically studying both the external and internal use of social media, particularly in the context of Facebook and Twitter. Previous studies on Facebook (Olmstead et al., 2011, p. 2; Gerlitz & Helmond, 2011; Newman, 2012, p. 15; Ma et al., 2012; Villi, 2012; Sasseen, Olmstead, & Mitchell, 2013) show how Facebook has emerged as an important channel for distributing media content, almost as a news medium or portal in itself. Research on Twitter (e.g. Engesser & Humprecht, 2012; Kwak, Lee, Park, & Moon, 2010) has been focused the external use of social media (i.e. Twitter accounts). Only a few studies (e.g. Messner et al., 2011) have focused both on the internal and external use of Twitter by media companies. Our contribution to the research on Facebook and Twitter is to provide an extensive amount of data on social plugins, Facebook pages and Twitter accounts gathered continuously during several months, instead of obtaining data only on a few individual dates.

3 Method and Data Collection

In order to study the various forms of UDC that are supported by the news media, three research questions are presented in the paper:

RQ1: How are social plugins used on news media websites (internal use of social media)?

RQ2: How active is audience participation in the social media platforms used by the news media (external use of social media)?

RQ3: Does the activity of the news media in social media, especially on Facebook and Twitter, impact the activity of the audience?

To address the research questions, we utilize data collected from the websites, Facebook pages and Twitter accounts of 15 different news media. The data consist of almost 50,000 news items released during the 4-month period between 14 January 2012 and 14 May 2012. All of the news items were put through statistical observation in order to reveal how and what were shared and how communication between the media and audience was promoted.

The 15 news media from seven different countries were selected based on their circulation and popularity. Concentration was made on major media in each country, as well as on internationally well-known news media. They consist of the following newspapers and broadcasters: Finland: *Helsingin Sanomat* (newspaper); *MTV3* (commercial television network); *YLE* (public service broadcasting company); Germany: *Die Zeit* (newspaper); the U.S.: *The Associated Press, CNN, The New York Times* and *Huffington Post;* Sweden: *Aftonbladet, Dagens Nyheter* and *Expressen* (all newspapers); the UK: *The Guardian*; Russia: *Argumenty i Fakty, Izvestia* (both newspapers); Qatar: *Al Jazeera.*

The study collected empirical data by using a data-harvesting program written in Python. The data-harvesting program was specifically created for the purposes of this study. The data was accumulated in an SQLite database. The data-harvesting program consisted of news and statistics harvesters. The news harvester required going through the list of the RSS feeds for each media and saving the data to the database.

A pilot study was carried out prior to the collection of the data. The purpose of the pilot study was (1) to test the method of data collection and the harvesting program and (2) to identify and eliminate the sources of errors when collecting data. In the pilot study, eight Finnish newspapers were chosen for testing the method and the program. During the test period, the newspapers released 4791 news items on their websites. The number of Facebook recommendations and Twitter shares for every news item was monitored for 10 days after they were released.

The data collection method proved to be suitable for the purposes of the study. However, one source of possible errors was identified. The newsfeeds of most of the media contained news released only within the last 2–5 days. The 10-day period of news monitoring was, therefore, a potential problem, as news items older than 5 days disappeared from the newsfeeds and could be found only by their URL. In order to eliminate the problem and decrease possible errors, it was decided to shorten the monitoring period. In addition, the pilot study revealed that almost all (up to 99 %) of the total Facebook recommendations and Twitter shares happen within 3 days after a news item is released.

After the pilot study, the 4-month long data collection was carried out. The statistics module retrieved and saved the data for each news item that was not older than 3 days (the length of the 'aging' period of news items was determined to be roughly 70 h, based on the statistics collected during the pilot study). The program was executed on a periodic basis as a single process on one computer. The length of the period of data processing varied from 1 to 7 h every day, as the number of news to be processed also fluctuated heavily.

The program checked 73 URLs daily, as several of the 15 selected media had newsfeeds in different categories, such as sport, entertainment and culture. The program scanned the news items released within 2 h and counted the recommendations, tweets and other shares for every news item from the moment it was released until it was 3 days old.

Along with monitoring the URLs, the study also used the online service Socialbakers[3] to collect additional data from the Facebook pages of the selected media. The focus was on the number of those who 'like' the page; the amount of 'people talking about it'[4]; the number of 'likes' for individual news items; the amount of own posts by the media outlets and posts made by users; and the number of comments and shares. The Socialbakers service allowed the data to be exported in the form of PDF files. These files were used as a source for further analysis and interpretation of the news media's external use of social media. The study also explored the Twitter accounts of the selected media, the focus being on the number of tweets and followers.

In general, the research methods were quite simple: the main target was to estimate the extent of both internal and external social media use. The statistics (frequency, average value, correlation) were calculated to compare the different news media. It is important to note that the sample is not representative of the media in each studied country. However, the large size of the sample enables the drawing of conclusions on the application of social media features by news media.

4 Findings

4.1 Internal Use of Social Media

In the first section, we address research question 1 (How are social plugins used on news media websites?). The empirical work began with the examination of the webpages of the selected media. A typical webpage of a news media contained a

[3] Socialbakers (www.socialbakers.com) is a global social media and digital analytics company that aids in the measurement of activities across all major social media services, such as Twitter and Facebook.

[4] 'People Talking About This' indicates the number of unique users who have created a story about a page in a 7-day period. On Facebook, users create stories when they, for example, post on the page wall, comment on a post or mention the page in a post (http://www.insidefacebook.com/2012/01/10/people-talking-about-this-defined/).

Table 1 The most common social plugins on the news media websites (Spring 2012)

	Facebook	Twitter	Google+
Aftonbladet (http://www.aftonbladet.se/)	X	X	
Al Jazeera (http://www.aljazeera.com/)	X	X	X
Argumenty i Fakty (http://www.aif.ru/)	X	X	
Associated Press (hosted.ap.org/dynamic/fronts/HOME?SITE=AP)			
CNN (http://edition.cnn.com/)	X	X	
Dagens Nyheter (http://www.dn.se/)	X	X	
Expressen (http://www.expressen.se/)	X	X	
Helsingin Sanomat (http://www.hs.fi/)	X	X	X
Huffington Post (http://www.huffingtonpost.com/)	X	X	X
Izvestia (http://www.izvestia.ru/)	X	X	X
MTV3 (http://www.mtv3.fi/)	X	X	
The NY Times (http://www.nytimes.com/)	X	X	X
The Guardian (http://www.guardian.co.uk/)	X	X	X
YLE (http://yle.fi/)	X	X	X
Die Zeit (http://www.zeit.de/index)	X	X	X
Total number of media outlets: 15	14	14	8

number of elements connected to the social media platforms, most notably the social plugins (e.g. the Facebook Recommend button). The social plugins enable the audience to spread the news further in their social networks. The layouts of the websites varied between the selected media outlets, but most of the websites embraced almost a similar set of social media related elements. They included:

- The set of social plugins.
- A counter indicating the number of recommendations and other shares. Information is also provided if somebody from the personal network of a user has recommended or otherwise shared the news item.

Table 1 does not include information on social plugins that were used by less than five of the selected media. For example, several media had attached to news items the plugins for LinkedIn (four media) and Delicious (two media). Eight media used the Google+ plugin. By contrast, all but one of the selected 15 media websites used plugins for both Facebook and Twitter. As such, in this paper, the focus is on these two social media services. According to Mitchell et al. (2012), Facebook and, to a lesser extent, Twitter, dominate the intersection of social media and news. As the one exception, *Associated Press* did not have any social media plugins on its website, but it did have a page on Facebook and a Twitter account. Consequently, *Associated Press* is examined only in the section focusing on the external use of social media.

Table 2 Statistics showing how many news items were published during the study period and how many times they were recommended and shared using the Facebook and Twitter Plugins

Media	News items	Tweets	Tweets/ news item	FB recommendations	FB recommendations/ news item
Aftonbladet	1639	16,867	10.3	190,075	**116.0**
Al Jazeera	502	34,385	**68.5**	72,361	**144.1**
CNN	417	5740	13.8	5231	12.5
Dagens Nyheter	3246	14,485	4.5	55,073	17.0
Expressen	753	2500	3.3	19,688	26.1
Izvestia	1514	11,024	7.3	6767	4.5
The NY Times	1699	14,941	8.8	224,206	**131.0**
The Guardian	912	57,526	**63.1**	248,148	**272.1**
Huffington Post	3949	116,360	**29.5**	1,514,814	**383.6**
Die Zeit	5019	9104	1.8	38,703	7.7
Argumenty i Fakty	2214	8334	3.8	2235	1.0
HS	9522	18,195	1.9	360,289	37.8
MTV3	6455	7599	1.2	44,251	6.9
YLE	11,406	16,929	1.5	253,632	22.2
Total	49,247	315,989	6.4	3,035,743	61.7

The top media in Tweets and Facebook recommendations per news item are marked in bold

During the data-gathering period, the harvesting program collected and analysed in total 49,247 news items released by the 15 news media. Those news items were tweeted 315,989 times on Twitter and recommended[5] 3,035,743 times on Facebook (Table 2). It is important to note that in some cases, the data-harvesting program covered only the top newsfeeds, as it was extremely difficult to analyse all news items released by such media giants as *CNN*, *The Guardian* or *Al Jazeera*. This preferential manner of monitoring the news explains the comparatively small amount of news shown in Table 2 for *CNN* or *Al Jazeera*. However, the amount of data collected is large enough to enable the evaluation of audience behaviour, particularly the ways in which the audience takes advantage of the social media elements on the news websites.

Table 2 shows how the Facebook plugin is more popular than the Twitter plugin in sharing the news online, except only in the case of *CNN* and the Russian newspapers *Argumenty i Fakty* and *Izvestia*. *The Huffington Post* is clearly the

[5] Facebook uses three terms (recommend, like, share) to refer to the sharing and distribution of online content items (Benton, 2011). The basic function behind these terms is the same; therefore, we have combined them under the term 'recommend', which is currently most commonly used in the news media that were studied for this paper.

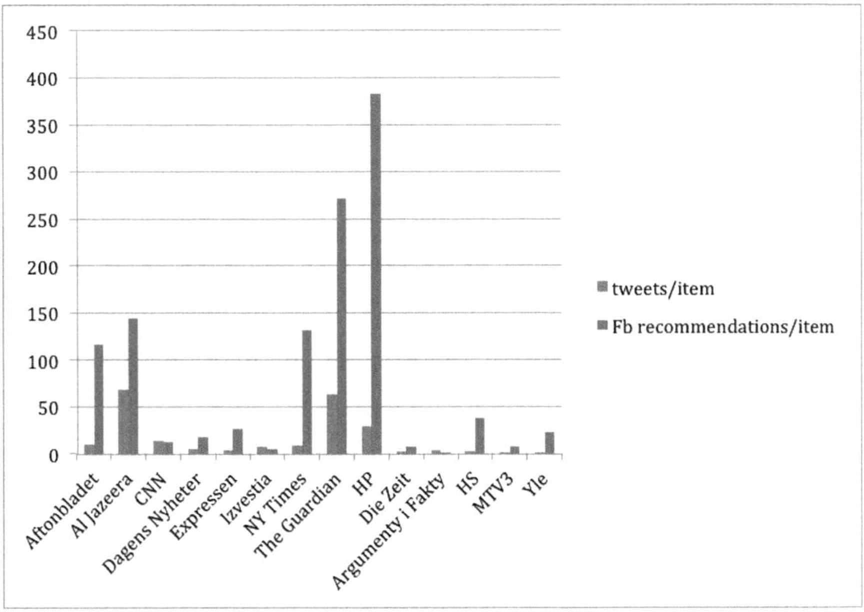

Fig. 1 The ratio of tweets and recommendations per news item

frontrunner in terms of the absolute number of recommendations and tweets (cf. Sasseen et al., 2013).[6] More importantly, when comparing the ratio between published and shared news items, *The Huffington Post* is still number one in Facebook recommendations and third behind *Al Jazeera* and *The Guardian* in tweets. These are in the top-three in both recommendations and tweets per news item. In Facebook recommendations, the *New York Times* and the Swedish newspaper *Aftonbladet* fare also well.

By studying the average ratio between the number of published and shared news items (Fig. 1), it is fair to state that Twitter is a less popular way to share news in audience communities than Facebook. In addition, the average ratio suggests that the amount of news items does not explain the number of tweets or Facebook recommendations. There is no correlation between these variables. The case of *Aftonbladet* especially indicates that the extent of the potential audience is not always the reason for a substantial ratio in Facebook recommendations. *Aftonbladet*—for a newspaper publishing in Swedish with a limited worldwide audience—is surprisingly successful in obtaining recommendations for its content. In order to understand the reasons for this better, we would need more quantitative

[6] In our study, the actual number of tweets and recommends is biased as the harvesting program collected a different share of news items of those published by each media. For *Huffington Post*, for example, there were almost 4000 news items collected during the reported time period while for *CNN* the program collected only slightly over 400 news items.

and especially qualitative data, such as interviews in media companies focusing on the practices regarding the use of social plugins.

4.2 External Use of Social Media

By the external use of social media, we refer to the social media activities run by the news media outside of their own domains. To address research question 2 on how active audience participation is on social media platforms used by the news media, we focus especially on the Facebook pages and Twitter accounts of the selected media. The Facebook pages and Twitter accounts were monitored for 1 month, between 1 March and 1 April 2012. The main focus was on the number of 'likers' (those people who have liked the page in Facebook) and 'people talking about', the amount of likes (individual news items liked), the number of posts made by the media and by the audience and the number of comments and shares. All the examined media had settled on Facebook and Twitter.

The Facebook pages of the selected media differed especially in terms of permitted audience engagement. Some of the news media (e.g. *CNN*) allowed users to post on their Facebook wall, whereas other media did not offer this possibility (*Huffington Post, The NY Times, Die Zeit, Al Jazeera*). Some of the news media updated their Facebook pages frequently, at least two to three times a day (e.g. *Helsingin Sanomat, Huffington Post*); others posted links to their news only a few times a month (*MTV3*). However, all news media had obtained likers on their Facebook pages, though at different rates (Table 3).

The *participation rate* given in Table 3 is calculated using the following formula: total number of interactions (comments, likes, shares) divided by the number of likers (their number counted at the end of the reported time period). The participation rate illustrates how much interaction there was on the Facebook pages on average for each liker.

The news media differ in terms of scale; therefore, it is not reasonable to draw a comparison between them based on the amount of likers, likes, comments and shares. For example, according to the data, the most successful media in terms of likers on its Facebook page is *CNN* with 3,481,636 likers. The smallest is *YLE* with just 6838 likers. Hence, it is appropriate only to make a comparison in the number of own posts (as a demonstration of the media outlet's willingness to communicate with the audience and maintain activity on its Facebook page) and the participation rate (as a demonstration of the success of communication between the media and its audience). Figures 2 and 3 demonstrate the differences in the number of own posts and the participation rate among the studied media.

The comparison of Figs. 2 and 3 shows that those media, which post most frequently on their Facebook pages, generally (with notable exceptions such as *Al Jazeera*) succeed in bringing about participation from the audience. Correlation points to the same direction as there is quite a strong correlation ($r = 0.58$ $p < 0.05$) between the amount of posts and the participation rate. According to our study, *Al Jazeera* and *Huffington Post* are media that make Facebook updates many times a

Table 3 News media activities on Facebook pages within the reported time period

Media	Likers	People talking about	Own posts	User posts	Likes	Comments	Shares	Total interactions	Participation rate
Helsingin Sanomat	23,358	1295	134	50	4332	1523	1120	6975	0.30
MTV3	41,811	438	1	29	0	1	0	1	0
YLE	6838	200	65	45	673	126	18	817	0.12
Zeit	89,627	3692	178	0	15,294	3222	6632	25,148	0.28
Huffington Post	562,999	71,298	941	0	57,3041	243,988	240,728	1,057,757	1.88
NY Times	2,133,141	43,019	226	0	166,870	25,769	58,696	251,335	0.12
CNN	3,481,636	49,806	189	10,611	145,965	82,526	32,627	261,118	0.08
Associated Press	75,351	3420	163	0	7923	5082	5000	18,005	0.24
Expressen	8699	922	153	0	2361	1238	90	3689	0.42
Dagens Nyheter	9577	329	60	0	1008	224	89	1321	0.14
Aftonbladet	42,627	2521	82	161	5756	3550	917	10,223	0.24
The Guardian	353,889	17,505	244	0	36,543	12,652	12,763	61,958	0.18
Argumenty i Fakty	10,036	476	265	49	2349	479	110	2938	0.29
Izvestia	10,350	481	353	28	2027	708	26	2761	0.27
Al Jazeera	995,003	21,570	1048	0	76,786	44,027	1016	121,829	0.12

The number of FB likers was measured on the last day of the monitoring period

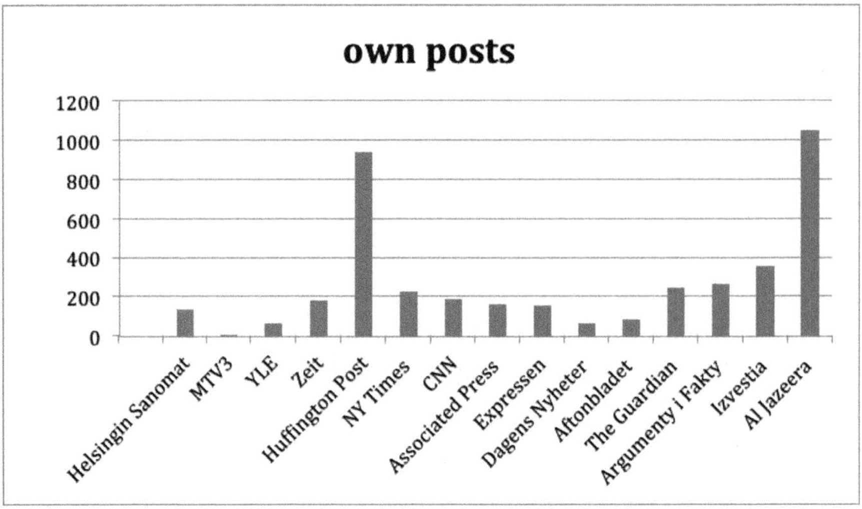

Fig. 2 Number of posts by the media on their Facebook pages

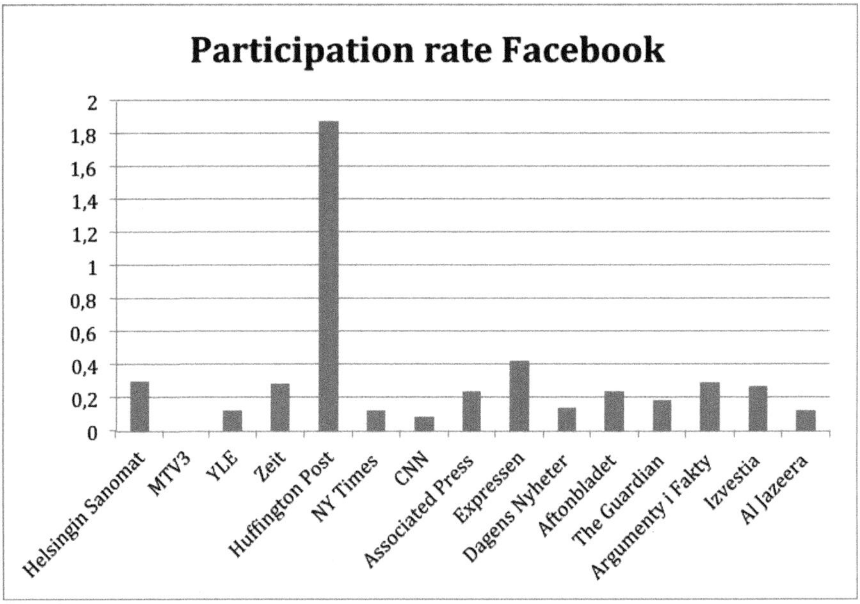

Fig. 3 Average participation rate on Facebook pages

day. Accordingly, *Huffington Post* has the best participation rate. In contrast, the rate of *Al Jazeera* is rather low. Investigating further the differences in Facebook use by *The Huffington Post* and *Al Jazeera* would be a worthy subject of supplementary study.

Table 4 Tweets and Followers on the Twitter Accounts of the Selected News Media

News media	Tweets (March 2012)	Tweets (all time)	Followers
Helsingin Sanomat	191	12,022	10,001
MTV3	2044	40,362	2177
YLE	986	60,991	3794
Zeit	293	42,768	139,899
Huffington Post	2778	90,399	1,728,014
NY Times	1559	80,995	4,778,956
CNN	629	44,780	289,286
Associated Press	1644	20,094	884,953
Expressen	586	7329	8757
Dagens Nyheter	2920	18,789	622
Aftonbladet	2370	29,490	4427
The Guardian	971	13,238	358,407
Argumenty i Fakty	1440	10,529	2562
Izvestia	1498	9031	5919
Al Jazeera	2990	57,158	926,896

The number of the Twitter followers indicates the amount of followers at the end of the monitoring period

The Twitter accounts of the selected media were monitored to find out the number of tweets and followers during the reported time period (Table 4). The main goal was to observe how seriously the news media take Twitter as a resource to reach the audience. All of the media have a Twitter account and also use their accounts. There is a strong correlation between the number of tweets and followers ($r = 0.66$, $p < 0.01$). This indicates that the more the news media tweet, the more followers they have.

An interesting observation is that all the selected media tweet significantly more often than post news on Facebook (Fig. 4). Nonetheless, most of them have remarkably fewer followers on Twitter than likers on Facebook, particularly those based outside of the United States. One explanation for this is that the popularity of Twitter among Internet users is behind that of Facebook in many countries. Or then, users consider it to be more adequate or more pleasing to 'like' a news media on Facebook than to begin following it on Twitter (see Gerlitz & Helmond, 2011; Olmstead et al., 2011, p. 2). Although other studies (Hermida et al., 2012; Ma et al., 2012; Mitchell et al., 2012) have also compared audience participation on both Facebook and Twitter or have focused on studying Twitter users (e.g. Humphreys, Gill, Krishnamurthy, & Newbury, 2012; Marwick & boyd, 2010), the question surrounding the reasons behind the popularity of liking versus following remains open. The response to research question 3 (Does the activity of the news media in social media impact the activity of the audience?) is that the activity of the news media does have a positive impact, but that the impact seems to differ among social media platforms.

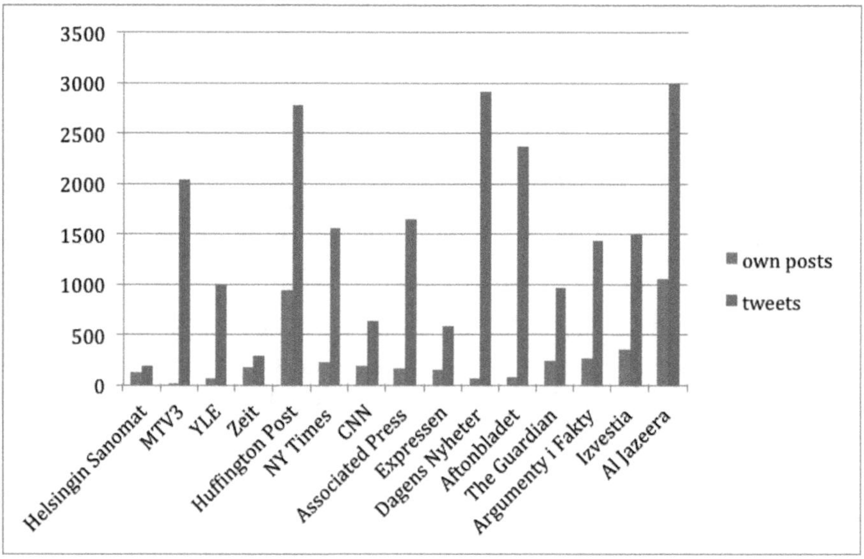

Fig. 4 Comparison of social media accounts: number of posts on Facebook pages vs. tweets on Twitter

5 Conclusions

The objective of this paper is to explore the potential of social media platforms in *converging the networks of communication of the news media and the audience* and promoting the distribution of media content in audience communities. Here, we have utilized user-distributed content (UDC) as a conceptual framework. The concept of UDC refers to the process by which the *mass media converge with the online social networks* through the intentional use of social media features and platforms to expand the number of content delivery channels and maximize visibility online.

Based on the results of the study made on the internal and external use of social media by the selected 15 news media outlets, the conclusion is that, although all of these media have established presence on the various social media platforms and have used various social media tools, some of them are still unsuccessful in actively inducing communication among audience communities. In 2012, *Associated Press* had not placed any social plugins on its website. Others, despite having incorporated social plugins, were not successful in creating bottom-up, audience-driven interest. For example, the news published on the websites of *Die Zeit* and *Argumenty i Fakty* earned, in general, less than five Facebook likes and Twitter shares per item. This indicates that content distribution in audience communities is a difficult and challenging process—it takes more thought, effort and time from the news media. As Bechmann (2012) suggests, the media need also to consider whether or not they have ownership and control of the channels and traffic generated by the content. Both the external and internal use of social media rely mainly on the platforms and services offered by commercial companies, such as Google and Facebook, which can be more or less regarded as competitors of the legacy media companies.

In general, according to the average number of social interactions garnered when using social plugins, *The Huffington Post*, *The Guardian*, *The New York Times*, *Aftonbladet* and *Al Jazeera* lead the pack when it comes to Facebook interactions. The news from *The Guardian* and *Al Jazeera* were tweeted more frequently than the news from the other news media. When looking at the use of the social plugins, the number of Facebook recommendations clearly surpassed the number of tweets.

Apart from the data concerning the number of news items that get tweeted and recommended through the use of social plugins (internal use of social media), our study shows how the news items create social buzz when posts by the media on their Facebook pages and Twitter accounts are liked, commented and shared (external use of social media). The main finding regarding the external use of social media is that the level of presence and activity of the news media on Facebook and Twitter impacts positively the participation rate of the audience. However, the news media are more active on Twitter than on Facebook in their external use of social media, even though the audience is often more active on Facebook (see also Mitchell et al., 2012).

One reason for this discrepancy might be in the different natures of mass communication and interpersonal communication. Mass media, as its name indicates, is grounded in one-way, one-to-many communication. Twitter suits well the conventions and preferences of mass media, with its content flowing to a rather anonymous audience (although Twitter also offers the possibility of two-way communication). Therefore, those working in mass media may eagerly adopt Twitter as their preferred social media communication platform. Facebook, by contrast, is a semi-closed and a more communal web service. Consequently, it might be better suited for the more interpersonal and intimate relations in the audience communities where the content is being distributed. Both Twitter and Facebook, as social media services and platforms, are technological instruments created for the production of sociality and connectivity (van Dijck, 2013). However, it is conceivable that Twitter produces sociality and connectivity that are closer to the conventions of mass media. Facebook, on the other hand, produces a more communal sociality.

A clear implication of our study is this: as the *news media have partly converged with social networks*, they can take advantage of the connections within the audience communities and gain by nurturing them. The degree to which media companies facilitate social sharing might, therefore, become an important factor in the consumption of their content. Professional media content can be regarded as a fuel or social glue for the sustenance of the social relations in social media (Villi, 2012, p. 620). As such, practices related to UDC should be more strongly incorporated into the management strategies of media companies, as peer-communication among the audience has a growing significance in the consumption and distribution of media content. It is not reasonable for media companies to assume that the users find their way to their sites only out of habit or by the help of search engines.

Media scholars have argued that for contemporary media companies, engaging, encouraging and assisting the audience in the circulation and distribution of media content is more important than having them participate in content production (Hermida et al., 2012; Singer et al., 2011). Pre-existing social networks are becoming fundamental to the sustenance of media. With this, media companies need to acknowledge the *convergence of media and communication*, where the social and communicative (interpersonal) dimensions have invaded, informed and mutated the media elements (Marshall, 2009). However, although social media are now an important source of media content for many users, the empirical study of how they engage this content is still in its infancy (Weeks & Holbert, 2012).

In this paper, we have studied the UDC practices of news media by means of quantitative data and analysis. By carrying out qualitative research among the news media, it would be possible to bring forward clarifications for the reasons why they use Twitter more eagerly than Facebook and why other media struggle when others thrive when it comes to attracting shares and recommendations via social plugins. A qualitative study in news media could also enable a more profound examination on how the media benefit from the social plugins as well as from their Facebook pages and Twitter accounts in terms of promoting the distribution of their content in the online social networks.

Acknowledgments This research has received funding from the Helsingin Sanomat Foundation. We would also like to thank Vladimir Grigor for his help in the programming of data collection.

References

Anderson, C. W., Bell, E., & Shirky, C. (2013). *Post-industrial journalism: Adapting to the present*. Columbia Journalism School, Tow Center for Digital Journalism. Accessed April 16, 2013, from http://towcenter.org/wp-content/uploads/2012/11/TOWCenter-Post_Industrial_Journalism.pdf

Banks, J., & Humphreys, S. (2008). The labour of user co-creators: Emergent social network markets? *Convergence: The International Journal of Research into New Media Technologies, 14*(4), 401–418.

Bechmann, A. (2012). Towards cross-platform value creation. *Information, Communication and Society, 15*(6), 888–908.

Benton, J. (2011). *"Like," "share," and "recommend": How the warring verbs of social media will influence the news' future*. Nieman Journalism Lab: Pushing to the future of journalism—A project of the Nieman Foundation at Harvard. Accessed October 18, 2012, from http://www.niemanlab.org/2011/02/like-share-and-recommend-how-the-warring-verbs-of-social-media-will-influence-the-news-future

boyd, d. (2008). Facebook's privacy trainwreck: Exposure, invasion, and social convergence. *Convergence: The International Journal of Research into New Media Technologies, 14*(1), 13–20.

boyd, d. (2011). Social network sites as networked publics: Affordances, dynamics, and implications. In Z. Papacharissi (Ed.), *A networked self: Identity, community, and culture on social network sites* (pp. 39–58). New York: Routledge.
Castells, M. (2006). *The theory of the network society*. Cambridge, MA: Polity Press.
Castells, M. (2009). *Communication power*. Oxford: Oxford University Press.
Chan-Olmstead, S. M. (2006). *Competitive strategy for media firms: Strategic and brand management in changing media markets*. Mahwah, NJ: Lawrence Erlbaum Associates.
Engesser, S., & Humprecht, E. (2012, May 24–28). How professional news media adopt Twitter: Findings of a cross-national comparison. In *2012 International Communication Association Conference*, Phoenix, AZ.
Gerlitz, C., & Helmond, A. (2011, January 24–25). *Hit, link, like and share: Organizing the social and the fabric of the web in a like economy*. In *DMI Mini-Conference*, University of Amsterdam.
Glynn, C. J., Huge, M. E., & Hoffman, L. H. (2012). All the news that's fit to post: A profile of news use on social networking sites. *Computers in Human Behavior, 28*(1), 113–119.
Heikkilä, H., Ahva, L., Siljamäki, J., & Valtonen, S. (2011, May 26–30). *What if everyday life is not that civic?* In *2011 International Communication Association Conference*, Boston, MA.
Hermida, A., Fletcher, F., Korell, D., & Logan, D. (2012). Share, like, recommend: Decoding the social media news consumer. *Journalism Studies, 3*(5–6), 815–824.
Himelboim, I., & McCreery, S. (2012). New technology, old practices: Examining news websites from a professional perspective. *Convergence: The International Journal of Research into New Media Technologies, 18*(4), 427–444.
Humphreys, L. M., Gill, P., Krishnamurthy, B., & Newbury, E. (2012, May 24–28). Historicizing new media: A content analysis of Twitter. In *2012 International Communication Association Conference*, Phoenix, AZ.
Jenkins, H. (2006). *Convergence culture: Where old and new media collide*. New York/London: New York University Press.
Jenkins, H., & Deuze, M. (2008). Convergence culture. *Convergence: The International Journal of Research into New Media Technologies, 14*(1), 5–12.
Jensen, K. B. (2010). *Media convergence: The three degrees of network, mass, and interpersonal communication*. Milton Park: Routledge.
Kontaxis, G., Keromytis, A. D., & Markatos, E. P. (2012, August). Privacy-preserving social plugins. In *Proceedings of the 21st USENIX Security Symposium*, Bellevue, WA.
Kwak, H., Lee, C., Park, H., & Moon, S. (2010, April 26–30). *What is Twitter, a social network or a news media? Categories and subject descriptors*. In *WWW 2010*, Raleigh, NC.
Lee, C. S., & Ma, L. (2012). News sharing in social media: The effect of gratifications and prior experience. *Computers in Human Behavior, 28*(2), 331–339.
Livingstone, S. (2004). The challenge of changing audiences: Or, what is the audience researcher to do in the age of the internet? *European Journal of Communication, 19*(1), 75–86.
Ma, L., Lee, S. C., & Goh, D. H.-L. (2012, May 24–28). Understanding news sharing in social media: An explanation from the diffusion of innovations theory. In *2012 International Communication Association Conference*, Phoenix, AZ.
Malmelin, N., & Villi, M. (2015a). Audience community as a strategic resource in media work. *Journalism Practice* (first published online).
Malmelin, N., & Villi, M. (2015b). Co-creation of what? Modes of audience community collaboration in media work. *Convergence: The International Journal of Research into New Media Technologies* (first published online).
Marshall, D. P. (2004). *New media cultures*. London: Arnold.
Marshall, D. P. (2009). New media as transformed media industry. In J. Holt & A. Perren (Eds.), *Media industries: History, theory, and method* (pp. 81–89). Chichester: Wiley-Blackwell.
Marwick, A. E., & boyd, d. (2010). I tweet honestly, I tweet passionately: Twitter users, context collapse, and the imagined audience. *New Media and Society, 13*(1), 114–133.

Matikainen, J., & Villi, M. (2012, August 30–31). Mediasisältöjen tuottaminen, jakelu ja kulutus sosiaalisessa mediassa. In *Finnish Conference of Communication Research*, University of Jyväskylä.

Messner, M., Linke, M., & Eford, A. (2011, April 1). Shoveling tweets: An analysis of the microblogging engagement of traditional news organizations. In *International Symposium on Online Journalism*, Austin, TX.

Mitchell, A., Rosenstiel, T., & Christian, L. (2012). *The state of the news media 2012: What Facebook and Twitter mean for news*. The Pew Research Center's Project for Excellence in News Journalism, an Annual Report on American Journalism.

Napoli, P. M. (2009). *Navigating producer-consumer convergence: Media policy priorities in the era of user-generated and user-distributed content*. Bronx, NY: The Donald McGannon Communication Research Center, Fordham University.

Napoli, P. M. (2010). Revisiting "mass communication" and the "work" of the audience in the new media environment. *Media, Culture and Society, 32*(3), 505–516.

Newman, N. (2012). *Reuters Institute Digital News Report 2012: Tracking the future of news*. University of Oxford: Reuters Institute for the Study of Journalism.

Newman, N., & Dutton, W. (2011, May 26–30). Social media in the changing ecology of news production and consumption: The case in Britain. In *2011 International Communication Association Conference*, Boston, MA.

Olmstead, K., Mitchell, A., & Rosenstiel, T. (2011). *Navigating News Online: Where people go, how they get there and what lures them away*. Washington, DC: Pew Research Center's Project for Excellence in Journalism.

Purcell, K., Rainie, L., Mitchell, A., Rosenstiel, T., & Olmstead, K. (2010). *Understanding the participatory news consumer: How Internet and cell phone users have turned news into a social experience*. Pew Research Center's Internet & American Life Project.

Sasseen, J., Olmstead, K., & Mitchell, A. (2013). *Digital: As mobile grows rapidly, the pressures on news intensify*. The State of the News Media 2013, The Pew Research Center's Project for Excellence in News Journalism, an Annual Report on American Journalism.

Singer, J. B., Domingo, D., Heinonen, A., Hermida, A., Paulussen, S., Quandt, T., et al. (2011). *Participatory journalism: Guarding open gates at online newspapers*. Chichester: Wiley-Blackwell.

Thurman, N. (2008). Forums for citizen journalists? Adoption of user generated content initiatives by online news media. *New Media and Society, 10*(1), 139–157.

van Dijck, J. A. G. M. (2013). Social media platforms as producers. In T. Olsson (Ed.), *Producing the internet* (pp. 45–62). Göteborg: Nordicom.

van Dijk, J. A. G. M. (2006). *The network society: Social aspects of new media*. Thousand Oaks, CA: Sage.

Villi, M. (2012). Social curation in audience communities: UDC (user-distributed content) in the networked media ecosystem. *Participations: The International Journal of Audience and Reception Studies, 9*(2), 614–632.

Villi, M., & Jung, J-Y. (2015). Accessing the audience community: How newspapers connect with audience communities in Finland, Japan, and Korea. *International Journal on Media Management* (forthcoming).

Wardle, C., & Williams, A. (2010). Beyond user-generated content: A production study examining the ways in which UGC is used at the BBC. *Media, Culture and Society, 32*(5), 781–799.

Weeks, B. E., & Holbert, R. L. (2012, May 24–28). Predicting dissemination of news content in social media: A focus on reception, friending, and partisanship. In *2012 International Communication Association Conference*, Phoenix, AZ.

Guerrilla Media: Interactive Social Media

Andrej Duh, Sebastian Meznaric, and Dean Korošak

1 Introduction

There is another way to look at a newspaper, and that is as an interface to news
N. Negroponte, Being digital (1995)

The Internet is the most important medium since the printing press (Gillmor, 2006). In today's digital renaissance the Internet is overturning so many things we have assumed about media and adequate business models that we can hardly follow. But one thing can be deducted from the history of media: old media never die. Genres and delivering technologies come and go, but media remain as layers within complicated information and entertainment system (Jenkins, 2001).

As today's consumers interact with Internet society using social networks or social blogs, there are unpredictable ways how new and traditional media collide. Consider for instance the story on Dino Ignacio posting posted his Photoshop collage of Sesame Street's Bert with terrorist leader Osama Bin Laden on his homepage as a part of a series "Bert Is Evil" in the fall of 2001. Several months later, the collage appeared in CNN coverage of anti-American protest following September 11 (Jenkins, 2006). Representatives from the Children's Television Workshop, creators of the Sesame Street, spotted the CNN footage and threatened to take legal action. It was not clear against whom they wanted to take this action: the young Dino Ignacio or the terrorist supporters. But at the end, amused fans produced a number of new sites, linking various Sesame Street characters with terrorists. This example showed how the circulation of media content across different media systems and across national borders depends heavily on consumers'

A. Duh (✉) • D. Korošak
University of Maribor, Slomškov trg 15, 2000 Maribor, Slovenia
e-mail: andrej.duh@um.si

S. Meznaric
Clarendon Laboratory, University of Oxford, Parks Road, Oxford OX1 3PU, United Kingdom

A. Lugmayr, C. Dal Zotto (eds.), *Media Convergence Handbook - Vol. 1*, Media Business and Innovation, DOI 10.1007/978-3-642-54484-2_17

pro-activity and interaction with media contents. Consumers perform in new media system where they are as well also at least partially producers.

Nowadays, there is more information on any given topic that anyone can consume. We like to talk among ourselves about the media we consume. The rise of consumer-generated media, such as: social networks, social blogs, weblogs, podcasts, pictures and videos, has even increased communication intensity between people leading to information consumption becoming a collective process through consumer-generated media. Such mass behaviour, also called collective intelligence (Lévy 1994), where we can put the pieces together if we pool our resources and combine our skills, can be seen as a new source of media power.

Conversation through the consumer-generated media creates buzz that is increasingly valued by the traditional media industry. For more than 50 years, television has been ruled by the Nielsen family whose TV habits provide statistical snapshot of nations' viewing habits (Vanderbilt, 2013). But nowadays we have complete new era of TV ratings using consumer generated buzz on the social channels. Companies like Bluefin Labs (recently acquired by Twitter), SocialGuide (recently acquired by Nielsen) and Trendrr, offer similar service: mining messages from Twitter, Facebook and other social networks to find out viewing habits of TV consumers.

The consumer-generated buzz is full of stories, from small local news to large stories that are still in their early stage of development but have not yet caught the attention of the traditional media and represents huge collection of extremely detail data on social behaviour. Using McLuhan's famous line: "The medium is the message." (McLuhan, 1964), we can state today that the social media is the message.

There are already solutions as Flipboard, News360, News.me, Prismatic, Pulse, Zite and many others that are focusing on how helping individuals to discover the messages they are interested in. Guerrilla media idea is not focused on individuals, but on using collective intelligence and detecting important messages from huge amount of social network accounts and spreading these messages through interactive social media news streams into the digital universe.

In their new book "Big Data: A Revolution That Will Transform How We Live, Work and Think" (Mayer-Schoenberger & Cukier 2013) Viktor Mayer-Schoenberger and Kenneth Cukier write: "Big data will be a source of new economic value and innovation. But even more is at stake. Big data's ascendancy represents three shifts in the way we analyze information that transform how we understand and organize society" (Wang & Huberman, 2012a). The three shifts caused by the Big Data are the emergence of ability to gather and analyze *all* the data in particular problems, loss of exactness as a result of increased scale, and settling for correlations as good enough—"Big data is about what, not why" (Wang & Huberman, 2012a).

In 2010, IDC said, the amount of information created and replicated in digital universe surpassed 1 zettabyte (zettabyte (ZB) = trillion gigabytes, 1,000,000,000,000,000,000,000 bytes = 10^{21} bytes) and more then doubles every

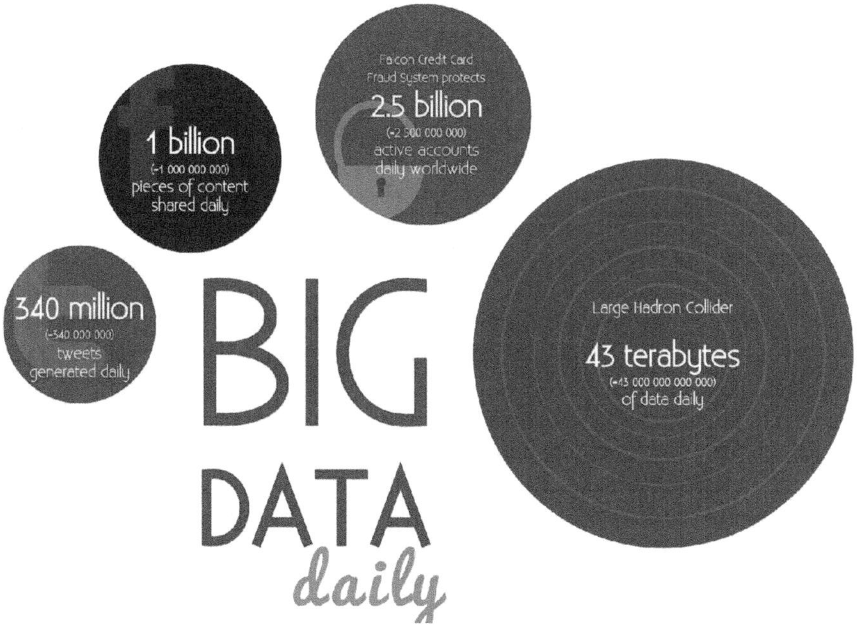

Fig. 1 Big Data production per day in the digital universe

2 years; by 2014, we are expected to generate about 7 ZB a year (Villars, Olofson, & Eastwood, 2011).

Where does all of this data come from? Companies produce huge quantities of data about their customers on a daily basis. For example, Falcon Credit Card Fraud Protection System protects 2.5 billion active accounts worldwide (FICO 2013). On the scientific side, Large Hadron Collider produced 15 petabytes of data annually (CERN 2013). And simply through exchange of e-mails and sharing of thoughts through social networks, users of computers and now also smartphones and tablets, contribute to an increasing data explosion. There are 30 billion pieces of data ready for "like", shared on Facebook every month and there are more than 1 billion tweets generated per week (Kissmetrics, 2013). Figure 1 shows the infographic of daily Big Data.

Data collections of this size that are beyond the ability of a typical database software tools to store, manage and analyse and are collectively referred to as Big Data. In such cases the size of data collections itself becomes part of the problem. Potential of Big Data is infinite in value as it is in size. Big Data is already a game changer in several different aspects of human life. Actually, Big Data is transforming business the same way as information technology did—transformation of retail has already occurred. User data driven pricing and recommendations is a "must have" on retail web portals. Users of web-based e-mails services, search tools, social applications are all exposed to and familiar with transformation of online marketing. Disruptive marketing and sales recommendations based on user or location data are present everywhere. Transformation of law enforcement is something that is not visible to a public user, but visible are the results such as for example crime hot-spot predictions. There are many other Big Data transformed businesses like IT, customer service, fraud management etc.

The global use of social media present huge collection of extremely detailed data on social behaviour. The ubiquitousness of Twitter as a micro-blogging service has reshaped how information spreads in digital space. An example of using Big Data analysis for prediction of a social event (Ciulla et al., 2012) is a demonstration of the usefulness of tweets as an open source data. An in depth analysis of the micro-blogging activity surrounding the voting behaviour on the contestant in American Idol has shown that Twitter activity during the TV show and the voting period following it, correlates with the contestants ranking and enables predictive voting results.

Prediction of activity of a single user and aggregate activity of topical tweets that Twitter users will generate within a time period is subject of the article in Ruan, Purohit, Fuhry, Parthasarathy, & Sheth (2012). Introduced framework for modelling and predicting topic-specific tweets that will be generated by Twitter users in the future is next step in prediction.

The Internet is full of stories, from small local news that large-scale media do not consider newsworthy due to the small number of potentially interested people, to larger and more important developments in remote places that are unlikely to immediately resonate with the audience. And finally, there are stories that are still in their early development but have not yet caught the attention of the large news networks, whose camera crews and journalists might still be on the way to the spot or in the process of producing the newsreel.

People notice the big developments in their locality simply as they go about their daily lives. And they tweet them, or post them on Facebook or another social network, often along with the newest images making interaction semantics, contextual and social aspects of media as important as the media content itself (Sundaram, Xie, De Choudhury, Lin, & Natsev, 2012).

The idea of detecting these news patterns, and to present those to interested audience, is the subject of this chapter. These news patterns are shaped by the structure and properties of the underlying social media. We explore how to detect news, content or media texts patterns, how content is spreading through digital universe. We briefly touch technological issues and discuss how to implement an application for predicting content spreading.

2 Related work

> Imagine a computer display of news stories with a knob that, like a volume control, allows you to crank personalization up or down.
>
> N. Negroponte, Being digital (1995)

There is a predicting power that lies in social media content that and can be used to detect news patterns using persistent dynamics of trends in social media. In example of using tweets for predicting box-office movies revenues, the attention was found to be strongly correlated with future ranking and sentiment analysis important in improving prediction (Asur & Huberman, 2010). A stochastic model for trend persistence, tested on Twitter data (Wang & Huberman, 2012a), predicted linear relationship between trend duration and resonant user activity on the topic. It is the resonance with the topic that determines the trend rather than general activity or number of followers. Interestingly, user's activity, defined as a sequence of possible states, have deterministic components which are less expressed when user acts in a group activity (Wang & Huberman, 2012b).

Social media are in a way a perfect disseminator of such news. A good example is Twitter. Study of information diffusion in the entire Twittersphere (Kwak, Lee, Park, & Moon, 2010) revealed that the number of followings and followers shows a power-law distribution (up to a limit because of the effect of few users with extremely large number of followers). Trends analysis disclosed Twitter as a breaking news media source with a fast diffusion of retweets after the first retweet irrespective of the number of followers of original tweet. The social structure of Twitter using mentions and retweets was investigated in (Grabowicz, Ramasco, Moro, Pujol, & Eguiluz, 2012) where the importance of organization of the Twitter in groups was discovered. There seem to exist three basic types of links in the Twitter network: strong ties represented by mentions are in-group links, weak ties correspond to retweets that link different groups and are crucial for information propagation, and intermediary links that combine in-group mentions and have large diversity due to linking members in different groups.

In Quercia, Capra, and Crowcroft (2012) real-life social networks were compared to Twitter network. Social brokers with both positive and negative tweets were detected in Twitter structure. The sentiment of users is shown to be correlated in geographical local networks (locality increases with the strength of links). Distance, language, country and measured such as number of flights all influences the network structure of Twitter (Takhteyev, Gruzd, & Wellman, 2012). More specifically, around 39 % of links are found to be localized to metropolitan area size clusters, where as the long-range links seem to be correlated with the migration activity expressed as the number of flights. The geographic proximity seems to importantly influence the structure of Twitter network and also other social media. In Kulshrestha, Kooti, Nikravesh, and Gummadi (2012) approximately one third of links were found to be transnational, existing between geographically and linguistically close countries. The importance of geography in LiveJournal blogging community is studied in Liben-Nowell, Novak, Kumar, Raghavan, and Tomkins (2005) where a new model based on ranking is introduced so that probability of linking nodes i and j is inversely proportional to rank defined as the number of people closer to j than to i. One third of links in these large social networks were geography independent and the findings were connected to navigability in spatially embedded networks showing that rank-based networks contain long-range links.

Multiple studies have shown that people prefer to read stories that they agree with, exhibiting the so-called confirmation bias and the prior attitude effect (Taber & Lodge, 2006). When confronted with evidence to the contrary of our pre-existing beliefs we become *motivated sceptics* (Kunda, 1990), trying to find any and all reasons for why the evidence might be false, examining and picking apart flaws in the study. We only consider changing our mind when we have exhausted all the attempts at convincing ourselves that the evidence is false.

In one such study (Taber & Lodge, 2006), after being divided into groups either favouring or disfavouring gun control, the participants were asked to even-handedly rate the arguments both in favour and against gun control. In addition to being instructed to evaluate arguments even-handedly, the participants were told they would have to explain those arguments to a group of students, something expected to make people act in a less biased fashion. However, despite these efforts by the researchers, those with strong opinions in favour of gun control were far more likely to rate highly the arguments in favour of gun control than they were those against gun control and vice versa.

News, especially when it concerns politics, is an extremely contentious subject with many opposing theories and opinions already developed in the minds of the people. It should therefore come as no surprise that these effects are amplified when consumers are confronted with news that do not fall neatly into their pre-existing world view, rating those with opposing views as less trustworthy, less credible and biased, while those with the same world view as neutral and rational. There are many possibilities of manipulating social media and network in politics using "Twitter bombs", spam, automatic content adjustment, promoted trends, etc. There is also a large discrepancy between Twitter volume and number of accounts in political conversations (Metaxas & Mustafaraj, 2012).

Since people will view information they consider credible and neutral, it is therefore a survival strategy for any media stream, be it social network based or traditional, to provide information their consumers agree with. This is the point where social networks can have a great impact. In a recent study (Morales, Gionis, & Lucchese, 2012), authors gathered data from users' Twitter profiles in order to construct a news stream composed of mainstream news stories from online media. They tackled the recommendation problem to find most relevant news from the news stream for particular user (at a particular time) given the tweet stream of users using recommendation ranking model linearly combining social and content relatedness, and new popularity based on entity.

Predicting popularity of news on Twitter has also been studied based on four content features: source, category-topic, language and named entities (Bandari, Asur, & Huberman, 2012). The most important indicator out of these four was the source indicating the crucial importance of good topic classification.

In *Spreadable Media: Creating Value and Meaning in a Networked Culture*, Jenkins, Ford and Green (2013) emphasize the importance of strong connections between media producers and their audiences in "spreadable media landscape": "Spreadability recognizes the importance of the social connections among individuals, connections increasingly made visible (and amplified) by social media platforms. This approach may still include quantitative measures of how frequently and broadly content travels, but it makes important actively listening to the way media texts are taken up by audiences and circulate through audience interactions" (Jenkins et al., 2013).

Several studies have already detected and analysed the interdependence of content transfer or spread and the properties of social networks providing several hints into the dynamics of content spreading.

For instance in spreading (contagion) in Facebook depends on the density of the network neighbourhood of users rather than on its size (Ugander, Backstrom, Marlow, & Kleinberg, 2012), while the weak ties in this social network and their abundance is crucial for new content transfer (Bakshy, Marlow, Rosenn, & Adamic, 2012).

The analysis of the structure of a blogosphere (Cha, Pérez, & Haddadi, 2012) showed that the blogs network is less connected than other social networks. The blogosphere also responds differently to the content type and spreads news and political topics fast but with a short lifetime, while contents around topics, such as music, spreads slowly in comparison to news but with a longer lifetime. Such media events seem to cause "seismic" waves in the blogosphere (Klimek, Bayer, & Thurner, 2011), with exogenous and endogenous events showing different scaling exponents of their size distribution. Also evidence of correlations of pre- and after-media events in the blogosphere was found.

In a blogs-Big Media interaction study (Leskovec, Backstrom, & Kleinberg, 2009) showed that there is a 2.5 h phase shift between Big Media and blogs. Furthermore, patterns in news cycles show that 3.5 % of memes diffuse from blogs to media. The detected differences in diffusion and response between blogosphere and mainstream media are caused by the underlying time-dependent network structure for information pathways and diffusion (Rodriguez, Leskovec, & Schölkopf, 2013).

An agent-based model of meme diffusion on social networks (Weng, Flammini, Vespignani, & Menczer, 2012) demonstrated that only the social network structure and the limited time and attention of agents are sufficient for rich and complex meme patterns. That content might also play an important role in meme acceptance is reflected in the spread of memes in Twitter social network using hashtag as an idea concept (Tsur & Rappoport, 2012). The lifetimes of Twitter memes were found to be power-law distributed, where the interaction with Big Media was suggested as a support mechanism for the long-lived memes (Weng et al., 2012). Similarly, phrase volume in the phrase graph is found to be power-law distributed (Leskovec et al., 2009).

3 Methodology and Approach

3.1 Quantifying Media Convergence

> Our message is simple and direct: if it doesn't spread, it's dead.
>
> H. Jenkins, S. Ford, J. Green, Spreadable Media (2013)

Jenkins, Ford and Green define spreadability as a combination of "technical resources that make it easier to circulate some kinds of content than others, the economic structures that support or restrict circulation, the attributes of a media text that might appeal to a community's motivation for sharing material, and the social networks that link people through the exchange of meaningful bytes" (Jenkins et al., 2013). They contrast spreadability to "stickiness", a model in which the media content is measured by counting the number of users and which "capitalizes on the easiest way companies have found to conduct business online—rather than the ways audiences want to and do experience material online." While it is clear in a stickiness model "who the 'producer,' the 'marketer,' and the 'audience'" are, "In a spreadable model, there is not only an increased collaboration across these roles but, in some cases, a blurring of the distinctions between these roles" (Jenkins et al., 2013).

Motivated by this shift of the focus from "stickiness" to "spreadability", we introduce and discuss a model for content flow in spreadable media space that allows to quantify media spreadability and convergence. We see the spreadable media landscape as a Big Data space with semantic metric (i.e. with defined distances between media texts), driven by social network of users. Figure 2 depicts the basic setup of the model. We have users interacting (create, change, update, comment, like, tweet, retweet, cite, etc.) with the media content denoted as "i". Content can be in any form of text, image, video, sound or convergent combination of these media including meta-data about the content itself. Users also form or belong to a social network through which they can communicate, link and share.

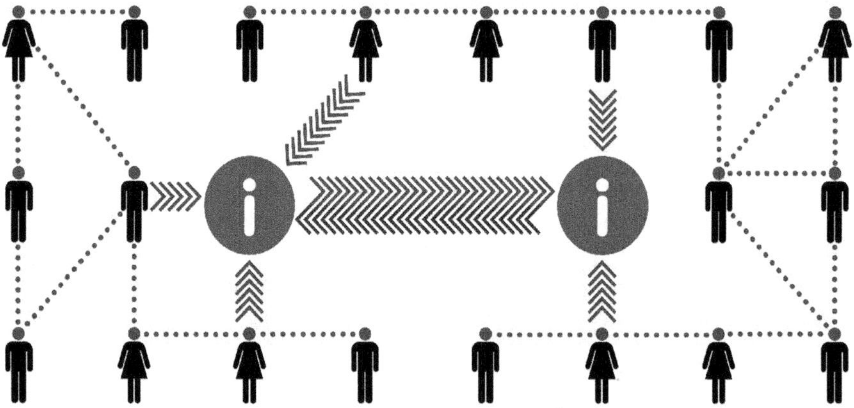

Fig. 2 Schematic view of the media text spreadability model. Around each media text (i) there is a social network of users some of which are generating and interacting with the content (indicated by *arrows*). The spreadability of the media text depends on the media content (stickiness) at emitter and absorber locations, at locations between them and the defined distance. Our suggestions for metric (see main text) include definitions based on number of active users interacting with media text and definitions using properties of the social network of users (with links indicated by *dotted lines*)

To quantify the media spreading and converging processes we follow the idea that was recently developed to predict mobility patterns (Simini, Gonzalez, Maritan, & Barabasi, 2012) where the basic underlying processes are emission and absorption of particles. A particle (in our case content or media text) is emitted from location i with certain absorption threshold (property of location i to spread media content) and it is absorbed at the location j where the absorbance (a property of location j to convergently combine content) is greater than the absorption threshold. The transfer from i to j depends on the number of particles m_i at location i, the number of particles m_j at location j, and the number of particles in the space between i and j (in the circle with the radius equal to distance d_{ij} from i to j) s_{ij}. The probability p_{ij} to transfer a particle from i to j is (Simini et al., 2012):

$$p_{ij} = \frac{m_i m_j}{(m_i + m_j + s_{ij})(m_i + s_{ij})} \tag{1}$$

When applying this model to predict content spreading we replace "number of particles at location $i(j)$" in our case with the quantified interaction of users with the content $i(j)$ expressed as user activity such as tweeting, retweeting, liking, commenting, citing or changing the content. The choice of using user activity in the model instead of for instance volume of the content data or number of users is motivated by recent discoveries of structural properties of social media such as Twitter and Facebook (Bakshy et al., 2012; Cha et al., 2012; Grabowicz et al., 2012). The weak ties, corresponding to retweets, are crucial for information spreading in Twitter, while the diffusion of retweets (after the first retweet) does not depend on the number of followers of the original tweet. Similarly, the weak ties are crucial for content transfer in Facebook and the transfer process depends on the structure of users's network and not on the size. The resonance of users with the topic of the content is seemingly more important in determining trends than sheer number of followers or total activity (Wang & Huberman, 2012b). That meme (viral information that spreads fast) absorbance importantly depends on content was shown using Twitter hashtags as spreading ideas (Tsur & Rappoport, 2012).

In addition to users' activity and interaction with the content (m), the spreading of the content depends also on the distance d_{ij} between content at i and j, so we also need to introduce a metric in this space. There are several possibilities for defining such distance: semantic similarity between contents i and j, network distances in the social network around i and j, or simpler measures that depend on number of users at i and j. Starting with the later choice, the distance can be defined using intersection of sets of users at i and j as introduced in the analysis of Twitter media landscape (An, Cha, Gummadi, & Crowcroft, 2011). The predictive power of the model might possibly depend on the choice of the metric, so each case needs to be experimentally tested.

The spreadable media network model can also be viewed as the coupling of two networks: there is a social network of users interacting with the content and with each other, and then there is a content network characterized by links between content given by the measure of spreadability. The two networks are coupled and interdependent—the changes of the contents influence the interaction of users with it, and user actions change the contents. It would be interesting to test whether these coupled networks behave similarly as some other interdependent networks (Buldyrev, Parshani, Paul, Stanley, & Havlin, 2010; Vespignani, 2010).

We illustrate the basic features of the model using a simple example: let's consider a social network of users with a known topology and let's take that the media content m_i at network location (node) i is given simply by the number of links of node i. The distance between i and j is equal to the shortest path between these two nodes, and the content s_{ij} between i and j is a sum of links of nodes along this shortest path. We pick a starting node randomly and then follow the spreading by jumping with probability p_{ij} from Eq. (1) to the next randomly selected node j. Arriving at node j, we select the next node and continue with the process. Spreading pattern is in this very simple example a path of all visited nodes in the network. Activity of users interacting with the content is equal to number of nearest neighbors at each visited node. In Fig. 3 we present spreading patterns computed for this example on two different network topologies: scale-free (left) and regular lattice (right). We see that in scale-free network the spreading pattern includes most of the nodes with large number of links (the hubs of the network). This is not surprising since the probability to spread is proportional to the product of the links of nodes i and j. In a regular lattice where each node has the same number of links, it is the distance that mostly governs the spreading pattern. We have also compared the spreading process given by Eq. (1) with random jump process for a network with scale-free topology. From results in Fig. 4 we can see that spreading governed by Eq. (1) becomes much more effective (measured by the number of influenced nodes) than random process as the number of visited nodes increases.

A more complex model that includes the interaction between users and the dynamic media content which perpetually changes the structure of the social

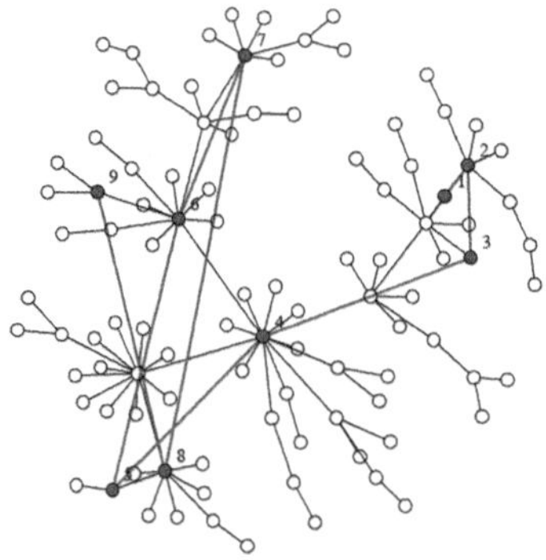

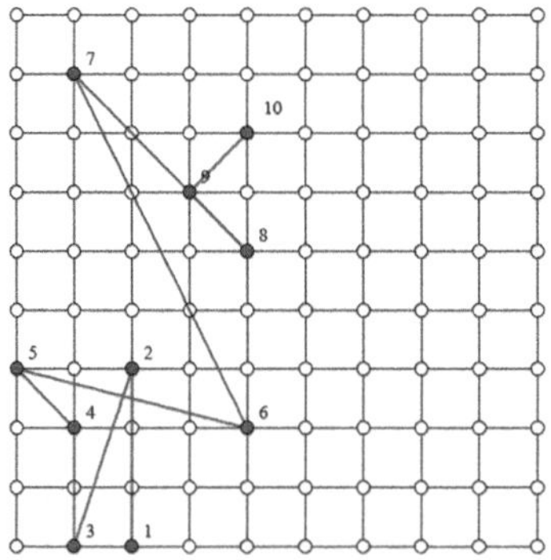

Fig. 3 Spreading patterns on a scale-free network (*left*) and on a regular lattice (*right*) computed for a simple example

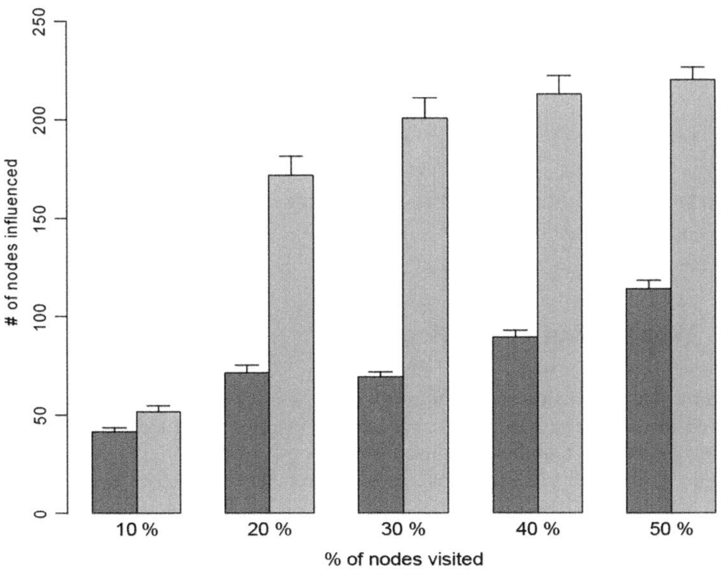

Fig. 4 Comparison of spreading process on a scale-free network governed by Eq. (1) (*light gray*) and random process (*dark gray*). The difference in number of influenced links in both processes increases with number of nodes visited

network is the basis of our Guerrilla Media news application (technology of implementation is described in the next section).

The benefit of building the application on spreadable media model is that it allows us to predict the spreading potential of media contents by inferring network links between media texts and to quantify the media convergence. We use the average flux $< T_{ij} >$ from location i to location j as a predictor for media convergence. Given the probability for emission and absorption process by Eq. (1), the average flux (Simini et al., 2012):

$$\langle T_{ij} \rangle = m_i p_{ij} \quad (2)$$

We connect convergence with absorption potential of content at location j for emitted news stream from the node i.

Let's consider two limiting cases for media convergence predictor: a Big Media emitter and a grassroot news generator. In the first limiting case of Big Media emitter (mainstream media outlet) and grassroot news absorber we have $m_i >> m_j$ and $m_i >> s_{ij}$. In this case the convergence is limited by the absorption potential of the receiving content:

$$\langle T_{ij} \rangle = m_j. \quad (3)$$

In the opposite limit with the grassroot news generator and Big Media receiver (for example social media news (blogs) that are adopted by mainstream media) we have $m_i << m_j$ and $m_i << s_{ij}$. The predictor for convergence is in this case given by:

$$\langle T_{ij} \rangle \approx \frac{m_i^2}{s_{ij}} \quad (4)$$

and depends on the content of the emitter and also on the content between the emitter and receiver.

3.2 What About Technology?

> The idea that the proprietary software social system-the system that says you are not allowed to share or change software-is unsocial, that it is unethical, that it is simply wrong may come as a surprise to some people.
>
> R. Stallman, Open Sources: Voices from the Open Source Revolution (1999).

Big Data requires significant new technical capabilities in dealing with new demands regarding data volume, velocity, variety and veracity. The entire modern analytical system must be prepared to manage immense amounts of data quickly. The use of parallel computing for tackling data scalability is natural solution already known from our brains. The new programming systems are designed to get their parallelism from large collection of commodity hardware. The unlimited linear scalability to several thousand of nodes is solution for very affordable price comparing to traditional symmetric multiprocessing shared memory server solution. To deal with unlimited scalable system of nodes, a new software stack has evolved. Central to the new software stack is a MapReduce programming model that simplifies parallel processing by abstracting away complexities involved in distributed systems.

The MapReduce model decomposes work into small parallelized Map and Reduce workers i.e. Map and Reduce functions. The input data format is application specific and the output data is a set of `<key, value>` pairs (Ranger, Raghuraman, Penmets, Bradski, & Kozyrakism, 2007). The Map function is applied on the input data. The Map function prepares a list of intermediate `<key, value>` pairs, which are processed by the Reduce function. The Reduce function is applied to all intermediate pairs with the same key. The Map function typically performs some kind of filtering and sorting and the Reduce function performs some kind of merging operation.

The pseudocode example that counts the appearance of each word in a set of documents is as follows (Ranger et al., 2007):

```
function map(document) {
      // document : document content
      for each word w in document
            emit(w, 1)
}
```

```
function reduce(word, partialCounts) {
      // word : a word
// partialCounts : a list of aggregated partial counts
      result = 0
      for each p in partialCounts
            result += p
      emit(word, result)
}
```

The main benefit of the MapReduce model is that allows the programmer to focus on functionality and not on parallelization. The parallelization and the details of concurrency management are left to the platforms such as Hadoop.

We want to test our ideas in the real application called Guerrilla Media (GM). The GM application is still under development, but the ultimate aim is to offer the GM application to the market and to test the ideas in a living laboratory.

The application is based on dynamic network of spreadable media contents that perpetually changes with time as new media content is created and shared, and as the social network evolves. We use our model for the media contents network where the nodes of the network (news items, *m*) are connected with links inferred from computed news spreading and convergence predictor (<*T*>).

Current version of the GM application provides data pre-procession, aggregation and categorization in a real time for English texts. The data is collected from several different data channels: social channels (Twitter and Facebook) and several news channels. Collection from Twitter is performed using Twitter stream API, which offers low latency access to Twitter's global stream of Tweet data. There is no such access to the Facebook data therefore Facebook data is obtained only for users that allow accessing their data through GM Facebook module. Through sharing their social channels within the GM application, the Facebook users will improve predictive output of the application and increase the spreadability of generated news. Different news channel data is obtained via RSS feeds and is used to follow news stream from traditional media.

Each obtained news is pre-processed, aggregated and categorized. News pre-processing is very important part in the GM application pipeline. As all data stream is either in XML or HTML, extracting meaningful text is the first obvious task. While parsing XML stream of tweets and extracting meaningful text is almost trivial task, identifying meaningful article inside original HTML document is complicated by the large amount of unrelated material such as navigation menus, forms, scripts and ads as discussed in several articles about this topic (Kohlschütter, Fankhauser, & Nejdl, 2010; Pasternack & Roth, 2009). We decided to use boilerpipe library for extracting articles from HTML documents. Boilerpipe library is the official implementation of the overall algorithm presented in the previously mentioned paper (Kohlschütter et al., 2010) and it is open source Java library integrated in the Apache Tika. Apache Tika is content analysis toolkit and is available separately or integrated in even more sophisticated toolkit called Lucene. Apache Lucene is search engine toolkit and contains several exiting libraries.

Lucene has powerful parsing library for next phase of article processing: extracting tokens from text. Before tokenizing process a part of clean text was used for news aggregation where first few sentences of article text are used. The document is tokenized to explore words in a sentence and to remove unimportant words like articles, conjunctions, pronouns, prepositions, adverbs and non-informative verbs. The next step in the pre-processing phase is called stemming and is a technique to reduce words into their roots. The stemming for the English language is quite simple as the endings are simple and similar. The result of reduction of the words may cause that some words are transformed into wrong words but for statistical interpretation of data is this practically negligible error. There are many stemming algorithms, for example Lovins stemming, Porter stemming (included in Lucene), Krovetz stemming, inflectional stemming and derivational stemming (Hull, 1996). As the Porter stemming algorithm has problems with endings "ed" or "es", we implemented the inflectional stemmer, where corrected word is iteratively checked in the dictionary.

The next step in the pre-processing phase is the recognition and identification of nouns, verbs, adjectives and prepositions using part-of-speech tagging (POS) (Brill, 1992). POS tagging is in general very complex task, but for current version of the GM application we use corpus-based approach using WordNet. WordNet is sophisticated thesaurus developed by the Princeton University for the English language (Miller, 1995). Because current version of the GM application uses only nouns as topic words, is the corpus-based approach is good enough.

To penalize common words, which appear very often in all documents, we use model called term frequency—inverse document frequency (tf-idf) (Somlo & Howe, 2001):

$$tfidf(w) = tf \cdot log\left(\frac{N}{df(w)}\right), \tag{5}$$

where *tf* is number of word occurrences in a document, *df* is number of documents containing that word and N is number of all documents. If a word appears many times in a document but also appears many times in the collection (for example training set) it will get a lower score.

The final step in the pre-processing phase is topic modelling. Because we manage large document archive, we have decided to use probabilistic topic model called Latent Dirichlet Allocation (LDA) (Blei, 2012). LDA is a statistical model of document collections that reflect the intuition that news exhibits multiple topics. Obtained topics for selected news represent part of aggregated data.

The aggregated news is now ready for spreading if it passes a lightweight document-matching method. The document-matching method is focused on certain parts of aggregated news: title, aggregated sentences and topics keywords against news in the library. Weather the aggregated news is delivered to the individual Facebook user, depends on the details of user engagement with media content.

4 Conclusion

Henry Jenkins wrote "You are now entering convergence culture" already in 2006 in his seminal book Convergence Culture where he defined convergence as "the flow of content across multiple media platforms, the cooperation between multiple media industries, and the migratory behavior of media audiences who will go almost anywhere in search of the kinds of entertainment experiences they want" (Jenkins, 2006). However, that convergence is "one of those particularly hard-to-handle concepts in media studies" warns Espen Ytreberg reviewing three recent contribution to the field in his paper aptly titled "Convergence: Essentially confused?" (Ytreberg, 2011). The concept of converged media that emerged in marketing is defined as an intersection of three media spaces: owned, bought and earned media (Goodall, 2013; Himler, 2013; Web-strategist, 2013). Bought media are broadcast advertising, banner ads, sponsored links and similar. Owned media refers to content controlled by a brand in websites, branded blogs and owned social channels. Earned media is content generated and shared by users. Converged media results from confluence of channels from bought, owned and earned media spaces.

In this work we have adopted Jenkins' definition of convergence that rests on two processes: flow of content and migration of users. We use media convergence to refer to a process of spreading and combining of the content in spreadable media landscape. Our question was: can we quantify media convergence in such a way that would allow prediction of content spreading? To answer this we have introduced a network model of spreadable media where the nodes of the network (media content) are given by the probability to spread content between the nodes. Linking depends on the details of users engagement with media content through which their social network structure is reflected in the spreadable media network.

We have demonstrated the dependence of content spreading patterns on the topology of the underlying network with a simple example and defined the flow of content as a predictor for media convergence. Considering the case of Big Media or bought media channel content generator, we showed that the predictor for convergence is given by the absorbing media content. In case of user generated content or one of earned media channels, the media convergence predictor depends on the properties of emitted content and the content in media space around the emitter.

The model will be tested in a real application called Guerrilla Media (GM) currently under development. In an article entitled "The rise of guerrilla media" Glenn Reynolds refers to guerrilla media as news generated "by people who are much closer to what's really going on and are much more closely connected to their audiences" (Reynolds, 2006). The GM application will allow ranking of content according to the predictions to spread and converge with other sources in the spreadable media landscape. The insight into users' social media data will allow us to measure and perpetually infer the dynamic structure of the network. Through sharing their social channels within the GM application, the users will improve predictive output of the application and increase the spreadability of generated news.

Acknowledgments The work that led to this paper was partially financed by the project SEETechnology "Co-operation of SEE science parks for the promotion of transnational market uptake of R&D results and technologies by SMEs" co-funded by South East Europe Transnational Cooperation Programme. The work of one of us (DK) was partly financed within the framework of the operation entitled "Centre for Open Innovation and Research of the University of Maribor". The operation is co-funded by the European Regional Development Fund and conducted within the framework of the Operational Programme for Strengthening Regional Development Potentials for the period 2007–2013, development priority 1: "Competitiveness of companies and research excellence", priority axis 1.1: "Encouraging competitive potential of enterprises and research excellence."

The authors wish to thank Marko and Urška Samec for the infographics design.

References

An, J., Cha, M., Gummadi, K., & Crowcroft, J. (2011). Media landscape in Twitter: A world of new conventions and political diversity. In *International AAAI Conference on Weblogs and Social Media*

Asur, S., & Huberman, B. A. (2010). Predicting the future with social media. In *Proceedings of the 2010 IEEE/WIC/ACM International Conference on Web Intelligence and Intelligent Agent Technology* (Vol. 1, pp. 492–499).

Bakshy, E., Marlow, C., Rosenn, I., & Adamic, L. (2012). The role of social networks in information diffusion. In *Proceedings of the 21st International Conference on World Wide Web (WWW'12)* (pp. 519–522).

Bandari, R., Asur, S., & Huberman, B. A. (2012). The pulse of news in social media: Forecasting popularity. In *Proceedings of the Sixth International AAAI Conference on Weblogs and Social Media (ICWSM '12).*

Blei, M. (2012). Probabilistic topic models. *Communications of the ACM, 55*(4), 77–84.

Brill, E. (1992). A simple rule-based part of speech tagger. In *Proceedings of the Third Conference on Applied Natural Language Processing (ANLC '92)* (pp. 152–155).

Buldyrev, S. V., Parshani, R., Paul, G., Stanley, H. E., & Havlin, S. (2010). Catastrophic cascade of failures in interdependent networks. *Nature, 464*, 1025–1028.

CERN. (2013). *Large Hadron Collider at CERN*. Accessed March 2013, from http://public.web.cern.ch/public/en/lhc/Computing-en.html

Cha, M., Pérez, J. A. N., & Haddadi, H. (2012). The spread of media content through blogs. *Social Network Analysis and Mining, 2*, 249–264.

Ciulla, F., Mocanu, D., Baronchelli, A., Goncalves, B., Perra, N., & Vespignani, A. (2012). Beating the news using social media: The case study of American Idol. *EPJ Data Science, 1*, 8.

FICO. (2013). FICO, Flacon Fraud Manager. Accessed March 2013, from http://www.fico.com/en/Products/DMApps/Pages/FICO-Falcon-Fraud-Manager.aspx

Gillmor, D. (2006). *We the media. Grassroots journalism by the people, for the people*. London: O'Reilly Media.

Goodall, D. (2013). Accessed July 2013, from http://danielgoodall.com/2009/03/02/owned-bought-and-earned-media/

Grabowicz, P. A., Ramasco, J. J., Moro, E., Pujol, J. M., & Eguiluz, V. M. (2012). Social features of online networks: The strength of intermediary ties in online social media. *PLoS One, 7*(1), e29358.

Himler, P. (2013). Accessed July 2013, from http://www.forbes.com/sites/peterhimler/2012/09/11/converged-media-rising/

Hull, D. A. (1996). Stemming algorithms: A case study for detailed evaluation. *Journal of the American Society for Information Science, 47*, 70–84.

Jenkins, H. (2001, June). Convergence? I Diverge. *Technology Review, 104*(5), 93.

Jenkins, H. (2006). *Convergence culture: Where old and new media collide*. New York, NY: New York University Press.
Jenkins, H., Ford, S., & Green, J. (2013). *Spreadable media*. New York, NY: New York University Press.
Kissmetrics. (2013). A blog about analytics, marketing and testing. Accessed March 2013, from http://blog.kissmetrics.com
Klimek, P., Bayer, W., & Thurner, S. (2011). The blogosphere as an excitable social medium: Richter's and Omori's Law in media coverage. *Physica A: Statistical Mechanics and its Applications, 390*(21–22), 3870–3875.
Kohlschütter, C., Fankhauser, P., & Nejdl, W. (2010). Boilerplate detection using shallow text features. In *Proceedings of WSDM 2010* (pp. 441–450).
Kulshrestha, J., Kooti, F., Nikravesh, A., & Gummadi, K. (2012). Geographic dissection of the Twitter network. In *International AAAI Conference on Weblogs and Social Media*.
Kunda, Z. (1990). The case for motivated reasoning. *Psychological Bulletin, 108*(3), 480–498.
Kwak, H., Lee, C., Park, H., & Moon, S. (2010). What is Twitter, a social network or a news media? In *Proceedings of the 19th International Conference on World Wide Web (WWW'10)* (pp. 591–600).
Leskovec, J., Backstrom, L., & Kleinberg, J. (2009). Meme-tracking and the dynamics of the news cycle. In *Proceedings of the 15th ACM SIGKDD International Conference on Knowledge Discovery and Data Mining (KDD '09)* (pp. 497–506).
Lévy, P. (1994). L'Intelligence collective. Pour une anthropologie du cyberspace. *La Découverte*.
Liben-Nowell, D., Novak, J., Kumar, R., Raghavan, P., & Tomkins, A. (2005). Geographic routing in social networks. *PNAS, 102*(33), 11623–11628.
Mayer-Schoenberger, V., & Cukier, K. (2013). *Big data: A revolution that will transform how we live, work and think*. London: John Murray.
McLuhan, M. (1964). *Understanding media: The extension of man*. Cambridge: MIT Press.
Metaxas, P. T., & Mustafaraj, E. (2012). Social media and the elections. *Science, 338*, 472–473.
Miller, G. A. (1995). WordNet: A lexical database for English. *Communications of the ACM, 38* (11), 39–41.
Morales, G. D., Gionis, A., & Lucchese, C. (2012). From chatter to headlines: Harnessing the real-time web for personalized news recommendation. In *Proceedings of the Fifth ACM International Conference on Web Search and Data Mining (WSDM'12)* (pp. 153–162).
Negroponte, N. (1995). *Being digital*. London: Hodder & Stoughton.
Pasternack, J., & Roth, D. (2009). Extracting article text from the Web with Maximum subsequence segmentation. In *Proceedings of the 18th International Conference on World Wide Web (WWW2009)* (pp. 971–980).
Quercia, D., Capra, L., & Crowcroft, J. (2012). The social world of Twitter: Topics, geography, and emotions. In *International AAAI Conference on Weblogs and Social Media*.
Ranger, C., Raghuraman, R., Penmets, A., Bradski, G., & Kozyrakism, C. (2007). Evaluating MapReduce for multi-core and multiprocessor systems. In *IEEE 13th International Symposium on HPCA 2007* (pp. 13–24).
Reynolds, G. (2006, Winter). The rise of guerrilla media. *The New Atlantis* (pp. 97–111). This article is an excerpt from the book "An Army of Davids" by G. Reynolds.
Rodriguez, M. G., Leskovec, J., & Schölkopf, B. (2013). Structure and dynamics of information pathways in online media. In *Proceedings of the Sixth ACM International Conference on Web Search and Data Mining (WSDM '13)* (pp. 23–32).
Ruan, Y., Purohit, H., Fuhry, D., Parthasarathy, S., & Sheth, A. (2012). Prediction of topic volume on Twitter. In *Proceedings of the Fourth International ACM Conference on Web Science*. ACM 978-1-4503-1228-8.
Simini, F., Gonzalez, M. C., Maritan, A., & Barabasi, A.-L. (2012). A universal model for mobility and migration patterns. *Nature, 484*, 96–100.

Somlo, G. L., & Howe, A. E. (2001). Incremental clustering for profile maintenance in information gathering web agents. In *Proceedings of the Fifth International Conference on Autonomous Agents* (pp. 262–269).

Stallman, R. (1999). The GNU operating system and the free software movement. In C. DiBona, S. Ockman, & M. Stone (Eds.), *Open sources voices from the open source revolution.* Sebastopol, CA: O'Reilly.

Sundaram, H., Xie, L., De Choudhury, M., Lin, Y.-R., & Natsev, A. (2012). Multimedia semantics: Interactions between content and community. *Proceedings of the IEEE, 100*(9), 2737–2758.

Taber, C. S., & Lodge, M. (2006). Motivated skepticism in the evaluation of political beliefs. *American Journal of Political Science, 50*(3), 755–769.

Takhteyev, Y., Gruzd, A., & Wellman, B. (2012). Geography of Twitter networks. *Social Networks., 34*(1), 73–81.

Tsur, O., & Rappoport, A. (2012). What's in a hashtag?: Content based prediction of the spread of ideas in microblogging communities. In *Proceedings of the Fifth ACM International Conference on Web Search and Data Mining (WSDM'12)* (pp. 643–652).

Ugander, J., Backstrom, L., Marlow, C., & Kleinberg, J. (2012). Structural diversity in social contagion. *PNAS, 109*, 5962–5966.

Vanderbilt, T. (2013, April). The Nielsen family is dead. *Wired.* http://www.wired.com/underwire/2013/03/nielsen-family-is-dead/

Vespignani, A. (2010). Complex networks: The fragility of interdependency. *Nature, 464*(7291), 984–985. doi:10.1038/464984a.

Villars, R. L., Olofson, C. W., & Eastwood, M. (2011). Big data: What is and why you should care. *IDC.*

Wang, C., & Huberman, B. A. (2012a). How random are online social interactions? *Scientific Reports, 2*, 633. doi:10.1083/srep00633.

Wang, C., & Huberman, B. A. (2012b). Long trend dynamics in social media. *EPJ Data Science, 1*, 2.

Web-strategist. (2013). Accessed July 2013, from http://www.web-strategist.com/blog/2012/07/19/altimeter-report-paid-owned-earned-converged-media/

Weng, L., Flammini, A., Vespignani, A., & Menczer, F. (2012). Competition among memes in a world with limited attention. *Scientific Report, 2*, 335.

Ytreberg, E. (2011). Convergence: Essentially confused? *New Media & Society, 13*(3), 502–508.

Info Traders, Innovation and Information Spread

Biser Zlatanov and Maya Koleva

1 Introduction

Media convergence is often viewed as a disruptive change that alters the traditional communication between industries, markets, audiences, genres and simultaneously enables consumers to articulate and repurpose content through converging platforms. However, convergence also leads to redistribution of influence in social, cultural and economic sense and eventually produces social divergence. These are the side effects of the multilevel spread of information and the (limited) ability of the median type of online consumers to process enormous amounts of data that convergence has made abundant.

It may sound as an anecdote, but the problem of information spread and the ability to process information is connected with the common viruses that make researcher's life not so easy. This is evident in the Apple Inc. (AAPL) case which was recently thoroughly researched by a student who presented her MA thesis on social responsibility and charity of famous thought leaders such as Steve Jobs. Interestingly, after coming back as CEO in Apple in the 1990s, Jobs ceased all social responsibility programs and never re-launched them. Even tough, the student was aware of this fact, she insisted on Jobs charity. As an argument the student cited popular sites, which should have asserted that "Jobs is probably the most charitable guy on the planet", because rather than resorting to corporate charity as usual he was "massively improving quality of life" through selling great products at lower prices. The student used as a reference the statements of Ms. I, managing director of an equity fund who also happened to be an author of the AAPL generous valuation of "1–3 trillion dollars", before the slump of the Apple Inc. shares this autumn (2012). The same position was echoed by a few media such as AOL Daily Finance (Altucher, 2011a), Huffingtonpost Blog (Altucher, 2011b) and Ms. I's personal blog.

B. Zlatanov (✉) • M. Koleva
Sofia University, Moskovska 49, 1504 Sofia, Bulgaria
e-mail: biserz@abv.bg; maya.fadeeva@gmail.com

A. Lugmayr, C. Dal Zotto (eds.), *Media Convergence Handbook - Vol. 1*, Media Business and Innovation, DOI 10.1007/978-3-642-54484-2_18

In this way this controversial analysis on being charitable without making charity gained credibility and representativeness and appeared in the MA thesis of the student. Furthermore, in support of her thesis the student cited (The Economist Blog 2011), which in fact opposed and criticized this reasoning, but unfortunately cited Ms. I's insights, from where the student rushingly picked them.

Since we are working in a higher education institution which prepares future Journalists and PR experts, this incident made us to inquire for the vast array of such individuals or organisations who are communicative, web-savvy and advantageous in terms of their access to technologies, social networks and who are able to "sell" the information in one way or another. We call them here info traders.

2 Definition and Literature Review

2.1 Info Traders and Info Trading in Networks: Definition

The prominence of info traders is based on the premises that drive the technological process which is here a dependent variable and the limited ability of individuals to process information which is understood as a parameter (independent variable). The convergence in its technological, cultural and social sense is challenging this through the phenomenon of info trading in the context of powerful networks and information-rich and time-poor online social networkers. Trading of information is an approach in directing content or data that is transformed in information online or offline with the sole intention to create demand for it or to "sell" it, i.e. to provoke someone to comment, share, discuss, cite or keep the information.

At least three possible definitions (dimensions) of info traders could be discerned: (a) an individual (a node in the social network), who trades information as a commodity or good, but who is not motivated to inspect its quality; (b) a node in the social network that possesses the ability to trade i.e. to influence (an)other node(s); (c) a node in the social network that influences the ability of the other (neighbouring) nodes to trade in the network.

Info trading is a specific feature of communication which occurs in the converging, (world wide) web-based networks where the relevant search does not occur in the same manner that it occurs in the bricks-and-mortar environment. One of the main points of difference is the quantity of available information the user could access in both these environments. The second point is the structure and the nature of the social and economic webs. The third point is the mediation process and structure of the network—the sources and the links which resend information between the communicating parties (Zlatanov & Koleva, 2014). These differences impact the way information is valued and eventually traded. While in the every day face-to-face communication individuals value information according to the prioritisation manner contingent on the social authority of the source (Katz & Lazersfeld, 1955) or closeness (reference groups), in the web-based communication the pattern is different.

2.2 How Info Trader and Info Trading Differ from Opinion Leader (ship)

Originally Katz and Lazersfeld defined opinion leaders as the individuals who are likely to influence other people (1955:3). Later Merton used the term "influentials" (1968) as their influence were prescribed to be based on the shared values, competence and information exposure (by media). The domain of the influentials or leaders in the scientific literature was further developed with the diffusion of innovation literature (Bass, 1969; Chan & Misra, 1990; Rogers, 2003).

The concept of influence thresholds where agents took binary decisions and interact (Granovetter, 1978; Miller & Page, 2004; Schelling, 1978; Watts & Dodd, 2007) was introduced with the spread of literature on social networks and game theory. The threshold rule (Granovetter, 1978) discusses the idea that for the spread of information or particular behaviour at least one of the following conditions should be fulfilled—availability of influentials and/or easily influenced individuals (nodes in the social networks). Watts and Dodd (2007) found that in most cases a set of activations of new information flow (cascade) was driven by "critical mass of easily influenced individuals". They introduced the term "initiator" which comes close to the idea of the info traders presented here.

The nature and the origin of the info trader are in many aspects quite different from that of the influential or opinion leader. However, this does not mean that an info trader could not be an influential. Inversely, it states that the result of her actions is not dependent on his influence in the network, but on the "quality" of the information, as perceived by the other nodes, and the nodes thresholds. Under perceived "quality," we do not assume that information is formally accepted and integrated into the complex cognitive system of opinions and biases, since nodes in the network could strongly oppose the information. In the light of the opportunity cost concept, while recipients are receiving the "traded" information they are missing an opportunity to use other sources of information. Thus, "quality" is the ability of info trading to create congestion in the information flow and provoke reaction (either positive, neutral or negative), which would create maximum cost for the recipient to switch the source (supplier) of information.

The info trader possesses the ability to trade such information in a sense that she is able to produce, distribute and eventually infect or influence the remaining nodes in the network. The susceptibility of the nodes in the network could be latent (they resist/oppose to the information, but involuntarily spread it) or sustainable (the nodes are influenced and/or infected). In the second place, one's opinion matters only because of the motives of exchange, called here latent approval. Latent approval comprises a mixture of behavioral (convenience and inertia-driven) motives, and attitudinal motives (e.g. positive or negative opinions). People buy in information driven by different motives—availability, ease, strong opposition, centrality of the source they use as a reference point. The last motive, centrality, is understood in a geographical, cultural, demographic sense or the sense which is given by network theory (Freeman Wasserman & Faust, 1994).

Additionally, for the info trader the structure of the network is an important "threshold". The behavioral experiments conducted by Kearns, Suri, and Montfort (2006) and Judd and Kearns (2008) proved that individuals' influence thresholds and decisions in terms of network trading could be dependent on the structure of the network and the network environment in which individuals have to react.

Thus, we could explore if the info trader's ability to spread information and activate cascade could depend on the structure of the network and eventually his influence or prestige in directed graphs, measured in these case studies by eigenvector centrality (Bonacich, 1979). However, even nodes with no or relatively low influence in the network could activate a cascade provided that network structure allows it.

2.3 Info Traders and the Diffusion of Innovation

On the grounds of research on the problem of consumption of new information and innovation in particular, models of innovation diffusion assume that individuals could value correctly innovation once it had passed sufficient threshold level or once they have tested it. However, what happens when the fabricated consent produces wrong impression and thus, wrong valuation of the innovation? What happens when neither of the competing innovations is perceived as superior compared to the other? Or does innovation brings positive changes as a whole? Does always technology adoption is based on consensus? Is it mostly coordination problem as it is presumed in the game theory?

The research on the habits of information consumption is connected with the cognitive ability of individuals to construct and process data and here we will focus on the main theories that have large impact on social understanding of new technologies and their adoption.

The most recognised theory is that of the opinion leadership defined by Katz and Lazersfeld (1955), who built the two step flow model later extended to the multistep flow of information model. The opinion leader is an active media user, who interprets the meaning of media messages for lower-end media users. Typically the opinion leadership implies high level of acceptance of his or her opinions. The leader concept lies on the foundations of authority established due to competences, expressed and shared values and nature of social work that are discernable by his followers (Katz).

This understanding of leadership was incorporated in the diffusion of innovations theory by Rogers (2003), who focused on the ability of an opinion leader to spread innovation throughout a community. In terms of innovation and technology adoption what is important is the degree to which an individual is able to influence informally other individuals' attitudes or overt behavior in a desired way. Rogers (2003) added the concept of change agent who attempts to influence clients innovation-decisions in a direction that is deemed desirable by a change agency.

The normalisation process theory focuses on the implementation of the new information (technology or process) which is routinely embedded in a social context. The application is determined by four social mechanisms that are expressed through sense-making (coherence) work, engagement work (cognitive participation), the work of enacting a practice (collective action), and the work of understanding and appraising its effects (May & Finch, 2009).

The actor-network theory uses an alternative approach to innovation and studies how new information is internalised and translated in the network of human and non-human actors through enrolment of allies, mobilisation of resources and translation of interests. The theory discerns intermediation and mediator roles in the process of innovation transition and learning through adoption (Callon, 1986; Latour, 2005). The actor-network theory pays attention to the structural constraints in the use of a technology and specifies the roles of mediators who bridge the gap between social and technical dimensions of the process of technological adoption.

Social construction of technology theory adds new dimension to the research on the spread of the consumption habits of information. It is based on the idea that technology is socially constructed and is determined by human actions. It opposes the core notion that innovation determines human behaviour inferring that technology adoption is dependant on interaction between relevant social groups making use of it. The existence of multiple social groups implies multiple perspectives on the new technology (Bijker & Pinch, 1984). Thus, the consumption of new information through adoption of the innovation depends predominantly on the variance of the capacity, compatibility and interoperability of different interpretation frameworks.

Researchers in this field deal with the evolution of an innovation network and try to bridge empirical findings and conceptual understanding (Wigand and Frankwick, 1989; Nowotny et al., 2001). Nowotny et al. (2001) present typologies of social knowledge in the evaluation of innovation networks. They recognise the role of the users as key actors in an innovation process participating fully in the knowledge production process—the generation of insights, knowledge and innovation in a more distributed way. The technological solution appears as a result of the process of interactions between users and suppliers.

Constructivist researchers such as Bijker and Callon, who define the principle of general symmetry, both denounced technical reductionism and social reductionism as unproductive to the understanding of technological process (Bijker, 1995; Callon, 1986). Bijker further asserted that the technological process is often more connected with network-building rather than "laboratory" work on technical invention (Bijker, 1995). The interpretation of technological process needs a complex framework that includes different domains of knowledge—technical sciences, economics, sociology.

The theories of innovation diffusion add different actors and adopter categories in the process. The classic version of Rogers (2003) defined five categories, based on the different time windows in the adoption of the technology: innovators, early adopters, early majority, late majority and laggards. Bass (1969) model of innovation which is intensively applied in marketing added some external heterogeneity to the classification of adopters and their needs.

There are, however, models that assume that potential consumers of the information—in the case of technology—actors/adopters, have discernable needs and are not classified on the grounds of the time of adoption. Such models are defined by the threshold that usually explains more tightly the categories of adopters. The so-called probit models assign different threshold values to the adopters who react to the information dynamics accordingly. It assumes that during the process of communication the individuals use cost-benefit analysis to approach new technology. The decision for adoption is taken when it satisfies the benefits threshold, which has different levels for each group of adopters.

Bayesian models of diffusion discuss the availability or lack of information as a constraint to diffusion. Potential adopters hold different beliefs regarding the value of the innovation, which they may revise according to the results of trials to test the innovation (Tidd, 2006). There is no supposition for the perfect information which is the dominant assumption of the earlier models. Under imperfect information, individuals receive private signals and it is highly likely that the imitation either cannot take place or it needs additional prerequisites in order to start. Individuals cannot learn from the trials. Information advantage does not suggest that better-informed potential adopters will adopt an innovation earlier than the less well informed, which was an assumption of earlier models.

3 Method

We use case studies to outline two main problems of the info trader concept—info trader as an influencer who changes the direction of the information flow and info trader as a cascade initiator.

In the first role we call the info trader information stream changer which is connected with ability to influence the direction of the information stream. This role is important in the study of phenomenon of path dependence which has profound application in the theory of innovation diffusion.

The second role is part of the research on the technology of rumours spread and in this case the rumour for the bankruptcy of a bank.

To study the problems presented above first we performed content analysis and then we apply social network analysis and its measures. Since we studied citations networks of influence we applied measures which are appropriate for directed networks. The eigenvector centrality mentioned above measured the influence in the network. Additionally we included degree, in-degree and out-degree measured which are connected with the measurement of other centrality indexes. We applied also closeness centrality and eccentricity to find how close, respectively far are the nodes who trade information in network from the rest. Modularity algorithm was used to detect the communities in the network and the role of influentials in them.

The calculations and visualisation of the network was performed using Gephi software. Additionally, we also used Google Trends online service.

4 The Information Stream Changer

The question of congestion, influence of changing information and information cascade (Easly & Kleinberg, 2010) is connected with the phenomenon of the path dependence. Since the path dependence is defined and realised as a "sensitive dependence on initial conditions" (Arthur, 1997; Liebowitz & Margolis, 1995) the critical part of this assumption lies on the initial conditions. Who controls the initial conditions he/she determines the trajectory of the path. This assumption is introduced in the body of "creative destruction" research: coordination games (Daidj, Grazia, & Hammoudi, 2010), social network analysis (Christ & Slowak, 2009).

The battle for establishing the winning standards is a part of the history of the so-called format or standard wars. Since the research based on the theory of path dependence and game theory gained popularity the battlefield populated by modern consortiums has been armed with refined scientific weapons. The rate of adoption and the formation and power of coalitions, in most cases consortiums of manufacturers and distributors of the new technology-related products, are the main topics that remain in the researchers' focus.

Through its Betamax adventure Sony was the company which discovered that the successful adoption does not only depend on the choice of consumers, but rather on the strength of coalitions to enforce it. This experience eventually marked its behaviour on its future strategic moves and eventually in the introduction of innovations. The same holds true in the Blu-Ray versus HD-DVD format war, where Sony played one of the leading roles.

The history of the rivalry is well known and started when Sony and Toshiba decided to introduce next generation technical standard to replace the old DVD. There was no evidence that either of the standards was superior to the other or performed better (Christ & Slowak, 2009). The adoption of a new technology or dominant design is often a problem of coordination in the sense implied by coordination games in the game theory and the Blu-Ray/HD-DVD rivalry was not an exception.

The history of the rivalry was covered by popular media and we identified a few events which presumably throw light on the Blu-Ray/HD-DVD controversy. These events have marked the role of players who were not suspected to take the leading part by the popular theories of leadership and innovation diffusion. In the wake of this, we were able to discern several critical points or levels in coalitions that could nominate its leaders—movie studios, electronic manufacturers, movie rental companies, large retailers, technical standard developers. Any of them could have introduced change in the battle and some of them did, however only one of them succeeded to change the balance of the system play.

The rivalry between the consortium led by Sony and the consortium led by Toshiba (and Microsoft) took place between February/March 2002 to February 2008, when Toshiba announced defeat. From our point of view this could be interpreted as an obvious and logical result of this rivalry after an important player such as Warner Bros had left the Toshiba coalition, known as the DVD Forum.

However, judging by the circumstances and the year(s)-long resistance of the DVD Forum to the superior in terms of both number of members and sales Blu-Ray Disc Association (BDA), the Toshiba decision could be viewed as sudden and surprising.

As the researchers of this format war concluded, the reasons were complex and could be sought in different directions. The study of Christ and Slowak (2009) showed that the HD-DVD consortium lost, because it was supported by small part of the content producers, while Daidj et al. (2010) found that we need a conceptual framework at different levels (studios and manufacturers involved) to understand the format war. The latter research supposed that the Warner Bros decision had changed the balance and gave the Blu-Ray consortium a dominant position that would have made the rivalry economically unreasonable. Additionally, there was a general impression that the adult movie industry and movie rental service providers were the key to the problem (Christ & Slowak, 2009: 13).

The proponents of Blu-Ray consortium structural advantage and the role of movie studios lose a sight of the view. There are, however, at least two topics that do not receive much attention: first, few months before Warner Bros's defection to Blu-Ray, two large studios the Paramount and Dream Works announced they would back exclusively Toshiba; second, the number of movie releases produced on Bly-Ray or HD-DVD during the standard war did not differ systematically (Christ & Slowak, 2009).

Interestingly, all the researchers presume that the time of Warner Bros' turnaround somehow coincided with the decision of Toshiba to accept defeat and was the key to the problem. Probably it was—however, this time we will not assume Warner Bros with its functional characteristics and economical role, but the role of the company in the information stream. When the announcement that Warner Bros would support BDA was made public, this produced congestion in the information stream, which was driven by surprise and novelty.

The information that Warner Bros had moved to the rival was surprising, because the company started as a supporter solely of the HD-DVD consortium. Later, in 2005 it changed its position and started to produce films in both standards. The role of Warner Bros is that of the information stream changer, because it "influenced" effectively the demand for the information received and searched for in the HD-DVD/Blu-Ray controversy. The unveiling of the decision changed the direction of the information flow with regard to the topics discussed in the diffusion of innovations' research, namely introduction of the dominant design and the rate of adoption. We could speculate that this had the so-called "butterfly effect" known from the chaos theory, which implies that we could hardly define the problem using models that rest on the logic which lie behind linearity.

Notwithstanding the aforesaid, through applying the demand-side view of the format war and information spread we could find how the decisions of the players in the controversy and the message produced shape the online behaviour of the (message's) consumer. Until the Warner Bros turnaround in January 2008, the spread of the diffusion of information demonstrated the dominance of HD-DVD standard. The line chart in Fig. 1 presents the proportions between the search terms

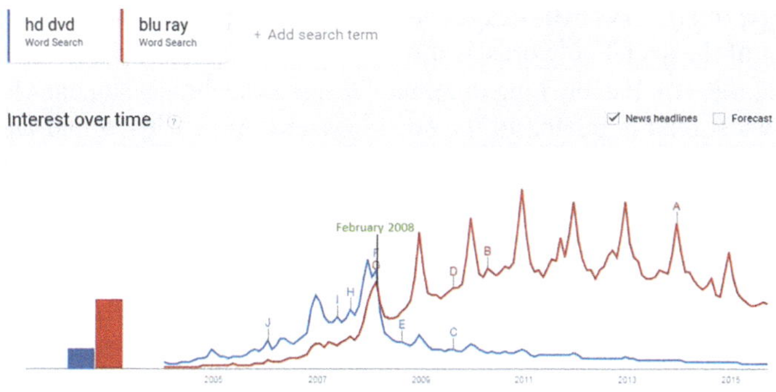

Fig. 1 Information search (by Google Trends)

"hd dvd" and "blu ray" in Google. The implied results by the chart, however, show only the online dominance.

From the Google trends chart it is easily visible that until February 2008 "hd dvd" searches outnumbered "blu ray" searches. Actually in December 2007 "hd dvd" search term reached its highest peak point, while in January it experienced a sharp drop and the spread between the two search terms is converging. After February, the trend changed in favour of "blu ray". Actually, the trend changed in March after other events of the technical standard controversy took place.

Admittedly, it could be argued that Warner Bros is not the information stream changer since it allegedly announced its intentions to back Sony in the beginning of January (BBC 2008). However, after the announcement the information cascade followed. In mid-February, Toshiba acknowledged its defeat, but before that the domino effect already had swept Blockbuster, FS Film, New Line Cinema. In February, the defections continued with retailers such as Best buy and Wal-Mart—Woolworth's did that a month earlier, movie rentals such as Netflix defected as well.

It should be pointed out that while playing HD-DVD supporter, the Warner Bros's share of the HD-DVD discs sales has become substantial and as a whole, the media company had the largest share of film revenues among all the other studios. It was obvious that the company would play important role among the studios in both consortia, especially in HD-DVD consortium, because from the inception of the latter the largest film studios such as Walt Disney, Twentieth Century Fox, as well as Sony Pictures and Metro-Goldwyn-Mayer supported Blu-Ray. Inevitably, this raises the question of Toshiba's sudden decision to accept defeat after succeeding for about 4 years to resist to the obvious dominance of Sony in the movie studios coalition. The cascade of events that followed Warner's decision to back Sony, implies that there should be internal instability in the HD-DVD coalition and low influence thresholds. Toshiba recognised the defeat fast in just 43 days. From Warner Bros announcement on January 4th until February 16th 2008, many companies ceased their support, including newly-adopted exclusive supporter Paramount and long-time backer Universal.

The case of HD-DVD/Blu-Ray is not only an example of strategic behaviour, but also a test of the social networks as it was pointed in Christ and Slowak (2009) and Daidj et al. (2010). Establishing dominant design or technical standard followed by innovations spread is related to the social system, driven by formal and informal institutions (North, 1981; Willamson, 1985). Even coalitions born to manage the process of diffusion from the initial steps, if we accept the non-cooperative approach and endogenous coalition formation (Bloch, 1995; Daidj et al., 2010; Hart & Kurz, 1983), are subordinates of the institutions in society. The introduction of change in the social systems, established in this case on the ground of external competition and internal coordination, is the way to test the robustness of the internal rules and institutions inherent to the network of relations.

Change usually comes through the diffusion of information and sometimes through players, evidently accepted as non-leaders. Thus, the information stream changers, such as Warner Bros, play crucial role not only in testing the stability of the coalitions, but also to recognize the internal or hidden properties and processes (Bijker, 1995) which keep the system together.

5 Influentials, Info Traders and Cascades in the Bank Bankruptcy Controversy

The opinion leadership could hardly be defined in the current state of variability of sources of information production and spread—common forms of participatory journalism and its practices such as citizen journalism, content farming, crowd sourcing. Currently everyone, no matter of their competence or position in the social network and using any distribution platform, could become an information stream changer, provided that the adopted threshold levels of influence are low enough and do not need source of information we perceive as established and credible. This is evident in the bank bankruptcy controversy in Bulgaria, which coincidentally happened to be in the same year as the exodus of the HD-DVD/Bly-Ray battle—namely, 2008. Probably, it proved that the online information market-place is mature enough to produce and spread its influence in the complex and dynamic offline environment.

The alleged bank crisis in Bulgaria started in May 2008 when a Bulgarian bank, namely First Investment Bank (FIB), changed its information system software in an attempt to introduce an innovative service in the bank market. This attempt started misfortunately for the bank's officials, since during the "trial" period they encountered problems with the new software and it took more time to process information of their clients. Simultaneously, the financial crisis had already started, but it barely found place in credible local public sources of information in Bulgaria. However, people close to bank and other financial circles were sensitive to this information and were concerned with the incoming news. Information about the impact of the financial crisis on the stability of the local financial institutions was scattered. In the deficiency of credible sources of information, the information demand found its own supply or sources which were formed in the online community forums and eventually produced community-born news. These were the premises for the inception of the rumour of the bank's bankruptcy, which started in several web portals. The rest was just the logic of the events that followed.

In May, public opinion was shocked by the alleged news that First Investment Bank is going to bankruptcy due to liquidity problems. More and more bank depositors rushed to the bank to withdraw their deposits, but met people queuing in front of bank desks, while bank officials nervously processed their accounts information. It took hours for lucky depositors, as they were thought to be, to withdraw their savings, creating an epidemic model of inverted innovation stream. These were the switching cost paid by both the customers and the bank officials. Bank officials, who albeit not helpless, were not as effective as they are supposed to be in processing the necessary documents in the new authorisation system, while striving to assist customers to close their deposit accounts.

5.1 The Inception of a Rumour

The public outcry against the introduction of the software innovation, which was wrongly perceived a as liquidity problem, produced the real danger of spreading to other banks and the bank's bankruptcy started to seem highly probable. The country's central bank took preventative measures to support the bank if it really experienced liquidity problems and to calm anxious customers. The Bulgarian Prime Minister at the time, Sergey Stanishev, made a statement to calm the public and then in the middle of May, the situation was under control. Afterwards, few Internet forums, such as the web portal BG-mama with target user groups Bulgarian housewives and young mothers, was accused for spreading the false rumour.

Although it could be stated that the birth of the rumour was in web portals, it is still not clear where exactly the rumour first started. What is more important in the story was the question which source of reference traditional media and eventually people as a whole tended to use. A blogger (Oggin.net) discussing the alleged liquidity problems of the FIB stated in his post that responsible for the rumour was the BG-mama online service, unambiguously called the Gossiper's Forum, which used to host information for young mothers—i.e. perceived as a place where they exchanged opinions, advices and gossips. This information was afterwards literally copied by traditional media citing that BG-mama.com has created and was responsible for the spread of the rumour. Media buzz was here the "push," which additionally provoked the dramatic rise of online searching of the terms "bankruptcy FIB" and "bgmama."

The information presented in the chart on Fig. 2 shows the changes in the Google searches of terms "FIB bankruptcy", "bankruptcy", "bgmama" and another site that allegedly took part in the rumour spread—"investor.bg". Between the first two terms—"FIB bankruptcy" and "bankruptcy" there is correlation and tipping point with a peak in May 2008. The demand for the two search terms is followed by a

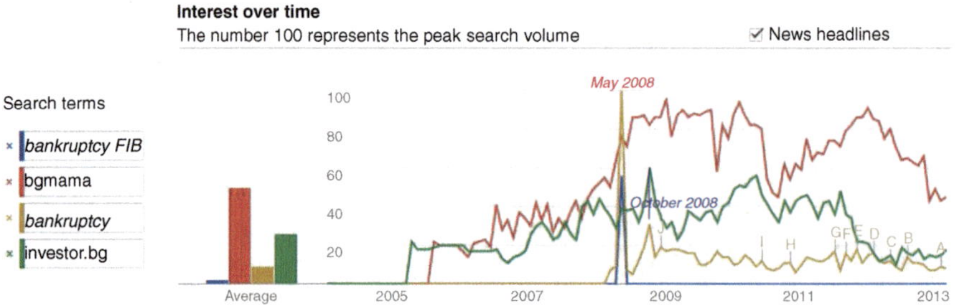

Fig. 2 Search term dynamics

sharp peak in the search for "bgmama", while search term "investor.bg" experienced a moderate rise. The difference is even more striking between "investor.bg" and the rest of the searches in the chart after May when the former experiences smooth decline in demand. By contrast, the drop in bankruptcy searches after the same month is sharp, as well as the web portal searches. It is interesting to point out that the demand for the search item "bankruptcy" in May outnumbers more than three times the demand in the second experienced peak of the same one in October, which coincides with the burst of the crisis after Lehman Brothers bankruptcy.

5.2 The Spread of a Rumour in a Closed Network

The role and the influence of the Bg-mama forum in the events that eventually caused the mass spread of the rumour was the main reason we considered it for the social network analysis. We focused on BG-mama for three reasons:

- the portal was connected with the rise in activity of people searching for information on the alleged bankruptcy, correlated with the demand in the bg-mama site;
- the importance of the site, the traditional media coverage and its effect that added representativeness and authenticity of its story even though media corrected their position;
- the widespread and resistant notion, which was not supported by enough evidence, that this Internet forum was responsible for spreading the rumour.

We used the archive of 295 posts in the Gossipers' Forum, which were created between 18 of April until the 19 of May when the administrators of the site closed the discussion thread. The rumour in the forum was "initiated" when a user with the alias "petiden" ("fifth day"), asked an ambiguous question and thus started the discussion titled "For the bankruptcy of a bank" (BG-mama.com, 2008):

> From the beginning of April there are rumours for a bank bankruptcy and I'm wondering if they are serious or it has been an April Fools' Day joke. Have you heard anything like that recently?

After few "gossipers" had posted their opinion the same user shared his thoughts just one more time:

> I don't wanna say which (bank) so as not to spread the alarm, but it had already started. . .Some people have already closed their deposits.

The discussion followed and a cascade was activated. User "petiden" literally disappeared from the forum, while more and more people entered to see what happened and what the vocal minority of gossipers shared in their opinions.

The directed graph on Fig. 3 reveals something more interesting: the discussion initiator disappears when we apply the influence measures, based on the citations in the forum.

The citation network is suitable for evaluating the dynamics of such discussion, taking place in an online social medium. The users who appear on this map were actively engaged in the conversation around the rumour: they were asking questions or suggesting answers. The most cited users sparked debates: their opinions were either found most authoritative, or the other way around—were most contested and controversial.

In the presented graph, the most powerful node in the Gossiper's Forum is called "Selindhzyr" and it is quite likely that the individual behind it is male, judging by the citations and posts of his neighbours in the network. Afraid of legal prosecution or any other troubles with local authorities, some of the users, including "Selindhzyr", changed their usernames while the discussion was active. Thus, in such cases in the archive there were two cited nicknames of one the same source, but we were able to detect them during the content analysis.

Most of the users in the forum (depicted by the nodes on the graph), who posted their opinion, were cited by other users or themselves cited someone from the forum. Examining their citations, we created a directed network (graph). 64 % of the users who posted opinion in the analysed discussion thread were presented in this network, where inbound edges (links) mean the user was cited, and the outbound links showed the user has cited someone else. Another 36 % of the users engaged in this discussion were not represented on the graph, since they were not cited by the rest of the users in the forum. These included also the discussion initiator "petiden".

The densely interlinked structure of the network suggested there was a core of users, who engaged more actively than the rest. Interestingly, most of the links were one-directional, meaning of each connected pair of nodes, only one of cited the other. The content analysis revealed that most of these one-directional citation links mean general agreement with the statement of the other user or a continuation of their line of thought.

Two strong bi-directional links appear in the margins of the network – between the pairs with usernames "eagle" and "Katiusha" and "asza" and "Oktopod s roklia". The content of their forum posts revealed that both pairs engaged in heated verbal

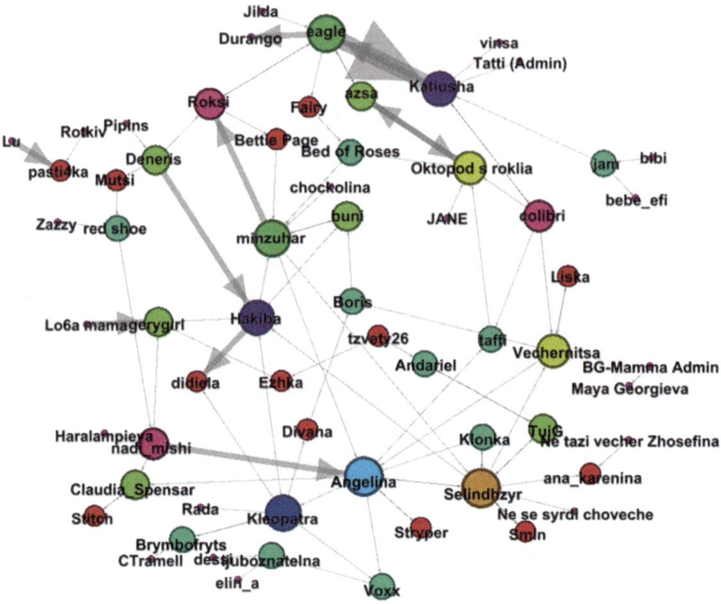

Fig. 3 The social network of influence (based on citations)

exchange: "Eagle" and "Katiusha" argued over whether anyone has ever been held responsible for spreading a bankruptcy rumour, while "asza" and "Oktopod s roklia" had disagreement over the financial literacy of the general public (keeping money under the mattress) and whether it was "right" or "just" to "throw dirt" on the bank.

The most cited and influential forum users in the network (measured through eigenvector centrality, Bonacich, 1979; Mizruchi & Bunting, 1981) were also the more "close" to the rest of the network—where closeness centrality is measured based on the out-degree of a node. There is correlation between the influence indicator (eigenvector centrality) and eccentricity/closeness centrality measure. This correlation is clearly visible through the data in Table 1, which lists the most powerful influentials in the Citation network.

The presented data has one main disadvantage—it does not give an idea of the influence of the forum users, who remained outside of the citation network. The more important implication is that we do not receive any information for the assumed influence of the node "petiden" which was not presented in the citation graph and metrics. Since the node has created just two edges in terms of posts (and no citations) its influence was not measured. The node received just two responses by one node with username "Lu", which was presented in the graph, but had 0.0 prestige and belonged to the community (modularity class) 6, where additionally belonged two other nodes. As it is seen from the graph on Fig. 3 none of these three nodes is among the influentials in the network.

Interestingly, the situation changes when after few posts the influentials entered the discussion. "Lu" received answer from the most powerful node in the network

Table 1 Measures of influence in the network of the seven most influential nodes

Id	In-degree	Out-degree	Degree	Eigenvector centrality	Modularity class	Eccentricity	Closeness centrality	Betweenness centrality
Selindhzyr	6	7	13	1	0	6	0.358	0.1132
Angelina	5	7	12	0.839	4	6	0.358	0.149
Klonka	2	1	3	0.66	0	7	0.266	0
Hakiba	4	3	7	0.527	3	5	0.304	0.0919
minzuhar	5	4	9	0.478	2	4	0.405	0.175
Buni	3	1	4	0.476	4	5	0.291	0.0201
Vechernitsa	3	3	6	0.468	4	7	0.293	0.06

and from here started the next shift in the discussion. “Selindhzyr” with prestige rank 1 was the one to first name the bank in question, but added that there was no evidence, and financial metrics do not support the allegations in the rumour. After him, the next two influential nodes “Angelina” (rank 2) and “Klonka” (rank 3) entered the discussion, presenting or pretending to be curious about the alleged situation of the bank. Simultaneously, a new rumour gained traction: that FIB was going to be bought by another financial institution. At this point, the discussion experienced a thematic shift. Users wondered what was going to happen with people with credit or debt exposure to the bank, namely mortgages and deposits, and the discussion entered into another cascade.

Some of the information stream changers who opened new information cascade were of relatively low prestige ranks in the citation network (“Lu”, “Claudia_Spensar”, “Oktopod s roklia”). This implies that influentials were not unimportant in the diffusion of rumours and information, but the nodes who could change the direction of the information stream in the network could be invisible or did not possess as high prestige, competence and verbally (or degree) affluent, as it is presumed.

Through content analysis we were able to identify and measure the influence of keywords, which served as indicators of the thematic aspects of the general conversation. In the presented keywords graph on Fig. 4, the most powerful keywords were “bank”, “rumour”, “money” and “bankruptcy”. Substantially less utilised were words such as “deposit” and “credit”, “mortgage” or “deposit withdrawal”.

Interestingly, the keyword analysis revealed that very few users mentioned terms with negative connotations such as “gossip” or “gossip place”/“gossip forum”, which suggests that most of the participants in this conversation were not particularly critical of the mode of the rumour spread in the information stream. Even though some aspects of offline customer behaviours did suggest there was “panic,” this keyword was also not among the most used terms in the analysed forum discussion.

Only two users mentioned the keyword “software”, showing that this rational explanation to the problem with slow processing of customer data remained on the margins of the discussion in this BG-mama thread.

The most central user on the keyword network was “eagle”, with eigenvector score of 0.3555, and with links to nine different keywords. This user was not among the most influential nodes on the citation network, but the keyword network clearly revealed that “eagle” touched upon different thematic aspect of the discussion. This made this user also important in the analysis of the rumour spread, and revealed how the two networks complement each other and inform the understanding of the rumour dynamics in this closed group.

“Eagle” actually attempted to break off the rumour by repeatedly highlighting that spreading rumours for bankruptcy was a criminal offence and it was not just harming a single financial institution, but all its customers, especially the ones falling for the rumours and withdrawing their deposits thus losing interest payments. Even tough user “eagle” was not successful in breaking off the rumour

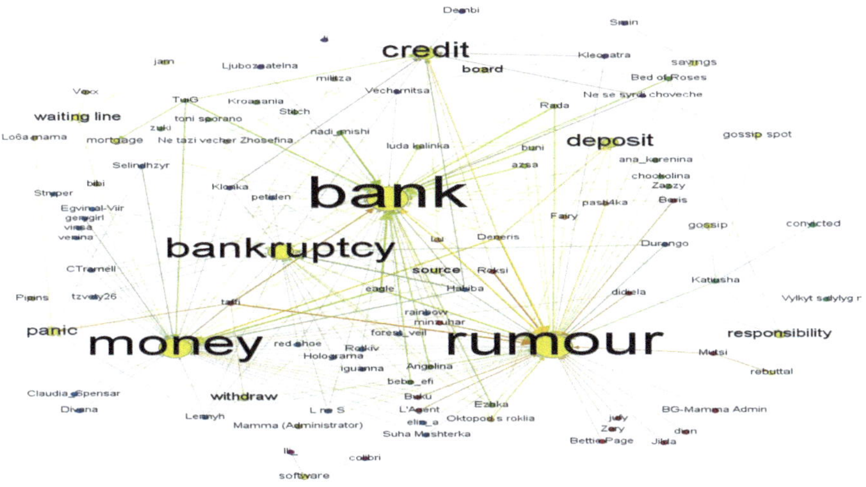

Fig. 4 The keyword network

flow, her activity suggests that the discussion participants were not united in their interpretation of the information.

This is confirmed by the second most central forum user in the keyword network—"Rada" with eigenvector score of 0.3528. She was among the discussion participants to clearly express an intention to withdraw her funds and a belief that the bank was in fact going down. In the presented example, the creator of the information streams was an unimportant node, but the information was provocative enough to receive the attention of the most influential nodes from the initial steps of the rumour spread. The influence analysis revealed that several participants in the discussion played a pivotal role in boosting the rumour by adding details or merely speculating. What made the rumour powerful within the limitation of this particular network, was that different details and different rumours were added gradually and were found compatible, which fostered the diffusion of the main message. In this sense, the role of the information stream changers was essential and accelerated both the rumour adoption and the diffusion rate. The rumours gave different pieces of information for different niche audiences which eventually led to the peak in the consumption of the overall information.

A key consideration in the analysis of the rumour spread and evolution in the 'bgmama' forum is the fact that the contents of the discussion were open. Our research focused solely on the "active" participants who posted in the forum thread, but could not encompass the potentially unlimited "passive" audience for this rumour and its development outside the online forum. However, much of the dynamics of the online rumour discussion could be applicable to rumour evolution models in real social networks.

It was observed during the study how rumours evolved, grew shorter and concise (Zhang et al., 2013) so as to be transmitted with a minimum effort in the network.

The active 'bgmama' forum participants were able to make the rumour more easy to comprehend by commenting on the key details, and an influential group of users put an effort in questioning the rumour altogether.

6 Conclusion

The technological changes that eventually led to convergence produce interesting phenomena online and introduced "segregation" rules in the online social networks. The convergence has led to redistribution of influence based on the modes of online transactions and the emergence of players with different network status who shaped the process of information spread. The models of technology adoption the mechanisms of network formation (Barabasi & Reka, 1999) are important determininants which affect the acceleration of the information diffusion and the rate of contagion and eventually the process of convergence.

The role of opinion leaders, influentials, brokers of information should be revised in the light of the impact of the often "invisible" actors in the communication process, or nodes in the social network, who could produce similar changes in the information stream. In this article we introduce the information stream changers—the nodes who are able with few actions to change the information stream or initiate discussion and thus evoke or prevent the diffusion of innovations.

In both presented case studies, the nodes in question grew in prominence not due to the qualities commonly ascribed to the opinion leaders, yet still successfully changed the process of innovation adoption and rumour spread. Their role was influenced by the network dynamics and the changing levels of trust thresholds within these networks. In the first case the consequences were irreversible, while in the second just the intervention of the government and central bank was able to prevent the probable crisis.

In the case of coalition collision during the Blu-Ray/HD-DVD format war the change was introduced by an influential company, which grew in prestige in the network with the decision to leave the coalition. Warner Bros' decision to support exclusively BDA led to domino effect in the rival coalition of HD-DVD supporters. This accelerated the process of defection among the supporters of the HD-DVD coalition which survived just 43 days after the Warner Bros's public announcement. It was presented by the numbers of searched terms that the decision eventually led to the change of the information stream. However, it should be noted that this decision itself could not produce the impact we prescribe to it, if it is separated from the network of relations and events that followed.

The second case study revealed a situation, in which the action of a literally invisible node in the influence network initiated discussion, which was then echoed in traditional media and diffused in different offline networks. While in the first case we could point that it was a coalition of companies and Warner Bros had influential position in the network, here we have different network. Interestingly, the most influential nodes in terms of prestige took part from the beginning of the discussion, but much of the nodes that produced information cascade and accelerated the rate of diffusion within that particular network did not possess power in the network.

The expression of opinion and activity does not create influence in the presented networks. The most of the nodes that eventually effectively transmitted the rumour, changing the direction of the discussion were not in the group of nodes commonly perceived as influentials.

Accidentally, during the study it was confirmed that contrary to the commonly held belief, men could become proficient gossipers, as the node which was on the peak of the gossipers' network in the presented story of bank bankruptcy was, in fact a man.

Both cases have to do with the coordination problems arising in dynamic systems. The introduction and diffusion of change in the social systems, established in this case on the grounds of external competition of network(s) and internal coordination, is possible to be initiated and in many steps performed by what we define as info traders. Info trading within a network is highly likely when external networks experience coordination problems and where demand of information is not saturated. Thus, the spread of information by the info trading networks is the way that demand finds its own supply—as it is the case with rumours. Simultaneously this gives the info trader the opportunity to test the robustness of the internal rules and institutions inherent to the network of relations.

The concepts presented here support the conclusion that online communities are rising in prominence and that they invade into different parts of the social life. It was argued that influence, opinion leadership or competence could not play a role in cases where influence thresholds are low. In such cases information diffusion is accelerated through different channels.

References

Altucher, J. (2011a). *10 unusual things you didn't know about Steve Jobs*. Available from www.huffingtonpost.com

Altucher, J. (2011b). *10 unusual things you didn't know about Steve Jobs*. Available from http://www.dailyfinance.com

Arthur, W. B. (1997). Competing technologies, increasing returns, and lock-in by historical small events. In B. Arthur (Ed.), *Increasing returns and path dependence in the economy* (pp. 13–32). Michigan: Michigan University Press.

Barabasi, L., & Reka, A. (1999, October 15). Emergence of scaling in random networks. *Science, 286*, 509–512.

Bass, F. (1969). A new product growth model for consumer durables. *Management Science, 15*, 215–228.

BBC News. (2008, January 7). *Warner backs Sony blu ray format*. Available from http://news.bbc.co.uk/2/hi/technology/7174591.stm

BG-mama.com (2008). *Discussion forum*. Available: http://www.bg-mamma.com/?topic=285851.0#

Bijker, W. E. (1995). *Of bicycles, bakelites and bulbs: Towards a theory of sociotechnical change*. Cambridge, MA: MIT Press.

Bijker, W. E., & Pinch, J. T. (1984). The social construction of facts and artifacts: Or how the sociology of science and the sociology of technology might benefit of each other. In W. E.

Bijker, P. T. Hughes, & J. T. Pinch (Eds.), *The social construction of technological systems* (pp. 17–50). Cambridge, MA: MIT Press.

Bloch, F. (1995). Endogenous structures of association of oligopolies. *The RAND Journal of Economics, 26*(3), 537–556.

Bonacich, P. (1979). The common structure semigroup: A replacement for the Boorman and White joint reduction. *American Journal of Sociology, 86*, 159–166.

Callon, M. (1986). Some elements of a sociology of translation: Domestication of the scallops and the fishermen of St Brieuc Bay. In J. Law (Ed.), *Power, action and belief: A new sociology of knowledge*. London: Routledge & Kegan Paul.

Chan, K., & Misra, S. (1990). Characteristics of the opinion leader: A new dimension. *Journal of Advertising, 19*, 53–60.

Christ, J., & Slowak, A. (2009). *Why Blu-Ray vs. HD-DVD is not VHS vs. Betamax: The co-evolution of standard-setting consortia*. FZID Discussion Papers, University of Hohenheim.

Daidj, N., Grazia, C., & Hammoudi, A. (2010). Introduction to the non-cooperative approach to coalition formation: The case of the Blu-Ray/HD-DVD standards' war. *Journal of Media Economics, 23*(4), 192–215.

Easly, D., & Kleinberg, J. (2010). *Networks, crowds, and markets: Reasoning about a highly connected world*. Cambridge: Cambridge University Press.

Granovetter, M. (1978). Threshold models of collective behavior. *American Journal of Sociology, 83*, 1420–1443.

Hart, S., & Kurz, M. (1983). Endogenous formation of coalitions. *Econometrica, 51*, 1047–1064.

Judd, S., & Kearns, M. (2008). Behavioral experiments in networked trade. In T. Sandholm, J. Riedl, & L. Fortnow (Eds.), *Proceedings of the 2008 ACM Conference on Electronic Commerce* (pp. 150–159). New York: Association for Computing Machinery.

Katz, E., & Lazersfeld, P. (1955). *Personal influence: The part played by people in the flow of mass communications*. Glencoe, IL: Free Press.

Kearns, M., Suri, S., & Montfort, N. (2006). An experimental study of the coloring problem on human subject networks. *Science, 313*, 824–827.

Latour, B. (2005). *Reassembling the Social: An introduction to actor-network-theory*. New York: Oxford University Press.

Liebowitz, S. J., & Margolis, S. E. (1995). Path dependence, lock-in, and history. *Journal of Law, Economics and Organization, 11*(1), 205–226.

May, C., & Finch, T. (2009). Implementation, embedding, and integration: An outline of normalization process theory. *Sociology, 43*(3), 535–554.

Merton, R. (1968). Patterns of influence: Local and cosmopolitan influentials. In R. Merton (Ed.), *Social theory and social structure* (pp. 441–474). New York: FreePress.

Miller, J., & Page, S. (2004). *Standing ovation problem*. Available from http://www2.econ.iastate.edu

Mizruchi, M. S., & Bunting, D. (1981). Influence in corporate networks: An examination of four measures. *Administrative Science Quarterly, 26*(3), 475–489.

North, D. (1981). *Structure and change in economic history*. New York: Norton.

Nowotny, H., Scott, P., & Gibbons, M. (2001). *Re-thinking science: Knowledge and the public in an age of uncertainty*. Cambridge, UK: Polity Press.

Rogers, E. (2003). *Diffusion of innovations*. New York: Free Press.

Schelling, T. (1978). *Micromotives and macrobehavior*. New York: Norton.

The Economist. (2011). *Steve Jobs: Beauty justifies wealth*. Available from www.economist.com

Tidd, J. (2006). A review of innovation models. *Imperial College London, 16*

Wasserman, S., & Faust, K. (1994). *Social network analysis: Methods and applications*. Cambridge: Cambridge University Press.

Watts, D., & Dodd, P. (2007). Influentials, networks and public opinion formation. *Journal of Consumer Research, 34*, 441–458.

Wigand, R. T., & Frankwick, G. L. (1989). Inter–organizational communication and technology transfer: Industry–government–university linkages. *International Journal of Technology Management, 4*(1), 63–76.

Willamson, O. (1985). *The economic institutions of capitalism* (pp. 55–60). New York: The Free Press.

Zhang, Y., Zhou, S., Zhang, Z., Guan, J., Zhou, S. (2013). Rumor evolution in social networks. *Physical Review* E87. Available from http://arxiv.org/pdf/1102.1487.pdf

Zlatanov, B., & Koleva, M. (2014). *Networks of collective power: (Non)movements and semantic networks*. Conference paper, ECREA Conference, Lisbon.

Between M-Internet Users and Stakeholders: Convergence or Divergence?

Manuel José Damásio, Sara Henriques, Inês Teixeira-Botelho, and Patrícia Dias

1 Introduction and Problem Discussion

The consequences of technological mediation on human social forms of interaction, has always been one of the concerns for researchers within the field of communication and human-computer interaction (HCI). Nowadays the internet is accessible through a multiplicity of mobile devices, such as mobile phones, tablets, mp3, GPS or others. The convergence between the internet and mobile phones has resulted in a new type of devices, the so-called smartphones. Using such devices to access the internet is part of what came to be named m-internet. These new devices largely exceed the functions of a phone, blending voice communication, internet access and computer like functions. The usage of these devices and the possibility of accessing internet features "on the go" contribute to the emergence of new practices of social networking and communication, as these devices make it easier to maintain networks of relationships and continuous patterns of interaction (Damásio, Henriques, Botelho, & Dias, 2013; Ling, 2008). In this context, convergence can be found at three levels: between devices—the phone and other devices used to access the internet; between activities—talking and browsing; and between the local and the global through the permanent dynamic articulation of communication and coordination.

Such convergence processes are in line with a techno-social process that signals the broader coming-together of technological networks that connect computers, and nowadays smartphones, and the social networks that connect human beings.

Such convergence process could be broadly summarized under the "m-internet" concept. This concept is still a debated issue, far from a clear and delimited

M.J. Damásio (✉) • S. Henriques • I. Teixeira-Botelho • P. Dias
CICANT-Centre for Research in Applied Communication and New Technologies, Lusófona University, Lisbon, Portugal
e-mail: mjdamasio@ulusofona.pt; sarahenriques.sh@gmail.com; inesbotelho@hotmail.com; patricia.dias@ulusofona.pt

A. Lugmayr, C. Dal Zotto (eds.), *Media Convergence Handbook - Vol. 1*, Media Business and Innovation, DOI 10.1007/978-3-642-54484-2_19

description. A general definition poses that m-internet concerns the use of internet on mobile devices (Gerpott, 2010; Wang & Wang, 2010). Given the too broader nature of this definition, several questions have arisen, particularly on what concerns the conceptualization and operationalization of the term. Some consider m-internet a synonymous of wireless internet that includes both going online with a laptop using a wi-fi connection or broadband card or going online via a mobile phone (Smith, 2010). For others, mobile devices are considered mobile phones, with two broad categories: smartphones and feature phones, notebooks and tablets (Townsend and Rios, 2011). Going online via wirelessly connectivity is one of the agreed features of m-internet (Wellman, Quan-Haase, Boase, & Chen, 2002), an important aspect being to perform activities on the go. Gerpott (2010) refers that m-internet involves packet-switched and internet protocol-based access to a vast assortment of data-applications via mobile communication networks, such as e-mail, web browser, video streaming. Wang and Wang (2010) relate m-internet to accessing wireless internet anytime and anywhere via palm-sized mobile devices, such as mobile phones, personal digital assistants (PDA) or smartphones. According to these authors, mobile activities are electronic activities, performed in a wireless environment through a mobile device, via the internet, the major difference between mobile and fixed internet being the mobility and accessibility. We understand m-internet as internet access and consumption performed via wireless, 3G, LTE or 4G platforms using mobile devices that allow for physical displacement during use operations, which means, to perform activities on the move, anywhere and at any time. On contrary, fixed internet concerns access to the internet via a laptop or desktop in a steady and fixed space, whether one uses wired or wireless connectivity.

Past research (Ling, 2008; Smith, 2010; Vincent, 2004) has pointed to an increase in connectivity for those who use smartphones and other mobile devices with internet access, though merely within small networks of close relationships, as well as an increase of coordination of activities between these subjects. On the one hand, these devices make it easier to maintain and expand peripheral networks of relationships, promoting cross-cultural communication and contributing to a sense of living in a global village. On the other hand, users also find these devices extremely useful to coordinate with those who are geographically close. The combination of these two processes results in the emergence of a behavioral pattern called "network individualism" (Wellman & Rainie, 2012). The emergence of these social interactions in this particular context points to a tension between the convergence processes we have mentioned above and this particular type of social divergence.

This chapter will approach the study of mobile internet adoption patterns and discuss the interdependencies between existing socio-economic environment and personal interactions occurring as a consequence of this technology adoption and use. Our main problem concerns the fact that industry stakeholders (such as mobile devices developers, telecommunications operators and content creators) and users perspectives on mobile technologies adoption drivers, depict a tension between technological convergence and social divergence, that emerging adoption drivers of mobile internet and the behaviors of use that characterize it, seem to confirm. Such process leads us to question the value of mobility for the emergence of social activities that lead to social cohesion. Our main arguments states that, while stakeholders regard this technology adoption and diffusion as being driven by the same principles that molded past technological convergence, namely for the PC era, users act differently and focus on closer and more similar people, though diverging from technological determined convergence. Such opposite visions are in-line with past questioning of a functionalist view of convergence and the need of more observational research that rules out technological determinism as the main explanation for convergence (Skinner, 1976).

Our research explores industry and users' standpoint on m-internet adoption drivers and whether the determinants thereof have an influence on the type of activities undertaken online. This chapter presents the results from two complementary studies performed with users and with mobile industry stakeholders. One study is based on a set of interviews performed with some of the main stakeholders of mobile industry. This study is based on more qualitative techniques of analysis and approaches the market perspective of mobile technology adoption drivers and patterns of use. On the other hand, the second study is based on a quantitative inquiry on mobile internet use and drivers of adoption, applied to a representative sample of a southern European country—Portugal—population. Both perspectives are compared in order to better understand their viewpoints on what concerns this technology's adoption, use and dissemination, intending to contribute to a deeper understanding of mobile technologies impact on technological and social convergence and divergence processes.

2 Literature: State of the Art

2.1 Technology Adoption and Media Convergence

The generalized adoption of a new technology is usually accompanied by apocalyptic predictions of the replacement and obsolescence of another existing technology. Marshall McLuhan's tetrad (McLuhan & Powers, 1986) argues precisely that every new media causes the obsolescence of another (or several) previous media. However, newspapers coexist and have articulated with news websites. We can listen to the radio and watch television online. We have internet on our smartphones and varied access to all our documents stored somewhere in the cloud. Thus, concepts revolving around the notions of convergence and articulation have been suggested and discussed in order to more adequately describe the contemporary media landscape and the practices and choices of the users (Bolter & Grusin, 2000; Cardoso, 2006; Castells, 2009; Jenkins, 2006).

Bolter and Grusin (2000) developed McLuhan's argument that previous media tend to become the content of new media in their concept of remediation. According to them, every new communication medium reconfigures pre-existing formats and contents as they follow a logic of permanent communication which feeds on repetition and reconfiguration, remixes and mashups, shares and retweets. Bolter and Grusin (2000) add that every media implies a mediation process, i.e., its technological features shape its content and impact, but users prefer immediacy and seamless experiences. Thus, the process of remediation tends to be imperceptible. They also highlight the role played by users—prosumers (Castells, 2004; Tapscott, 2008; Toffler, 1980) or producers (Bruns, 2009)—in the remediation performed via digital media.

Jenkins (2006) addressed the issue of convergence suggesting that it is a process that goes beyond technology and that does not correspond to the integration of several digital media in the same "black box". Convergence is both technologically-enabled and culturally-performed, corresponding to the possibility of having access to contents via alternative platforms and devices and to the user-driven and collaborative creation of those contents.

Cardoso (2006) studied the Portuguese media landscape having Castells (2005 [1996]) concept of network society as theoretical framework, concluding that users do not replace one media with another nor do they divide their time in using different media. Users who are more active and sociable tend to use more media, and the effective use is often simultaneous, i.e., the same user may be texting while surfing a website on a laptop while having the TV on. Hence, users articulate the choice and use of different media according to their intentions and goals. Drawing on these findings, Cardoso (2006) suggested the concept of synthetic networked communication to describe the way that users articulate and find synergies among the different media available, namely mass media and digital media. This new type of communication is synthetic because it tends to integrate and articulate different types of media—text, images, sound, animation, and video—and it is supported by infrastructural media which are networked.

Shirky (2009) observes that digital media tend to promote group communication while previous media promoted interpersonal or mass communication. Castells (2009) considers that these categories no longer apply to describing the contemporary way of communicating supported by digital media, which he considers a self-mass communication. With this concept, the author describes a new type of communication which can assume different formats and modalities, that has potential global reach, and that is selected, organized, configured, received, interpreted, and often produced, by each individual user. (Wellman & Rainie, 2012) explores this notion in his concept of networked individualism. According to Wellman, networks are as important to contemporary society as an operating system is to a digital device. He adds, focusing on communities and digital capital, that these networks are created, maintained and managed by each individual, as social interaction is the main driver for using the technologies that facilitate it.

We suggest these concepts as a relevant theoretical framework for analysing the adoption of m-internet and its uses. We argue that the internet is remediated in smartphones and that these devices represent a specific form of convergence that supports network individualism and in doing so promotes at the same time social divergence. In addition, each user mass self-communicates in his/her individual networks, synthesizing and articulating different media and contents according to his/her needs, intentions, preferences and goals. These concepts are useful for exploring m-internet's drivers of adoption and also for comparing the activities performed online via fixed and mobile access. They are also relevant to understanding the contradictions found between the industry stakeholders' and users' perspectives on the adoption and use of m-internet.

2.2 Mobile Phones, M-Internet and the Online Experience

Relationships are core aspects in human life and interaction is considered an essential human need. Social interaction has been the basis for the dissemination and usage of mobile phones and m-internet (Lasen, 2004; Ling, 2004). The mobile phone was firstly regarded as an elitist technology, a device for businessmen that was progressively spread among the population for various purposes as users found out new ways to use mobile phones regarding interpersonal communication. As it enables personal and private communication, it became common for users to carry it everywhere (Vincent & Harper, 2003) and to use it in public. This allowed people to access anywhere to their buddy space (Vincent, 2005) and to have a nomad intimacy (Fortunati, 2002). Mobile phones also represent, in a symbolic way, the virtual presence of friends and family, as it allows connectedness (Vincent, 2005) and connected presence (Licoppe quoted by Rettie, 2006). In addition, it can also be a symbol of social integration and a way of displaying an active, filled, happy and desirable way of life—stage phoning (Plant, 2002).

As mobile phone use increased, the emergence of specific usage patterns became the main object of study. Researchers focused on teenagers as pioneers and trend setters. Teenagers also enhanced the symbolical dimension of the mobile phone by personalizing their devices, which expressed identity and life style and became a symbol of group integration and/or exclusion (Lobet-Maris & Henin, 2002; Plant, 2001).

Some studies (Vincent, 2005; Vincent & Harper, 2003) showed that people are willing to pay for new services that allow them to communicate with the ones they like the most. In addition, the mobile phone has changed relationships and work: people maintain a perpetual contact with their close network of interactions (Katz & Aakhus, 2002); relationships have become more intense as users are always in touch with their network of close relationships, although physical encounters are indispensable within these type of relationships (Dias, 2009; Vincent, 2005); the mobile phone enables people to express their emotions and feelings more often (Lasen, 2004; Levinson, 2004) and always being in touch results in a development/deepening of these relationships (Ling, 2007); the mobile phone is a social lubricant as it contributes to a more spontaneous and less formal communication (Federman, 2006; Intel, 2005; Plant, 2002; Utz, 2007), similar to what happened in small villages through the garden fences (Fox, 2001). Using a mobile phone is a *sine qua non* condition for social integration and coordination (Lasen, 2004; Levinson, 2004; Taylor & Harper, 2001, 2002).

At the same time, this growth in social connectivity is also associated with changes in the nature of the ties between subjects, with a tension emerging between the strong ties we maintain with our close relations now in perpetual contact and a the weaker and weaker ties we endure with the "outside" world (Hampton, Sessions, & Her, 2011).

Ling and Yttri (2002) distinguish between mobile phone mediated communication that is merely instrumental or which has symbolic meaning, proposing the concepts of coordination and hyper-coordination. Coordination is task-related and goal-oriented whether hyper-coordination is more evident in communication almost without content: for instance, the exchange of messages before going to sleep between teenagers in romantic relationships expresses love independently from the specific content of each message. Fox (2001) adds that the mobile phone enhances the frequency of communication and also transparency within social groups, thus increasing the social capital within groups. Other researchers argue that the mobile phone, by being so convenient and easy to use, facilitates the expression of emotions (Mobile Life Report, 2007).

Research on connectivity and relationships reveals that the mobile phone contributes more to enhancing the frequency of contact within networks of close relationships (mainly via calls and messages) than to expanding these networks (mainly via mobile social networking) (Katz & Aakhus, 2002; Ling, 2004).

The concept of perpetual contact, suggested by Katz and Aakhus (2002), sums up the effects of the mobile phone's mediation on relationships: the mobility and accessibility afforded by the mobile phone enables the constant possibility of instantaneously reaching and/or being reached, a possibility which has nowadays become the norm. However, this perpetual contact is merely desired and maintained with those whom we love, i.e. within close relationships. Research relating the mobile phone to social capital reveals that, although technological mediation has traditionally been related to social capital decrease, the mobile phone has the inverse effect (Damásio et al. 2013). Lasen (2004) argues that it actually enhances two types of social capital: bonding (the reinforcement of relationships, personal and group identities) and bridging (connecting around common interests). Hence, the mobile phone not only reinforces close relationships, but it also expands social networks, as long as they are dense ones. Literature thus points to the conclusion that the mobile phone and m-internet reinforce dense close relationships but can also facilitate the maintenance of peripheral social networks—i.e. perpetual contact (Katz & Aakhus, 2002), bonding and bridging (Lasen, 2004), selective sociality (Ito, Okabe, & Matsuda, 2006), bounded solidarity (Ling, 2008).

3 Methodology and Approach

3.1 Field Research

This section presents and discusses the results from empirical research that intended to contribute to the understanding and analysis of the relationship between m-internet adoption and diffusion and the influence and impact the type of internet access—i.e. on the go/mobility—has on the activities users perform online. Our research examined two complementary views—one based on stakeholders' perspective and the other based on users' perspective—that approach the problem of adoption and use of internet via mobile devices and the impact of the type of access and related variables, such as mobility, on activities performed on the go, on user online experience, and on social capital. Within this study we intended to assess both industry and users perspectives about m-internet usage, main activities performed, main added-value perceived, benefits and main differences in regard to online experience. We consider this data to be relevant for the problem at hand, since, as we will discuss, stakeholders clearly represent a view that claims that the technologically supported processes of convergence that are promoted via these technologies adoption and use generate a convergence of users' social relations with other people, while what users views actually depicts, is a process whereby m-internet use seems to mostly reinforce existing relations while at the same time promoting divergence with others outside the existing network.

In order to analyze stakeholders' views, interviews were conducted with some of the main stakeholders from the mobile industry at a national and international level. Stakeholders were asked their opinion with regard to m-internet adoption, dissemination and usage in the present and near future. Qualitative techniques of analysis were used to examine the data achieved. Each interview consisted of 18 predetermined questions. The questionnaire developed for this study was based on previous studies that also analysed stakeholders' views on technology adoption (Quico et al., 2012; The World Internet Project, 2012). All interviews were performed face-to-face, transcribed and analysed with the help of the software NVIVO.

Complementarily to this approach, a quantitative approach was developed in order to assess users' perspective of the same problem. This approach was based on the application of an inquiry about m-internet usage, with the main goal of analysing user's view of m-internet services and possibilities, identifying the factors that contribute to its growth and the activities performed via m-internet in comparison with fixed access to the internet. The inquiry was developed based on previous measures (Zickuh & Smith, 2011—Pew Internet & American Life Project; Smith, Aaron, 2010—Pew Internet & American Life Project; Lejnieks, 2008). The inquiry has in total 20 questions.

3.2 Results

3.2.1 Stakeholders

This stage was very important to explore and search for relevant research questions of how the industry faces the changing technological environment and how they perceived m-internet dissemination, key drivers, social and individual consequences and network effects to access to m-internet. Also, these interviews allowed us to identify pertinent and relevant views, data and questions to explore in the further quantitative stage.

The following stakeholders of mobile manufacturing companies were interviewed: Sony Ericsson's Key Account Manager; Nokia's Communication Manager; LG Marketing Manager; market research companies, Marktest's Internet Director; GFK's Business Group Director; Netsonda's partner; mobile marketing companies, TIMWE's product manager, and mobile network operators, Optimus' Internet Mobile Multimedia Services Manager; Internet Services Director at Vodafone; and ZON's voice product manager. It is highly relevant to state that stakeholders understood m-internet as internet on mobile phones and, or smartphones.

In our interviews, stakeholders pointed that mobile phones' sales are decreasing in Portugal and worldwide but, on the other hand, smartphones are having an exponential growth due to the price being so similar to the ordinary mobile phone or feature phone. As consequence of the smartphone widespread, stakeholders claim that the access to m-internet is growing, as smartphones naturally need internet for full extent using of their features.

According to stakeholders, the age group between 25–34 years is the one that uses m-internet the most in Portugal. The main activities are social media and email access, news reading and meteorology checking.

Vodafone believes that the use of m-internet will soon exceed the use of internet on a PC. In addition, their data shows that m-internet access is more utilitarian and frequent along the day but has a shorter duration. Also, there are peaks of m-internet usage at lunch time, between 6 p.m. and 7 p.m. and a primetime between 10 p.m. and 11 p.m. Friday is usually the day with more traffic.

According to all interviews, m-internet users are becoming multiplatform, which means that they can access it everywhere, anytime and through different devices. This characteristic allows mobility and consequently a better exploitation of the internet and its possibilities, the creation of a more appealing experience and spending less time on a normal computer. Thus, the stakeholders' perspective points to the possibilities afforded by technological convergence as the main adoption driver for smartphones and m-internet.

For the majority of the stakeholders, with exception of those from mobile networks operators, the price plans available are the main obstacle to the expansion of m-internet, as they still have high prices and very limited traffic. Other obstacles listed were the lack of information and literacy of Portuguese people regarding the existent services.

Among the motivations to access m-internet, stakeholders also referred that social networks represent a very high share of internet traffic through mobile phones. Market research companies highlighted that a high percentage of young people aged 15–17 access social networks on mobile phones, with the groups aged groups 18–24 and 25–34 also reaching significant values. Mobility is also one of the main reasons to use m-internet as it gives access to new services and possibilities (e.g. geolocation) and new types of consuming, sharing and communicating. However, in a near future, stakeholders expect that people will perform the same tasks via m-internet than they do now via fixed internet, like editing documents and store data, thus pointing to a more profound remediation of previous technologies on smartphones. Geo-location is another motivation mentioned by the stakeholders as geo-location services are among the most popular apps.

Stakeholders referred to an interesting topic about the impact of m-internet on social practices, as they considered that m-internet access makes it easier to communicate and share information in real time. This fact contributes to social closeness and public life involvement. Stakeholders remembered some of the social movements that were initiated via the internet and mobile access e.g. Arab Spring and London riots. These references show that, for the stakeholders', smartphone convergence is not only a technological process, but also a cultural one, that results in new social practices that "open" the users to the outside world.

3.2.2 Users

In this section, we present the results attained during the quantitative inquiry applied to a representative sample of the Portuguese population ($n = 1107$), during the first semester of 2012. The individuals from this sample are aged between 14 and 64 years (mean age = 38.69), 49 % being male and 51 % being female. The vast majority of the sample (81 %) has elementary and high school education, 42.7 % frequented elementary school, 38.0 % frequented high school, 9 % have a degree, 1.4 % have a master, 0.3 % have a PhD and 7.1 % have less than primary school. The sampling was conducted in a random way in all regions of the country and the inquiry application was performed face-to-face in the individuals' households, with no pre-requisite regarding mobile phone possession or internet usage having been considered.

Given the length of the questionnaire, only a partial set of the results will be presented here. Results indicate that almost all participants have a mobile phone device (96.6 %), however only 19.1 % own smartphones. From the total sample, only 15.7 % answered they use m-interned (84.3 % do not use m-internet), 14.5 % use it via mobile phone, 1.5 % use it via tablets, 3.3 % use it in their iPods, 1.4 % use it via iPad and finally, 2.1 % use it via portable playstation.

The vast majority of Portuguese users of m-internet state that they have been using this type of access for more than 1 year (73.1 %) and only 11.5 % for less than 6 months. Regarding frequency of use, most participants (30.5 %) use mobile internet two to three times per week, 32.2 % use it sometimes per week, 21.3 % replied they use it rarely and 16.1 % replied they are always on. Referring to the moment during the day in which individuals most use m-internet, participants

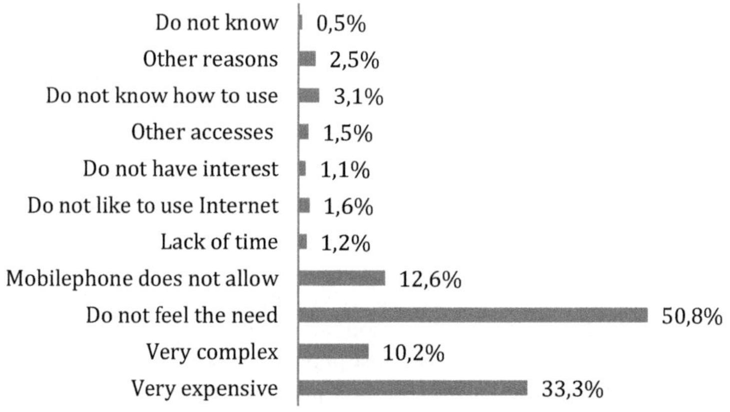

Fig. 1 Reasons for not using m-internet

responded they mostly use m-internet between 12 a.m. to 4 p.m. (43.1 %) and between 8 p.m. to 12 p.m. (41.4 %).

The inquiry also asked participants about their behaviors using mobile internet, if they normally use the browser or apps when accessing—internet, offering them a scale from 1 to 7 in which 1 meant they never use and 7 they always use it. Most participants answered they use the browser (mean = 4.77, sum = 830), when compared with apps (mean = 3.62; sum = 630). Also, from the total sample, 85.1 % of the individuals that use m-internet participate in virtual communities, such as social networking sites, Facebook, Hi5, Linkedin, collaborative games, discussions groups, interest groups, fans groups, among others.

Participants that replied they do not use m-internet (84 %) were asked their reasons. The main reasons presented were: they do not feel the need of using it (50.8 %) or it is very expensive (33.3 %). Figure 1 shows other reasons mentioned.

Thus, there is a divergence between the drivers for m-internet adoption considered by stakeholders and users actual statements. Stakeholders believe that mobility is the main added-value brought by smartphones and m-internet, as users point as main reason for not adopting, not feeling the need of having a technology that provides that mobility. Clearly, current prices also do not reflect the value perceived by users.

Regarding the activities performed via m-internet, the inquiry posted a list of activities performed online, both via mobile or via fixed access to the internet (the same online activity was always questioned twice, one regarding the mobile access and one regarding the fixed access to the internet), and asked participants their agreement with the sentences on a scale from 1 to 7, where 1 meant "I never do that" and 7 meant "I always do that". The more frequently conducted via mobile are ordered in Fig. 2.

Online activity	Mean	SD
Reading emails	4,17	2,264
Accessing social networking sites	4,24	2,313
Using chats	3,89	2,352
Searching information about maps, paths or roads use geo-location applications	3,74	2,293

Fig. 2 Activities more frequently conducted online

In order to explore if there is a divergence between actions performed online when fixed or mobile internet access was considered, we performed a multidimensional scaling analysis evaluating the proximity (similarity/dissimilarity) between the activities analyzed in the inquiry, intending to examine groups or types of activities that are more frequently performed via mobile. The values of proximity were calculated based on the answers to this item in the inquiry, using the algorithm Proxcal (PROXimity SCALing) with the support of SPSS (v14, SPSS Inc. Chicago, IL). The selection of the minimum number of dimensions to keep in the model in order to reproduce parsimoniously the similarities/dissimilarities between the activities (via mobile and fixed access) was held respecting the screen-plot criteria and the proximity versus transformed distance graphics criteria. The quality of the model was analyzed through the STRESS-I index and DAF index (Indexes that represent a measure of goodness of fit—STRESS-I $= \sqrt{\frac{Raw\ Stress}{\sum_{i=1}^{n}\sum_{j=i+1}^{n-1} d2ij}}$ which measure the difference between the observed similarity matrix and the estimated one, DAF = 1-normalized Raw Stress which measures the Dispersion Accounted For. The lower STRESS the better the fit) using the reference values defined in Marôco (2007). According to these assumptions, three dimensions should be retained in order to reproduce properly the perceived similarities between the activities (STRESS-I = 0.129, DAF = 0.983). Figure 3 illustrates the three-dimensional perceptual map.

When we look at this perceptual map, we can verify that the activities conducted online either via fixed or mobile access to the internet are distributed spatially according to its similarities (based on the respondent's answers) in a spatial map with three dimensions that we defined as: dimension 1—interaction; dimension 2—time; dimension 3—effort. Each point in the perceptual map represents the positioning of each activity according to these three dimensions, allowing to represent perceptually and to analyze the proximity and distance between the activities, considering the dimensions mentioned.

Examining the graphic, two main groups of activities emerge, while the remaining activities are distributed randomly within the graphic space. The two main groups are located at the superior left part and at the superior right of the perceptual map. The superior left part includes the following activities: using the e-mail via mobile access, using social network site via mobile, using geo-location applications via mobile, using online services via mobile, searching information about maps, paths or roads, using chats via mobile. These activities score higher

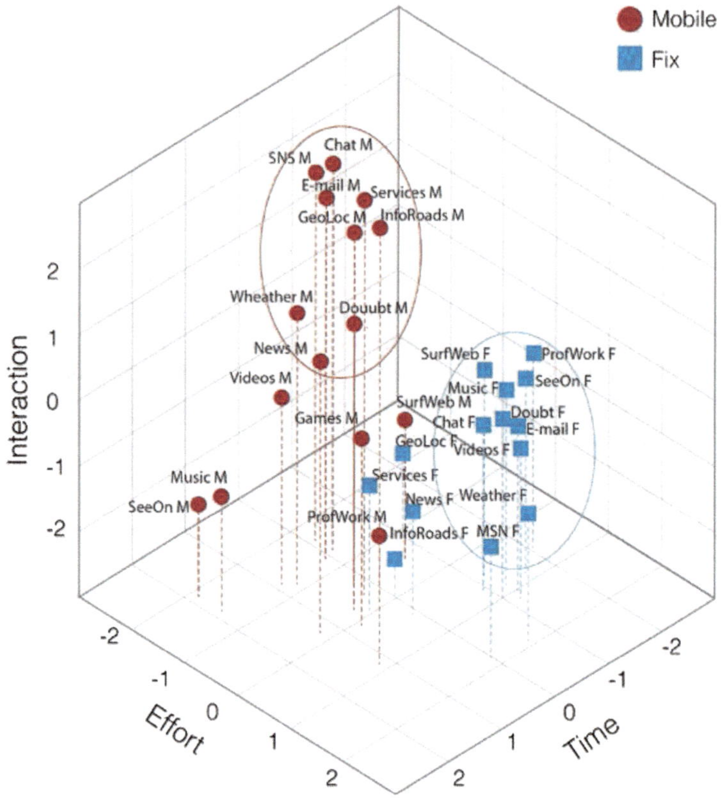

Fig. 3 Three-dimensional perceptual map of online activities (Euclidean distance model: STRESS-I = 0.129; DAF = 0.983)

values on mobile access. Regarding the superior right part of the perceptual map, we can find another group of online activities, such as: downloading music, settling a doubt, looking for someone online, performing professional/work searches online, watching videos online, which present higher values on the fixed access.

The perceptual map distinguishes between two main groups of activities: one more related to the fixed access and the other one more related to the mobile access. Interpreting this data and considering the dimensions' conceptual meaning, one can say that the activities more frequently done via mobile access commonly imply communication and participation, such as using chats, e-mailing, using social network sites, and activities that need to be done on the go, such as searching information about streets, roads or maps, used when someone is lost or is trying to find a new place. The activities more frequently done via fixed access are normally activities that take more time to be performed, imply higher levels of attention or knowledge. This distinction is thus another divergence identified in our research, this one concerning the activities performed online according to the type of access. These results are consistent with our previous argument of smartphones remediating the internet through the convergence process of these two technologies.

One last and highly relevant aspect, concerns the fact that all activities that involve the reinforcement of close ties within our dense personal network, are much stronger in mobile technologies. This shows this technology potential to increase high levels of connectivity with existing social networks—bonding type of social capital—and lower levels with outside networks—bridging type of social capital.

4 Viewpoints on Convergence

Our research on the social consequences of technological convergence between the internet and the mobile phone is consistent with previous research, as it shows that smartphones are associated with new use practices that mark the convergence of activities previously conducted in another way or with another resource. Our results also emphasize divergences resulting from this convergence process. On the one hand, there is a divergence between stakeholders' and users' perspectives on the advantages that smartphones provide and on adoption drivers. There is a strong divergence ion the expectancy values associated to this technology, as stakeholders and users perceive it and its role in daily life differently. However, these perspectives converge in a dynamic interaction process between the technologies and services pushed by the industry and the way they are presented to users and the users' actual adoption and appropriation. On the other hand, there is a divergence between the activities performed online via fixed and mobile access. Mobile access favors both utilitarian activities such as obtaining directions and social activities that extend the perpetual contact already provided by the mobile phone. In addition, smartphone adoption may also emphasize another divergence between users and non-users, creating a digitally-supported social divide between those who are able to maintain perpetual contact with their social circles and those who are not. This is in our view the most relevant tension coming out of this process and one that confirms our initial argument, showing that while convergence is positively regarded by both industry and users, is has particular consequences, namely in this case by reinforcing a divergence at the social level. This divergence happens between the level of social connectivity the technology allows for and the actual limited uses people do of that possibility, since in most cases they solely conduct activities that reinforce existing relations and connections.

5 Conclusions

The adoption and appropriation of new technologies has traditionally been studied focusing whether on the technology or on the users. More technology oriented research usually focuses on technology's effects, while more socially oriented research seeks to understand the consequences and outcomes of ongoing processes. Our research contributes to a more holistic understanding of m-internet adoption and appropriation by contrasting the perspective of relevant industry stakeholders with the users' point of view, emphasizing the convergent and divergent points among them. Such approach also intends to surpass both deterministic and cultural view by probing links between the individual level of use and adoption and the collective levels of dissemination practices.

Our findings reveal divergences both at a cognitive and a social level. At the cognitive level, expectancy values vary between industry stakeholders and users. Their understandings of what is m-internet or of what is a smartphone is different, a gap that constitutes a barrier to the development of technologies more suited to the users' needs, expectancies and preferences. Such a divergence is followed at the social level by a similar one. While industry stakeholders' believe subjects will use the technology connectivity potential to enlarge their social networks, users are actually conducting a set of activities that reinforces their existing dense social networks. Such a divergence can potentially highlight this technology's ability to prompt bonding social capital instead of bridging social capital.

Research results indicate that the industry' view appears to be more concerned with technological determinants such as screens, platforms, type of connection, speed of the browsers, quality of the devices, price plans, among others. On the other side, users' view seems to be more related to which social activities are allowed by the technology, what is the added-value of this kind of technology over other previous technologies specifically for social interaction, what value mobility brings to their daily-life, whether it is practical, useful and easy to use or what are the costs it involves. Regarding to this question, it seems that maybe we are facing a push movement whereby operators push m-internet consumption into the market and focus mainly of technology determinants when users not always feel totally engaged by using this technology and usually focus more on the added-value concerning the social dimension and the relationships and interactions allowed via this technology. In any case they all seem to agree the technology depicts a set of convergent features and possibilities that promote its adoption. Though, it is not in the early adoption stages that we find the above mentioned tension, but only later on when the technology reaches greater social diffusion. We can thereby conclude that the early adoption stage is marked by a positive regard towards convergence as a driver of adoption, while later stages point to social divergence as a consequence of the adoption and usage process.

Outlined divergence between the activities performed online via fixed and mobile access show that the main advantages perceived by users in smartphones and m-internet do not concern technological features, but the ways these technologies remediate the mobile phone and the internet through the convergence of these media, thus further facilitating the preferred uses of the internet and the mobile phone. It is these features they embrace in the adoption stage. By joining mobility and connectivity, smartphones enhance both perpetual contact and coordination, thus facilitating the performance of utilitarian and social activities, which in later stages result in social divergence and in the reinforcement of existing dense social networks. This outcome characterizes what some have called "network individualism". Users results indicate that the possibility of a more continuous connection with others, with higher levels of interaction and social activities the technology allows for, is embrace by users as a limited possibility of performing activities online anywhere at any time, with their already existing relations, thus reinforcing bonding social capital in close relationships.

Our research emphasizes the importance of perceptions, expectations and anticipations concerning the adoption and appropriation of technologies, and concludes that only a multifaceted approach that apprehends both industry and users can grasp a full understanding of these phenomena and surpassed the outlined tension. Future research is needed to further explore the individual and social consequences of m-internet adoption and use.

References

Bolter, J. D., & Grusin, R. (2000). *Remediation: Understanding new media*. Cambridge, MA: The MIT Press.

Bruns, A. (2009, September 3–4). *From prosumer to produser: Understanding user-led content creation*. Presented at transforming audiences, London.

Cardoso, G. (2006). *Os Media na Sociedade em Rede*. Lisboa: Fundação Calouste Gulbenkian.

Castells, M. (2004). *A Galáxia Internet: Reflexões sobre internet, negócios e sociedade*. Lisboa: Fundação Calouste Gulbenkian.

Castells, M. (2005 [1996]). *A Era da Informação: Economia, sociedade e cultura: Vol. 1—A Sociedade em Rede*. Lisboa: Fundação Calouste Gulbenkian.

Castells, M. (2009). *Communication power*. Oxford: Oxford University Press.

Damásio, M., Henriques, S., Botelho, I., & Dias, P. (2013). Social activities and mobile internet diffusion: A search for the holy. *Mobile Media and Communication, 1*(3), 335–355.

Dias, P. (2009). *The Message is 'I Love You': The role of mobile phone mediated communication in close relationships*. Presented at 1st autumn conference of interpersonal communication and social interaction section of ECREA, Tampere.

Federman, M. (2006). Memories of now. *Receiver, 15*, Vodafone.

Fortunati, L. (2002). The Mobile Phone: Towards new categories and social relations. *Information, Communication and Society, 5*(4), 513–528.

Fox, K. (2001). *Evolution, alienation and gossip: The role of mobile telecommunications in the 21st century*. Social Issues Research Center. [online] http://www.sirc.org/publik/gossip.shtml

Gerpott, T. (2010). Communication behaviors and perceptions of mobile internet adopters. *Info, 12* (4), 54–73.

Hampton, K., Sessions, L., & Her, E. (2011). Core networks, social isolation, and new media: How internet and mobile phone use is related to network size and diversity. *Information, Communication and Society, 14*(1), 130–155.

Intel. (2005). *Intel digital lifestyles report*. United Kingdom: Intel Corp.

Ito, M., Okabe, D., & Matsuda, M. (2006). *Personal, portable, pedestrian: Mobile phones in Japanese life*. Cambridge, MA: The MIT Press.

Jenkins, H. (2006). *Convergence culture*. New York: NYU Press.

Katz, J., & Aakhus, M. (Eds.). (2002). *Perpetual contact, mobile communication, private talk, public performance*. Cambridge, MA: Cambridge University Press.

Lasen, A. (2004). *Emotions and digital devices—affective computing and mobile phones*. Surrey: Surrey University.

Lejnieks, C. A. (2008). *Generation unplugged*. Research report, Harris Interactive, September 12, 2008. Accessed July 21, 2011, from http://files.ctia.org/pdf/HI_TeenMobileStudy_ResearchReport.pdf

Levinson, P. (2004). *Cellphone: The story of the world's most mobile medium and how it has changed everything!* New York: Palgrave MacMillan.

Ling, R. (2004). *The mobile connection, the cell phone's impact on society*. San Francisco: Elsevier.

Ling, R. (2007). *Mobile communication and mediated ritual*. Norway and USA: Telenor & University of Michigan.

Ling, R. (2008). The mediation of ritual interaction via the mobile phone. In R. Ling (Ed.), *Handbook of mobile communication studies* (pp. 165–176). Cambridge, MA: The MIT Press.

Ling, R., & Yttri, B. (2002). Hyper-coordination via Mobile Phones in Norway. In J. Katz & M. Aakhus (Eds.), *Perpetual contact, mobile communication, private talk, public performance* (pp. 139–169). Cambridge, MA: Cambridge University Press.

Lobet-Maris, C., & Henin, L. (2002). Hablar sin Comunicar o Comunicar sin Hablar: del GSM al SMS. *Revista de Estudios de Juventud, 57*, 101–114.

Marôco. (2007). *Análise estatística com utilização do SPSS*. Lisboa: Edições Sílabo.

McLuhan, M., & Powers, B. (1986). *The Global Village. Transformations in world life and media in the 21st century*. New York: Oxford University Press.

Plant, S. (2001). *On the Mobile: The effects of mobile telephones on social and individual life*. Motorola [online] http://www.motorola.com/mot/doc/0/234_MotDoc.pdf

Plant, S. (2002). On the mobile. *Receiver*, Vodafone.

Quico, C., Damásio, M. J., Henriques, S., & Veríssimo, I. (2012). Digital terrestrial TV switchover process in Portugal: Viewers between a rock and a hard place. In H. Sousa & S. Denicoli (Eds.), *Digital communication policies in the information society promotion stage*. Braga: LASICS, Universidade do Minho (eBook).

Rettie, R. (2006). How text messages create connectedness. *Receiver*, 17, Vodafone.

Shirky, C.(2009). *Here comes everybody*. London:Penguin

Skinner, R. (1976, January). Technological determinism: A critique of convergence theory. *Comparative Studies in Society and History, 18*(1), 2–27.

Smith, A. (2010). Mobile Access 2011. *Pew Internet & American Life Project*. [online] Accessed July 21, 2011, from http://pewinternet.org/Reports/2010/Mobile-Access-2010.aspx

Tapscott, D. (2008). *Grown up digital: How the net generation is changing your world*. New York: McGraw-Hill.

Taylor, A., & Harper, R. (2001). *The gift of the Gab? A design sociology of young people's Use of 'MobiliZe'!* Surrey: Surrey University.

Taylor, A., & Harper, R. (2002). *Age-old practices in the 'New World': A study of gift-giving between teenage mobile users*. Surrey: Surrey University.

The Mobile Life Report. (2007). *The Mobile Life European Report - Exploring how mobile phone usage reflects our national cultures*. London, UK: Mobile Life - The Carphone Warehouse, London School of Economics and Political Science. www.mobilelife2007.co.uk

The World Internet Project. (2012). *The world internet project international report* (4th ed.). USC Annenberg School Center for the Digital Future.

Toffler, A. (1980). *The third wave*. EUA: Bantam Books.

Townsend, L., & Rios, H. (2011). *Defragmenting mobile research*. Kinesis Survey Technologies, LLC. Accessed from http://www.kinesissurvey.com/wp-content/uploads/2013/06/Defragmenting MobileResearch_KinesisWhitepaper.pdf

Utz, S. (2007). Media use in long distance friendships. *Information, Communication and Society, 10*, 693–712.

Vincent, J. (2004). *Are mobile phones changing people?* Surrey: Surrey University.

Vincent, J. (2005). Emotional attachment to mobile phones: An extraordinary relationship. In L. Hamill & A. Lasen (Eds.), *Mobile World: Past, present and future*. London: Springer.

Vincent, J., & Harper, R. (2003). *Social shaping of UMTS: Preparing the 3G customer* (UMTS Forum Report 26). Reino Unido.

Wang, H., & Wang, S. (2010). User acceptance of mobile internet based on the unified theory of acceptance and use of technology: Investigating the determinants and gender differences. *Social Behavior and Personality, 38*(3), 414–426.

Wellman, B., Quan-Haase, A., Boase, J., & Chen, W. (2002). *Examining the internet in everyday life. Nijmegen*. [online] http://homes.chass.utoronto.ca/~wellman/publications/index.html

Wellman, B., & Rainie, L. (2012). *Networked: The new social operating system*. Cambridge, MA: The MIT Press.

Zickuh, K., & Smith, A. (2011). 28 % of American adults use mobile and social location based services. *Pew Internet & American Life Project*. Accessed July 21, 2011, from http://pewinternet.org/Reports/2011/Location.aspx

How Does Social Media Shape Media Convergence? The Case of Journalists Covering War and Conflict

Vittoria Sacco

1 Introduction

The role of the journalist during war is to relate a firsthand experience of conflict to distant audiences. In a globalised, digital and online world, the importance and relevance of journalistic work during war and conflict has not changed; global audiences still demand objective and verifiable information about the nature of the conflict. However, the ways in which journalists produce and disseminate this information and the way audiences engage and interact with it has changed dramatically. Through digital and social media, global audiences can interact directly and instantaneously, both with journalists and other diverse groups of people interested in the dissemination of news and information during war and conflict. The spread of news and information and the number of different voices creating and delivering content in the public sphere has been fostered by the improvement in technological devices and online social networks (Tumber, 2009, p. 386). These changes in news production, dissemination and consumption have both challenged and invigorated journalistic practices and the environment in which journalists do their work. Journalists who cover war and conflict have had to adapt to a different status in the online and social media age, mediating much more information through various networks, interacting with many more diverse sources of information with various vested interests and finding new ways to produce and disseminate news for global audiences.

Journalism during war and conflict has been investigated by several key researchers in disciplines such as communication, media and journalism fields (e.g. Carruthers, 2000; Clarke & Bromley, 2011; Cottle, 2006, 2008, 2011; Hallin, 1986; Kellner, 1992; Lynch, 2013; Sambrook, 2010; Sylvester, 2011). The interest in this topic stems from the dramatic nature of war and conflict, their relevance to

V. Sacco (✉)
Academy of Journalism and Media, University of Neuchâtel, Neuchâtel, Switzerland
e-mail: vittoria.sacco@unine.ch

A. Lugmayr, C. Dal Zotto (eds.), *Media Convergence Handbook - Vol. 1*, Media Business and Innovation, DOI 10.1007/978-3-642-54484-2_20

nations and publics on a global scale and the investment of both money and time by media organisations to cover these events (Tumber, 2009, p. 386). However, scholars in these disciplines (e.g. Cottle, 2006; Robinson, Goddard, Parry, & Murray, 2009) have suggested that investigation of how professional journalists cover war and conflict should continue because reportage of war and conflict involves complex dynamics including economic, political, social and cultural impacts at both a local and global level (Beck, 1997; Castells, 1997; Giddens, 1994).

Social media is part of the media digitization process which can be considered a central component of the convergence phenomenon (Latzer, 2013, p. 8). Indeed, this study was motivated by the potential for substantial change to war reportage that digital, online and social media are bringing. This is especially important at a time when the use of social media during war and conflict, such as the so-called "Arab Spring", has signalled changes in the way news and information is reported and disseminated. Taking convergence as the main framework of analysis, the outcomes and implications of social media in promoting convergence or divergence in journalistic practices during the coverage of war and conflict are explored and empirically analysed. Through a first content analysis based on 100 *Storify* "stories" written by both amateurs and media professionals covering the Arab Spring, changes in news production and dissemination occurring due to the integration of social media into the coverage of war and conflict will be highlighted. A second content analysis of 234 *Storify* "stories" about the Arab Spring written by media professionals with seven structured interviews with journalists who used *Storify* to cover the events of the Arab Spring will allow examining how journalistic practices have been affected by the use of social media, in particular in terms of newsgathering. Moreover, 22 semi-structured interviews with journalists covering war and conflict, editors and online digital media professionals were then collected to understand how journalists position themselves and their practices within this new media environment.

The chapter firstly reports existing research on major transformations that have affected journalists and their working environment when covering war and conflict. This is because convergence mainly describes processes of change (Latzer, 2013). Then, the methodology and the collection of data to show evidences of changes and challenges due to social media are discussed. Finally, the findings speculate on social media broader impact of introducing a new dynamism. Because social media increases the different available sources and the way to produce and distribute news content, people who were previously voiceless in the media coverage can now bring new points of view. This implies a "divergence" of views through which the information coming to the audience can differ in opinions, contribute to the formation of a more diversified public sphere and thus promote a democratic society. Moreover, these different available sources suggest a more complex and transparent coverage of today's war and conflict which has the potential to bring the audience closer to reality. However, findings pointed out that social media can also become a channel of misinformation, manipulation and subjective reportage. This is due to the different agents present on social media that could try to impose their narratives and thus influence public opinion. Because social media presents both positive and negative aspects, it can be considered as a complementary channel to traditional ways of reporting news. This is where the profession of journalism can also take its relevance. Indeed, the use of social media also leads to the convergence of different functions in the single role of a journalist. This contribution is a summary of the main findings and conclusions of a PhD dissertation on the influence of social media on media coverage of war and conflict.

2 Literature Review

The phenomenon of media convergence has a great impact on journalism. Media convergence redefines the role of journalists and newsrooms as well as affects the business environments of media companies (Russ-Mohl, Nienstedt, & Wilczek, 2013, p. 3). The introduction of networked platforms such as social media, where different technologies converge, has brought new possibilities for news distribution and exchange but also challenges for journalists covering war and conflict. The aim of this literature review is to present a summary of the current understanding of these changes, which relates mainly to three major transformations that have affected journalists and their working environment (Hamilton & Jenner, 2004, p. 303). The first change is related to economic convergence that increases economic pressures on traditional news coverage and engenders modifications on media ownership. This implies also some consequences on news organisations' routines. The second change is a form of social-cultural convergence. It is linked to global interdependence at the community level that has some consequences on the expectations of audiences in term of news coverage. The last change concerns technological convergence. The media sector has experienced technological innovation in line with the advent of a digital era.

2.1 Media Ownership Concentration

While coverage of war and conflict continues to be a major and important aspect of news reporting, the globalisation of media markets has affected the ways in which news organisations are able to bring war and conflict news to their audiences. Changes in the media landscape—including economic pressure, shrinking in editorial autonomy, changes in media technologies—have brought deep budgets cuts to newsrooms as well as the closure of major foreign news bureaus (Sambrook, 2010). Journalists also face professional uncertainties that weaken rather than improve their working conditions due to the above mentioned changes (Paulussen, 2012). While newsrooms rely more and more on freelance stringers and local journalists for news, these journalists are often being put into dangerous situations to meet editorial requirements; some remaining underpaid and living in precarious conditions to do so (Cottle, 2008).

Another important external factor that affects professional journalism is the emergence of global media conglomerates. The pressure to survive in a harsh economic climate has meant that some media companies have merged or collaborated with other media organisations to produce news (Hjarvard, 2000, p. 684). The necessity for media companies to generate profit can undermine the work of journalists. Researchers have been concerned by the dominance of Western global news agencies that might support westernised discourses to the detriment of diversity in the coverage of geo-political and economic issues (Boyd-Barrett & Rantanen, 1998; McChesney, 1999; Thussu & Freedman, 2003). The ever-increasing cost of foreign bureaus, coupled with the availability of cheaper, mostly online news wires has meant that relying on global conglomerates for news content tends to be cheaper (Hamilton & Jenner, 2004, p. 305). In this context, Thussu and Freedman (2003) have spoken of "CNNisation", or the homogenisation of the types of news and information disseminated due to less reporting diversity and by extension, an increased "infotainment" value in the coverage of news driven by profits, ratings and 24/7 content requirements (Henry, 2007; Thussu, 2007; Cottle, 2006, p. 85). Other scholars (e.g. Volkmer, 1999) position themselves at the other extreme, suggesting that globalised media and public environments increase opportunities to obtain multi-faceted information. Volkmer (1999, 2003) argues that today's "mediated global interconnectedness" supports cultural differences and allows audiences around the world to learn about war and conflict because of the emergence of a mediated "global public sphere" and the creation of cosmopolitan citizenship. Others argue that although global media conglomerates can, to some extent, lead to a homogenisation of news they can also sustain deeper and high-quality reporting through the leading position that they enjoy in the market (e.g. Archetti, 2013). These divergent opinions are probably due to the early life of this phenomenon and that analysis has been run only in the short term.

Another external factor affecting the coverage of war and conflict is the financial crisis that hit some media organisations, leading to job cuts. For instance, in the print sector the decreasing number of journalists is explained by the fact that newspapers prefer to cover local news rather than international news as it represents a more stable source of revenue (Adams & Ovide, 2009). In order to survive, newspapers started to focus on niche news markets, offering content that cannot already be accessed for free online (Archetti, 2013, p. 420). The availability of news online that covers war and conflict for global audiences can explain why some scholars have claimed that war correspondents become "redundant" (Kester, 2010, pp. 51–52; Sambrook, 2010; Barton, 2009). However, more recent studies (e.g. Archetti, 2013; Heinrich, 2012) have argued that the profession of war correspondents is evolving, but not at its end.

In this section, it has been shown that the transformations that media organisations went through and the strategies they followed had an impact on the coverage of war and conflict through a globalised and standardised representation of news. Through two content analyses, this study seeks to investigate whether the standardised coverage of war and conflict persists even when journalists rely on social media as a source of information. Social media content comes from many different sources and could present several different perspectives, disrupting the tendency towards homogenisation of news and information about war and conflict.

To confirm or refute this hypothesis, the frames and the sources will be examined in order to compare recent coverage of war and conflict with past ones.

2.2 News Organisation and Routines

While the institutionalised news environment has an impact on the way in which journalists are able to report on war and conflict, more pertinent to the way in which news content is represented is the impact of news routines on journalists. News routines have been defined as "those patterned, routinized, repeated practices and forms that media workers use to do their jobs" (Shoemaker & Reese, 1996, p. 105). These routines emerge partly because of the scarce resources of media organisations compared with the huge amount of information that could become news (Shoemaker & Reese, 1996, ibid). This scarcity is also related to technology, financial constraints, deadlines, space and norms (Lacy, 2000; Reese, 2001). In theory, the establishment of these routines should ensure the production and distribution of a quality product to the audience within the constraints of time and space and without a loss of efficiency (Shoemaker & Reese, 1996, pp. 108–109). With these routines—and their representation as professional journalistic conduct—journalists attempt to follow particular standards in the creation of content, protecting their reputation and credibility and being relied upon as trustworthy sources of news information.

The routine work and tasks of journalists can lead to events receiving broadly similar coverage because journalists rely on the same sources (Archetti, 2013; Boyd-Barrett & Rantanen, 1998; Paterson, 1997). Indeed when it comes to representations of conflict, journalists' professional norms show a bias towards the use of official sources (e.g. Dimitrova, Kaid, & Williams, 2004, 2005; Griffin, 2004; Hall, Critcher, Jefferson, Clarke, & Roberts, 1978). It has been shown that official sources (e.g. the government and/or the military) can manipulate coverage through biased or one-sided accounts of events (Bennett, 1990; Carruthers, 2000; Hallin, 1986; Robinson, Goddard, Parry, & Murray, 2010). The vested interests of official sources might also lead to censorship of what might otherwise be deemed important information.

Another limit in terms of newsgathering is that journalists working in situations of war and conflict are often dependent "on the host-country media for information and ideas" (Morrison & Tumber, 1985, p. 466). Journalists working in conflict situations often work on tight budgets and on their own (Wu & Hamilton, 2004, p. 527; Morrison & Tumber, 1981, 1985, p. 23, p. 458). Therefore, journalists are often time poor and reliant on sources with vested interests for content. Some journalists may even rely on content from wire services (Archetti, 2013, p. 422). Furthermore, they may be tempted to reuse without re-working or further researching what has already been disseminated online—a phenomenon called 'churnalism' (Atkins, 2011; Davies, 2008).

These issues can explain why many academic studies have concluded that most coverage of war and conflict is standardised to some extent and representation of war and conflict has recurrent themes (e.g. Campbell, 1997; Dimitrova et al., 2004; Dimitrova & Neznanski, 2006; Griffin, 2004; Harris, 1994; Kellner, 1992; Morrison, 1994). The study will further investigated to see whether social media might help to overcome some of these limitations and provide a more complete and balanced coverage of war and conflict.

2.3 From "Compassion Fatigue" to Globalised Audiences

Other scholars have focussed their attention on the effects that reportage of war and conflict might have on the audience that demand this news and information. Some scholars (e.g. Altmeppen, 2010; Robinson, 2007) have claimed that war correspondents will disappear because the audience is no longer concerned with war and conflict, a decline explained by what Moeller (1999) and Tester (2001) call "compassion fatigue". Some Non-Government Organisations (NGOs) refer to the same phenomenon as "media fatigue" (Cottle, 2008), corresponding to the decreasing audience engagement with news reporting war and conflict or other types of violence.

Through the notion of "compassion fatigue", scholarly research has encapsulated "the failures, practices and forms of international news reporting to audience-based questions about levels of news engagement and interests" (Cottle, 2009, p. 350). According to Cottle (2009, ibid), this concept is not exhaustive and studies on the news media's "spectacle of suffering" (Chouliaraki, 2006) and discourses of global compassion (Höijer 2004) might compensate the lack of the work on "compassion fatigue" and better acknowledge the complexities of this phenomenon. The current discussions about "compassion fatigue" are addressed by focussing on how the audience discern and react to mediated images of human pain—rather than how media should select, represent and explain the content of these kinds of images (Cottle, 2008). This calls for the need for studies that investigate how media approach this problem, so this contribution presents some discussion and research results about how journalists position themselves in overcoming this issue.

However, in a networked era "compassion fatigue" cannot fully describe the relation between a distant audience and journalists presenting news about war and conflict. Instead, it appears that the impact of an event in one part of the globe is much more significant (Sambrook, 2010, p. 59) because people have private and professional connections in many more countries than ever before. In a world where international trade and markets as well as human migration are growing, the line between foreign and domestic news is blurring (Hamilton & Jenner, 2004). People affected by war and conflict are more likely than ever to be dispersed and thus exceedingly dependent on media coverage. Nonetheless, the issues of news in a globalised world are discussed less compared with other sectors such as finance, governance or culture (Hjarvard, 2000, p. 686). While domestic and foreign news agendas are becoming uniquely aggregated as the news reporting tends towards much more global news perspectives, the events are still framed to support national political interests (Hjarvard, 2000, ibid). Indeed, everyday coverage of international and foreign politics is still debated at a national level by the media and political institutions (Hjarvard, 2000, ibid). According to Heinrich (2012, p. 769), this divide is still noticeable in the coverage of the recent protests that shaped the Middle East countries. She reported that mainstream media often lack the "intertextuality" that could provide a multidimensional understanding of the events covered on the ground.

This section highlighted that the problem of "compassion fatigue" cannot be the only source of the decline of the coverage of war and conflict (Hamilton & Jenner, 2004). It is too simple to blame the audience as the main source of this "crisis". Even if the audience tends to be more global, the coverage of war and conflict struggles to follow this pattern. This study further discuss if social media can increase audience engagement and concern about war and conflict.

2.4 Journalists Reporting on War and Conflict in the Digital Era

Social media is part of the media digitization process which can be considered a central component of the convergence phenomenon (Latzer, 2013, p. 8). Within the present online and social media enabled environment, audiences have increasingly been turning to alternative sources of news and information. While some academic discussions have focussed on why this turn to alternative online news providers might spell the end of professional journalism, others have shown that new forms of information gathering and dissemination based on online networks have brought journalists and users closer together. Recent research (Clarke & Bromley, 2011, pp. 4–5) has indicated that traditional and innovative journalistic practices are merging in reportage. Coverage of war and conflict has therefore taken on a "hybridism", relying on components such as "orthodox journalism, social media, civic expression and public assembly" (Clarke & Bromley, 2011, p. 5). The following are examples of these developments:

- The emergence of 24/7 reporting. Although the emergence of 24/7 real-time news demands more time, effort and resources, it also opens more channels for covering events around the world (Sambrook, 2010, p. 98; Cottle, 2008).
- Decrease of the time of news delivery. The Internet has reduced the time of news delivery through features instantaneously provided by hand-held devices or social media. Digitisation has increased the speed of news and it has removed the distance between producer and receiver, creating more immediacy but also the risk of superficiality (Kester, 2010, pp. 53–54).
- Unlimited space. The unlimited space in the online environment enables the emergence of a plurality of news providers that threatens the monopoly of delivery of major media companies. The flow of information becomes thus a many-to-many model where all news providers can be at the same level (Rheingold, 2002). Coverage of war and conflict can thus gain in complexity

and in addition, forgotten war and conflict areas and the long-term evolution of those events can be covered.

- New forms of user contribution. With the advent of the Internet, new forms of user contribution in the news cycle have arisen (Lewis, Kaufhold, & Lasorsa, 2010). As an example, platforms such as *Indymedia* (Platon & Deuze, 2003) and *OhMyNews* (Kim & Hamilton, 2006) allow amateurs to actively participate in the news cycle. Once users have an Internet connection and the right software, it is easy to create content and share it with people around the world. This plurality of available voices introduces alternative points of view on war and conflict and gives people who were previously voiceless the opportunity to be heard.
- New available voices in the media coverage. In their content analysis of Flemish newspapers Paulussen and Harder (2014) found that social media had become a common source of information for journalists. Social media does not seem to be subject to the typical "hierarchy of credibility". Journalists' primary sources on social media are ordinary citizens, celebrities and sports people rather than politicians, official institutions or experts. Broersma and Graham (2013) argued something similar; supporting that social media does not weaken the power of elite sources but has the potential to increase the diversity of voices in the news. Therefore, it can contribute to bringing balance in the coverage of war and conflict.

In this study, issues related to the ways in which forms of amateur or citizen journalism are used are debated. It discusses the way in which source material taken from citizen journalists is used by journalists to cover war and conflict. Thus, it provides an overview of the use of user generated content by journalists and the issues that may arise from its use.

3 Methodology

The main argument of this study is that use of different social media has brought a new dynamism to the coverage of war and conflict through increased possibilities in sourcing, producing and disseminating news and information. However, this dynamism has also meant that journalists covering war and conflict must also negotiate new and complex professional and social issues that affect reportage of war and conflict. In other words, concepts of convergence hold both blurring traditional boundaries between old media and novel diversification and differentiation of new media (Latzer, 2013). In forwarding this argument, the overall research question was:

Is social media promoting convergence in journalism practices when journalists cover war and conflict?

To answer to this question, the study firstly uses two content analyses to explore whether the frames used in traditional coverage of war and conflict persist with the introduction of social media to reportage. Indeed, scholars have shown that changes by convergences can impact on the content of a product (Kolo, 2010; Potts, 2011).

The frames identified were then compared with results from previous studies on the coverage of war and conflict to see if they held the same importance. The social media platform examined within the content analyses was the *Storify* website. The Arab Spring was selected as the topic for the content analyses because this event contributed more recent examples of collaboration between those who witnessed and those who reported on the events. Finally, it was considered more appropriate to investigate a group of stories, rather than an individual account. This choice can be supported by the fact that some biases could have been introduced according to the professional and social category of the selected account. In addition, the main findings that emerged from the second content analysis were supplemented by some interviews with journalists who used *Storify* to cover the Arab Spring. This additional data helped to assess if findings from both analyses (i.e. the content analysis and the interviews) matched, increasing the reliability of the results.

To investigate how journalists position themselves and their practices within this new media environment, other 22 interviews with journalists covering war and conflict, editors and online digital media professionals were performed. In collecting the information about journalists' routines and on how they organise their work, it has been possible to acquire some solid background about how they may rely on fact and be limited by use of social media in their work.

3.1 Content Analysis

Content analysis was selected as the best method to investigate frames and sources of stories because it allowed examining the media messages and assessing communication content (Riff, Lacy, & Fico, 2014). The selected frames where chosen from previous content and textual analysis on the coverage of war and conflict (Dimitrova et al., 2005; Neuman, Just, & Crigler, 1992; Schwalbe, 2006; Valkenburg, Semetko, & de Vreese, 1999). The most common frames were: the military conflict (emphasis is on the military conflict/action); human interest (emphasis is on the human participants in the event); economic consequences (emphasis is on the economic repercussions of the event, the costs of the event); morality (emphasis is on moral messages and religious tenets related to the event); attribution of responsibility (emphasis is on the party/person responsible for the event); diagnostic (emphasis is on the reasons for the event) and prognostic (emphasis is on the prognosis, the outcome of the event). It was noted that frames are not necessarily static and that they can change after time (Snow & Benford, 1988; Snow, Rochford, Worden, & Benford, 1986). As the Arab Spring was recognised as a time of major civic engagement, the "solidarity" frame—which represents messages of solidarity from people not directly involved in the war or conflict—has been added to the analysis.

The first content analysis was focussed on the changes in frames and sources due to social media. Therefore, it was important to have both amateurs' and professionals' stories included in the sample because with social media both groups can create and deliver content. In the second content analysis, the authors tried to examine the changes in journalistic reporting practices, thus the sample of the second content analysis was slightly different. There were more stories to analyse and they were exclusively written by media professionals. Moreover, the unit of analysis was not a specific single contribution (i.e. photo, video or hyperlink) but the coders considered the whole *Storify* "story". This means that the coders have not focussed on the frame(s) of each single contribution included in the *Storify* "stories" but on the narrative and storyline that the curators gave to their stories. The *Storify* platform signals when the final story was posted, as well as the time post of each single contribution that was used to build the story, therefore it is possible to measure the period between the original source publication and the story publication. Thus the analysis was also run according to *time* as a factor representing the length of time between the stories publication and the original contributions dissemination on other social media platforms. Three time frames were retained; the "hourly" coverage means that the story was published within 1 h of the original contribution, the "daily" coverage refers to the story publication within 24 h of the oldest curated source used in the story and finally, the "days" coverage represents when the story was published later than 24 h after the first original source. It was important to consider this factor because according to journalists deadlines it can influence the choice of sources and thus the frames used in the story. Journalists will rely only on the sources that they have the time to consult, as they must respond quickly to the information flows on social media.

3.2 Interviews

For the first round of interviews, structured interviews were preferred because all the participants were from different countries. Furthermore, because of their work engagements it was easier for them to answer the questions in a limited time frame. This was not restrictive in collecting the data because the questions were structured according to the goals of the second content analysis. In particular, the authors were interested to learn how these journalists have used *Storify* to cover the events of the Arab Spring. Combining these two methods creates a fuller picture of journalistic use of social media during war and conflict. Moreover, it overcomes some of the limitations of both methods. For example, one limitation of content analysis is that some bias can emerge when the coder is determined to find specific frames. On the other hand, the answers of interview respondents may be self-serving. Indeed, both methods need to be supplemented with other research sources or other participants' answers. This explains why the two methods have been used together.

The second round of interviews utilises semi-structured interviews, focussing on each respondent's experience rather than centring on data collection (Brennen, 2012, p. 28). The journalists interviewed had different backgrounds, length of work experience, ages and incomes, as well as different positions in mainstream media, which created some contrast in views about social media (see Table 1). Data from different interviews that refer to the same issues enable certain patterns to be observed; additionally, this approach improves the quality of the interviews,

Table 1 Interviewees' profile

Swiss Media	Print	Radio/Television	Others
French	ESH Media Group (1): Head of multimedia and Online	TSR (National Public broadcaster) (3): Head of foreign news & Head of Multimedia and Online & Social Media Specialist	UN Correspondent (1)
	La Liberté (1): foreign news correspondent		
	Le Temps (Tamedia Group) (2): Head of foreign news & Multimedia journalist		ATS (National Press Agency) (2): Assistant Editor of foreign news & Head of foreign news
	La Tribune de Genève (Tamedia Group) (1): Head of foreign news		
German	The Blick (Ringier Group) (1): Senior Editor of foreign news	SRF (National Public broadcaster) (3): Head of multimedia and Online & Quality and Market Director & Assistant Editor of foreign news	
	NZZ (AZ Medien Group) (1): Community Manager		
Italian	Corriere del Ticino (5): Head of foreign news & Multimedia and Online Team (4 journalists)	RSI (National Public broadcaster) (1): Head of multimedia and Online	

enabling the interviewers to crosscheck and to compare the data (Diefenbach, 2009). Semi-structured interviews present the greatest flexibility in administrating pre-defined questions while also permitting follow-up questions (Brennen, 2012, p. 28). Finally, semi-structured interviews allow the researchers to make adjustments in relation to the questions asked to the interviewees; this is important as it allows the researcher to follow up points of interest that are raised during the interview.

4 Results

In the first content analysis, the authors used existing frames and sources from past research on the coverage of war and conflict to build a suitable coding process. The aim was to investigate the changing practices in the sourcing and presenting of reportage about war and conflict. Previous research on the coverage of war and conflict found that the sources journalists relied on were "official sources" such as governments and military personnel (e.g. Carpentier et al., 2006, p. 161; Lee, 2004;

Bennett, 1990). It was speculated that these sources might have changed slightly with the advent of social media. The results supported the assertion that the advent of social media promotes some alternative frames and sources. The main findings of this content analysis suggested that the human-interest frame had greater coverage than the other frames, followed by the military conflict frame (see Table 2). The main sources of information used in the stories were media companies and citizens (see Table 3). When analysing the sources used by journalist Andy Carvin on his *Twitter* account, Hermida, Lewis, and Zamith (2012) also found that non-elite sources were preferred to other media or elite sources. Social media is thus seen to be more "people" oriented rather than "elite" oriented. The human-interest frame is mostly related to the coverage of the suffering of victims of conflicts, which is likely to sensitise public opinion on war and conflict realities (Shaw, 2012, p. 94). Stereotypes and clichés representing children and women that incite strong emotions in the spectators are associated also with this frame (Philippe, 2001). However, it can be assumed that through the human-interest frame, social media can "humanise" all sides of war and conflict. The positive implications of this representation of crisis are that social media might have the potential to allow the audience to understand the complexities as well as the realities of war and conflict. The audience might develop critical awareness of war and conflict because they are more exposed to the human face of events of war. Cottle (2008, p. 147) came to a similar conclusion, seeing alternative media as channels where the visualisation of suffering of distant "others" is humanised as a positive aspect of reportage on war and conflict.

Table 2 Frames of Web-specific features (1st content analysis)

Frames	Photos (%)	Multimedia (%)	Hyperlinks (%)
Human	89	85	68
Military	27	48	68
Diagnostic	13	14	24
Responsibility	–	17	18
Prognostic	–	2	7
Economic	–	–	4
Morality	–	2	3

It means that the frame was not present at all

Table 3 Sources of Web-specific features (1st content analysis)

Sources	Photos (%)	Multimedia (%)	Hyperlinks (%)
Media Professionals	36	50	33
Citizens	22	23	40
Arab People	6	10	4
Non-Arab People	4	10	8
Unknown	32	7	15
Sum	100	100	100

Previous research on the coverage of war and conflict has found that people are generally not represented in reportage and when they are, their portrayal is often stereotypical or ethnocentric (Wall, 2006, p. 12). Mobile devices and social media however have allowed citizens to use the power of images as "a screenshot of reality" (e.g., Pantti & Bakker, 2009; Williams, Wahl-Jorgensen, & Wardle, 2011). Therefore, social media such as *Storify* can be considered a more egalitarian media in which different sides of a conflict can be heard, lending to a more balanced reportage. The reportage has the potential to be more diverse and include a higher number of different sources and more complexity through the presence of a variety of frames, avoiding dualistic types of reportage. It should also to be added that social media allows for extreme images of violence and suffering to be more accessible to audiences and weaken the barriers of censorship in terms of sensitive images and videos.

Social media platforms such as *Storify* both perpetuate and disrupt traditional news frames and sources, as the content analysis underlined the presence of both old and new frames and sources (Sacco, Giardina, & Stanoevska, 2013). The principal advantage of platforms such as *Storify* seems to be that end-users can develop a more complete impression of events because the information comes from a variety of different sources, thus providing diverse points of view. It allows gathering content in different forms (e.g. photos, videos, text and hyperlinks), giving more options for personalisation and for supplementary information. End-users can have access to archived stories and read original source materials.

However, in the analysis of *Storify* it emerges that reporters sourced material from unknown sources and the identity of the original contributors could not always be determined (see Table 3). This highlights a problem with social media platforms such as *Storify* presenting insufficient ways to filter and verify sources. It also appears that the role of verification agents is still not well determined in these platforms. Although one can presume it should be up to the curator to check for the veracity of content and sources, end-users might actually end up with the responsibility of the verification of sources. The authenticity and the legitimacy of user-generated content that is integrated into reportage remains a major issue. Indeed, the introduction of user-generated content in newsrooms has raised some concerns among scholars, especially with regard to journalistic norms of impartiality, detachment and balanced coverage (Sambrook, 2010, p. 98; Cottle, 2008). Others (e.g. Halavais, 2002, pp. 26–32) worried about whether amateur journalists introduce misinformation, through the spread of rumours and ideological statements in place of "news". There could be other undesirable effects including the publication of manipulated or unverified information, lack of original investigation and poor information quality (Quandt & Singer, 2009, p. 139). Through this content analysis, the author concludes by asking whether there have been changes in journalists' practices of reporting since the advent of social media and if these changes have indirectly affected traditional journalistic principles and norms. The second content analysis completed by interviews with journalists that have used *Storify* to report the events of the Arab Spring attempts to answer these questions.

Table 4 Media professionals' choice of frames according to time coverage (2nd content analysis)

Frames	Hourly (%)	Daily (%)	Days (%)
Military	72	79	73
Human	72	76	87
Diagnostic	24	25	47
Responsibility	20	21	25
Solidarity	20	25	15
Prognostic	13	3	4
Morality	9	6	7
Economic	1	4	4

The major findings of this second investigation indicate that the dominant frames and sources were similar to those identified in the first content analysis. In particular, Table 4 shows that in the hourly and daily coverage, media professionals privilege the use of the *military* and *human interest* frames. After several days' coverage all the frames percentages increase. This is probably due to time passing, permitting more information to be published (Table 5).

In line with the results of this second content analysis, the interviews indicated that curators found *Storify* provided a more "global" perspective of events because it is possible to weave together different points of view. This is likely to escape limitations presented to traditional journalists as it was discussed in the literature section. In this regards, one of the participants said: "It's an extremely useful way to collect material from a plurality of sources and pull it together" (Participant D, 10 January 2013). Similarly, another participant observed that: "*Storify* is good for analysis of trends in the media and stories in the social media sphere because of the way I can aggregate information and show from first-hand sources whatever topic I am studying" (Participant A, 20 January 3013). The presence of many different sources might have the potential to help to recognise and legitimise the political views and democratic aspirations of activists or opposition groups, as well as "sustaining new forms of social conviviality in everyday life and thereby sustaining, both virtually and physically, pluralised identities and interaction in emergent civil societies" (Cottle, 2006). This form of citizen journalism can thus push individuals to behave as witnesses during crises (Allan, 2009; Dahlgren, 2005). An example of these developments was seen in the Iranian elections in June 2009. This event highlighted the opposition to the government-determined to restrict access to information—and the co-operation between newspapers, broadcasters and Iranian citizens to inform global audiences about the conflict occurring there (Newman, 2009). Interviewees further suggested that the advantage of this access was the pursuit of more "realistic" and "transparent" reporting. In this regard, one of the participants emphasized: "If realistic means the transparency of the reader to check and follow the sources, then yes" (Participant B, 14 January 2013). Therefore, social media has the potential to add complexity (as opposed to simplicity) to journalists' coverage of war and conflict.

However, the interviews highlighted also some limitations of *Storify*'s software and design, in particular, the limited appeal that the format has for curators and

Table 5 Media professionals' choice of original contributors according to time coverage (2nd content analysis)

Original contributors	Hourly (%)	Daily (%)	Days (%)
Media Professionals	87	94	87
Citizens	57	56	67
Non-Arab People	56	52	67
Arab People	47	38	35
Curators own content	23	28	31
NGOs	13	13	30
Government	11	6	9
Others	8	4	13

end-users. One of the participants was quite radical: "It's ugly. The user experience is still to be ameliorated" (Participant F, 5 October 2012). Similarly, another participant stated that: "The format is hard to embed in a standard news outlet" (Participant A, 20 January 3013). Finally, checking the original source of information is still challenging on *Storify*, as in other social media. According to one of the participants: "The limitation is 'how to know what's true' like in other social media" (Participant E, 14 November 2012).

Storify does not necessarily change processes of news reporting, but rather, points to increased user participation in news events, and increased collaboration with journalists through the use of social media (Sacco & Bossio, 2014). Social media can be considered as a complementary media channel, more suitable for documenting than for creating news content. For example, one of the participants liked: "The way I can use it to impose order, whether chronological or topical" (Participant D, 10 January 2013). Another participant thought that "*Storify* is better for collecting all the leads 'post-breaking' and [to] create an overview to be able to analyse news" (Participant A, 20 January 3013). The platform enables curators to store single social media contributions within a logical narrative and provide an overall context to individual user-generated content. The platform is flexible with several possibilities for news presentation, so curators can use their creativity and storytelling abilities to provide quality content. In addition, on *Storify*, curators are able to emphasise relevant information that otherwise would have gone unnoticed.

As this second investigation presents some advantages and disadvantages of social media use for journalists who cover a crisis, further analysis is required to investigate fully the complementary relationship that can exist between journalists and social media. The second round of interviews with Swiss media professionals moves the analysis forward by investigating this complex relationship. In particular, the interviews focused on how journalists deal with the complexity of social media and how their cultural and institutional values, as well as the news products have been affected by changes in newsgathering and news reporting due to the advent of social media.

Social media can give access to more remote geographic areas that would otherwise not be reached, due to location, the cost of getting there or the suddenness of events. According to one of the participants this opens new options to foreign correspondents that can include new perspectives in their reportage: “these tools allow foreign correspondents to follow and be followed by NGOs, news agencies, groups’ pressures etc. It is a tool for them to improve the visibility of their work as well as to have direct access to sources and be aware of unfolding events. Through social media, they can reach more people”. (Participant P, 10 April 2013). It seems that social media can thus compensate the lacks of traditional investigative practices. Therefore, social media can be considered as a complementary to traditional journalism practices. Moreover, its role as an alert can also explain their supplementary nature. This is extremely important as a supplement to journalists traditional work routines in an age where lack of resources, smaller newsroom and 24/7 deadlines mean they cannot always be on location or covering each event during a conflict or war situation (Sacco & Bossio, 2015). In this regard, one of the participants illustrated how social media can be integrated into traditional sourcing of news: “For example, after the assassination of the opposition in Egypt it would have ended there, except that suddenly we see the entire blogosphere that started rustle in a way that might result into a protest. Here, there is no need to check the info, it is a mood. So we switch to the modality on alert” (Participant F, 12 February 2013).

Although social media brings new possibilities in terms of reportage, journalists interviewed have also highlighted potential issues that range from technical limitations, especially in terms of access to quality content, to the overall verification of information, which can lead to misinformation, manipulation or subjective reportage (Sacco & Bossio, 2015). In this regards, one of the participants showed the importance to have still journalists on the ground: “I went in the summer of 2011 to Hamas to see with my eyes the young people who placed videos online during the protests. I established a link with these individuals to understand how they work and gain confidence in the content that they uploaded online. The trusted sources have become sources of trust because I went to see them and I know who they are. That does not mean that all of their content is good. However, I obtained a better view of the scheme and, thus, a better analysis of what they produce and distribute online” (Participant F, 12 February 2013). Despite the decrease use of “elite” sources, journalists continue to face issues of trust and reliability (Paulussen & Ugille, 2008, p. 34). The best practice in order to provide accurate coverage appears to be finding the balance between the information spread on social media and that collected “on the ground”. This is another way to look at the complementary relation that can exist between social media and traditional types of reportage, showing that both forms of reportage can co-exist in an online digital environment (Sacco & Bossio, 2015). As an extension of spreading misinformation, social media can be also a fertile ground for pressure groups and governmental organisations, not only to disseminate propaganda and manipulated information, but also to introduce censorship. One of the participants gave an example of how social media can be used for propaganda. He described a specific event in Syria during the Arab Spring: “On the Internet, everything is not clear because behind there is a machine of opinion that is not visible and that is extremely efficient. It is almost a laboratory where people, from pressure groups to governments, know how to make a buzz. [...] The Assad's regime used social media to promote his football championship, to show that there was a real life; however, this action was surreal because typically cities such as Homs were at war. The dictatorial regimes wanted to sell an image to the audience. Assad is very powerful because he quickly understood that, on *Facebook*, there were people in the revolution. Consequently, he has created fake accounts of the revolution to make them lose credibility. He exaggerated the information, and then, he blamed them as liars" (Participant K, 12 March 2013). Nonetheless, opposition groups also understood the potential of social media during the protests in the Middle East. Social media becomes channels for activism and for spreading messages about state abuses. Real communications strategies were implemented for reaching wide audiences. In this context, the same participant illustrated how the protests were organised in Libya during the Arab Spring: "Before posting a video, there was really an organisation, and there was real activism, which means that the communication was never performed superficially. There were people who had satellite phones and other foreign phone chips. Among these people in Libya, there were those who mastered image and video, and others were just repeaters, which mean that they delivered the content., *Facebook* pages emerged very quickly with well-organised information that has become real archives" (Participant K, 12 March 2013).

This last round of interviews supports a "dualistic" nature of social media which allow conceiving social media as a complementary way of reportage to traditional journalistic practice and point out there were the profession of journalism is still relevant in today's society.

5 Conclusions

Although this study is not extensive, it provides a sound framework for further research into fully understanding the role of social media during war and conflict. In particular, it underlines the key factors leading to an increasing acceptance of social media use by both newsrooms and audiences in the context of the coverage of war and conflict. For academics, it proposes some guidelines on how to analyse social media as a disruptive phenomenon in a continuously developing media sphere. For practitioners, it highlights the advantages and disadvantages of social media as well as possible strategies to more effectively and efficiently implement social media in newsrooms. It also provides some new perspectives on how journalists can regain their place in this changing online digital environment.

The chapter began by arguing that use of different social media has brought a new dynamism to the coverage of war and conflict through increased possibilities in sourcing, producing and disseminating news and information. However, this dynamism has also meant that journalists covering war and conflict must also face new and complex professional and social issues that affect reportage of war and conflict. In forwarding this argument, the results illustrate the advantages and disadvantages of social media use by journalists during war and conflict. The main findings suggest that social media might have the potential to enhance the reportage of war and conflict in a number of ways. Social media was seen to provide (a) journalistic access to remote areas; (b) additional sources of information that can promote alternative frames in reportage; and, (c) more complex (as opposed to simplistic) coverage of war and conflict. Therefore, social media enriches the news reporting, (d) allowing the audience to develop critical awareness of war and conflict; (e) increasing public engagement due to the more human-centred narratives produced by social media accounts of war and conflict; and, lastly (f) engaging with more innovative news formats. These statements imply a "divergence" of views through which the information coming to the audience can therefore differ in opinions and contribute to the formation of a public sphere more diversified than before which has the potential to promote a democratic society.

However, the use of social media also presents challenges to journalists. These drawbacks illustrate the complexities of the use of social media. In particular, it was shown that social media can also become a source of (a) misinformation, (b) manipulation and (c) subjective reporting. Due to its capacity to spread information and engage debate, social media is often used by diverse agents to influence public perception including understanding, feelings and values. Many actors with many different interests can seek to impose their own narrative on social media. As social media engages and creates new communities, these communities can become "social system[s] containing radical inequalities and conflicts of interests" (Williams, 1974, p. 149). Despite the advantages brought by social media, journalists who cover war and conflict are confronted with these limitations and must negotiate new professional methods for reporting in this new media environment.

In conclusion, social media use for reportage incarnates a "dualistic" nature which emphasised where the profession of journalism is still needed in this dynamic online digital environment. Platforms such as *Storify* can be seen only as a "starting point" for news formats that merge social media and traditional journalistic practices. It is a platform where converging information comes from various sources. It is in this convergence that emerges the decisive role of journalists in ensuring their editing and gatekeeping activities to be ruled by their professional ethics. In line with Jenkins' definition of convergence (2006), *Storify* and other social media platforms can be viewed as an ongoing process that should not be considered a displacement of the old media, but rather as an interaction between different media forms and platforms. The interesting point of these platforms lies in the fact that journalists can play the role of curators of social media content—a role that the results suggest is very important in the new media age. Similarly, Newman (2009, p. 10) envisioned the role of journalists in this digital online environment as "multi-media writers and curators, who can synthesise events in real-time with the best of the backchannel from social media". These platforms can help journalists to initiate relationships with communities and individuals that provide information about events of war and conflict. Mediating these relationships and the content that they bring, they can help journalists become the bridge between truth and audience. In this social environment, journalists can pursue both their professional and their social responsibilities by providing accurate coverage of the reality of war and conflict as a base for debate and dialogue and therefore "contribute to the prospects for monitory democracy by enabling the scrutiny and chastening of power" (Lynch, 2013, p. 20).

The study supports the importance of the journalism profession in delivering complex (as opposed to simplistic representations) and balanced coverage of the causes and consequences of war and conflict—even through social media platforms. The limits of social media platforms highlight the need for journalists to act with the same kind of gatewatching attitude they bring to traditional reportage. Relying on their professional principals, journalists can be the gatekeepers of social media information, verifying the information spread on these channels, adding value and pinpointing relevant and accurate content, as well as ensuring quality (Sacco & Bossio, 2015). The findings are in line with Friend and Singer (2007, p. 41) who presented the journalism profession as an unchanged profile no matter the media under scrutiny. Similarly, Picard (2013, p. 21) argues that there will always remain a need for news and journalists, even if journalism experiences pressures due to media convergence. Journalists' profession will continue to rest on the delivery of general information, safeguard of freedom of expression and reporting verified facts. As suggested by Cornu (2013, pp. 159–160), journalists should help citizens to find and understand information they may believe. The journalist is no longer a gatekeeper but an indicator, a sense-maker. The journalist is not there anymore to deliver the information, which is already available in the digital and online space; he/she should care about its reliability in reporting facts and its validity of meaning. Finally as already pointed out by Scherer (2011, p. 15), the added value provided by journalists would also be the fast process of putting the information into context. He explained that it would be necessary to save time to readers; it can be added that it is necessary to protect readers from possible manipulation, censorship or misinformation as well as present the complexity of today's war and conflict. Therefore, the use of social media leads to the convergence of different functions in the single role of a journalist. In order for journalists to be able to fulfil those functions, further research is finally needed to address (1) the internal tension that media organisations constantly face between their public service mission and the economic imperatives dictated by their firm nature and, (2) the economic opportunities and challenges linked to an uprising phenomenon called entrepreneurial journalism.

This study presents some limitations. The timeframe during which the chapter was written needs to be taken into account. In particular, all the discussions started from a specific event—the Middle East revolutions. Although the author tried as much as possible to draw parallels or underline differences with the coverage of previous war and conflict, in the meantime other protests and/or conflicts emerged, such as those in Ukraine. In the literature ethnic conflicts are represented only marginally. Further research could include regional conflicts in developing countries (e.g. protests in Venezuela) and developing nations (e.g. Darfur and Central Africa) as well as countries that have been victimised by terrorist attacks or continue to suffer of insecurity and political instability of past conflicts (e.g. Israeli-Palestinian, Afghanistan or Iraq). A comprehensive and worldwide vision would allow researchers to understand if the role of social media is more important in specific settings or even if it changes role depending on location (i.e. cultural values) or on the types of conflicts (e.g. ethnic conflicts). Another limitation concerns the sample. It is recommended performing some comparison between journalistic social media usage in different countries, to examine a diversified sample that included the varying degrees of freedom of expression and the use of the Internet (e.g. countries such as Africa or China). Regarding limitations that concern the methodology, it can be added that completing a network analysis could have been useful in particular for determining the relationship between the different sources of information. Further research could also focus on the economic and regulation (e.g. governmental and institutional) aspects of media companies. The author is aware that in this evolving digital and online ecosystem journalists have also to face important pressures that come from economic and regulation policies (i.e. political convergence).

Acknowledgments The authors would like to thank her two thesis supervisors Dr. Dal Zotto and Dr. Bossio as well as the scholars that have evaluated her Ph.D. dissertation Dr. Paulussen, Dr. Dubied and Dr. Grevisse. Many thanks also to the unknown reviewers of the chapter and the two editors of this book who have helped to improve and wisely valorized this study.

References

Adams, R., & Ovide, S. (2009, January 15). Newspapers move to outsource foreign coverage. *Wall Street Journal*. Retrieved August 24, 2012, from http://online.wsj.com/articles/SB123197973917183829

Allan, S. (2009). Histories of Citizen journalism. In S. Allan & E. Thorsen (Eds.), *Citizen Journalism: Global perspectives* (pp. 17–32). New York: Peter Lang.

Altmeppen, K. D. (2010). The gradual disappearance of foreign news on German television: Is there a future for global, international, world or foreign news? *Journalism Studies, 11*(4), 567–76.

Archetti, C. (2013). Journalism in the age of global media: The evolving practices of foreign correspondents in London. *Journalism, 14*(3), 419–36.

Atkins, C. (2011, February 23). Churnalism: When press releases masquerade as news stories. *The Guardian*. Retrieved June, 2014, from http://www.guardian.co.uk/media/video/2011/feb/23/churnalism-press-releases-news-video

Barton, V. (2009). *Killed in line of duty. Who is killing foreign correspondents and why?* Master thesis, Auckland University of Technology, NZ.

Beck, U. (1997). *The reinvention of politics: Rethinking modernity in the global social order*. Cambridge: Polity Press.

Bennett, W. L. (1990). Toward a theory of press state relations in the United States. *Journal of Communication, 40*(2), 103–127.

Boyd-Barrett, O., & Rantanen, T. (1998). *The globalization of news*. London: Sage.

Brennen, B. S. (2012). *Qualitative research methods for media studies*. New York: Routledge.

Broersma, M., & Graham, T. (2013). Twitter as a News Source. How Dutch and British Newspapers Used Tweets in Their News Coverage, 2007–2011. *Journalism Practice, 7*(4), 446–64.

Campbell, F. (1997). Journalistic construction of news: Information gathering. *New Library World, 98*(2), 60–4.
Carpentier, N., Pruulmann-Vengerfeldt, P., Nordenstreng, K., Hartmann, M., Vihalemm, P., & Cammaerts, B. (2006). The internet and the second Iraqi War: Extending participation and challenging mainstream journalism? In B. Cammaerts & N. Carpentier (Eds.), *Researching media, democracy and participation: The intellectual work of the 2006 European media and communication doctoral summer school. The research and teaching communication series* (pp. 159–177). Tartu, Estonia: Tartu University Press.
Carruthers, S. L. (2000). *The media at war*. Basingstoke: Macmillan Press.
Castells, M. (1997). *The power of identity. The information age: Economy, society and culture* (Vol. II). Cambridge, MA; Oxford, UK: Blackwell.
Chouliaraki, L. (2006). *The spectatorship of suffering*. London: Sage.
Clarke, J., & Bromley, M. (2011). *International news in the digital age: East-west perceptions of a new world order*. London: Routledge.
Cornu, D. (2013). *Tous connectés! Internet et les nouvelles frontières de l'info*. Genève: Labor et Fides.
Cottle, S. (2006). *Mediatized conflict*. Maidenhead: Mc Graw-Hill Education/Open University Press.
Cottle, S. (2008). *Global crisis reporting. Journalism in the Global Age*. Maidenhead: McGraw-Hill Education, Open University Press.
Cottle, S. (2009). Journalism and globalization. In K. Wahl-Jorgensen & T. Hanitzsch (Eds.), *The handbook of journalism studies* (pp. 341–356). New York: Routledge.
Cottle, S. (2011). Afterword – Media and the Arab Uprisings of 2011. In S. Cottle & L. Lester (Eds.), *Transnational protests and the media* (pp. 293–304). New York: Peter Lang.
Dahlgren, P. (2005). The internet, public spheres and political communication: Deliberation and dispersion. *Political Communication, 22*(2), 147–62.
Davies, N. (2008, February 4). Our media have become news producers of distortion. *The Guardian*. Retrieved June, 2014, from http://www.theguardian.com/commentisfree/2008/feb/04/comment.pressandpublishing
Diefenbach, T. (2009). Are case studies more than sophisticated storytelling? Methodological problems of qualitative empirical research mainly based on semi-structured interviews. *Quality & Quantity, 43*(6), 875–94.
Dimitrova, D. V., Kaid, L. L., & Williams, A. P. (2004). The first hours of online coverage of operation Iraqi freedom. In R. D. Berenger (Ed.), *Global media go to war: Role of news and entertainment media during the 2003 Iraq War* (p. 255). Spokane, WD: Marquette Books.
Dimitrova, D. V., Kaid, L. L., & Williams, A. P. (2005). War on the Web: The immediate news framing of Gulf War II. *The Harvard International Journal of Press/Politics, 10*(1), 22–44.
Dimitrova, D. V., & Neznanski, M. (2006). Online journalism and the war in cyberspace: A comparison between US and international newspapers. *Journal of Computer Mediated Communication, 12*(1), 248–63.
Friend, C., & Singer, J. B. (2007). *Online Journalism Ethics: Traditions and transitions*. ME Sharpe.
Giddens, A. (1994). *Beyond left and right: The future of radical politics*. Oxford: Stanford University Press.
Griffin, M. (2004). Picturing America's "War on Terrorism" in Afghanistan and Iraq. *Journalism, 5*(4), 381–402.
Halavais, A. (2002). Part 3: The Rise of Do-it-yourself Journalism After September 11. *One Year Later: September 11 and the Internet*. Pew Internet & American Life Project. Available from http://www.pewinternet.org/
Hall, S., Critcher, C., Jefferson, T., Clarke, J. N., & Roberts, B. (1978). *Policing the crisis: Muggng, the State, and Law and Order*. New York: Palgrave Macmillan.
Hallin, D. (1986). *The "Uncensored War?" The Media and Vietnam*. New York: Oxford University Press.

Hamilton, J. M., & Jenner, E. (2004). Redefining foreign correspondence. *Journalism, 5*(3), 301–21.
Harris, R. (1994). *The media trilogy*. London: Faber & Faber.
Heinrich, A. (2012). Foreign reporting in the sphere of network journalism. *Journalism Practice, 6* (5–6), 766–75.
Henry, N. (2007). *American carnival: Journalism under siege in an age of new media*. London: University of California Press.
Hermida, A., Lewis, S. C., & Zamith, R. (2012, April 20). *Sourcing the Arab Spring: A case study of Andy Carvin's sources during the Tunisian and Egyptian Revolutions*. Paper presented at the international symposium on online journalism, Austin, TX.
Hjarvard, S. (2000). News media and the globalization of the public sphere. *Kommunikations Forum*. Retrieved May 7, 2014, from http://www.kommunikationsforum.dk/artikler/news-media-and-the-globalization-of-the-public-sphere
Höijer, B. (2004). The discourse of global compassion: The audience and media reporting of human suffering. *Media, Culture & Society, 26*(4), 513–31.
Jenkins, H. (2006). *Convergence culture. Where old and new media collide*. New York: New York University Press.
Kellner, D. M. (1992). *The Persian Gulf TV War*. Boulder, CO: Westview Press.
Kester, B. (2010). The art of balancing: Foreign correspondents in non-democratic countries: The Russian case. *International Communication Gazette, 72*(1), 51–69.
Kim, E. G., & Hamilton, J. W. (2006). Capitulation to capital? Ohmynews as alternative media. *Media, Culture & Society, 28*(4), 541–60.
Kolo, C. (2010). Online medien und wandel: Konvergenz, diffusion, substitution. In W. Schweiger & K. Beck (Eds.), *Handbuch Online Kommunikation* (pp. 283–307). Wiesbaden: VS Verlag.
Lacy, S. (2000). Commitment of financial resources as a measure of quality. In R. G. Picard (Ed.), *Measuring Media content, quality and diversity*. Approaches and issues in content research, *Media Economics, Content and Diversity Project and Media Group, Business Research and Development Centre* (pp. 25–50). Turku, Finland: School of Economics and Business Administration.
Latzer, M. (2013, April). *Media convergence*. Working paper of the Media Change and Innovation Division, IPMZ, University of Zurich, Switzerland.
Lee, C.-H. (2004). *News coverage of US war with Iraq: A comparison of the New York Times, the Arab News, and the Middle East Times* (Doctoral dissertation). Retrieved from https://www.lib.utexas.edu/etd/d/2004/leec042/leec042.pdf
Lewis, S. C., Kaufhold, K., & Lasorsa, D. L. (2010). Thinking about citizen journalism: The philosophical and practical challenges of user-generated content for community newspapers. *Journalism Practice, 4*(2), 163–79.
Lynch, J. (2013). *A global standard for reporting conflict* (Vol. 7). New York: Routledge.
McChesney, R. W. (1999). *Rich media, poor democracy: Communication politics in dubious times*. Illinois: University of Illinois Press.
Moeller, S. D. (1999). *Compassion fatigue: How the media sell disease, famine, war and death*. New York: Routledge.
Morrison, D. E. (1994). Journalists and the social construction of war. *Contemporary Record, 8*(2), 305–20.
Morrison, D., & Tumber, H. (1981). *Foreign media correspondents based in London*. ESRC-funded project, final report.
Morrison, D., & Tumber, H. (1985). The foreign correspondent: Dateline London. Media. *Culture & Society, 7*(4), 445–70.
Neuman, W. R., Just, M. R., & Crigler, A. N. (1992). *Common knowledge*. Chicago: University of Chicago Press.
Newman, N. (2009). *The rise of social media and its impact on mainstream journalism*. Oxford: Reuters Institute for the Study of Journalism. Retrieved August 24, 2012, from https://

reutersinstitute.politics.ox.ac.uk/sites/default/files/The%20rise%20of%20social%20media%20and%20its%20impact%20on%20mainstream%20journalism.pdf
Pantti, M., & Bakker, P. (2009). Misfortunes, sunsets and memories: Non-professional images in Dutch news media. *International Journal of Cultural Studies, 12*(5), 471–89.
Paterson, C. (1997). Global television news services. In A. Sreberny-Mohammadi, D. Winseck, J. McKenna, & O. Boyd-Barrett (Eds.), *Media in global context: A reader*. London: Hodder Arnold.
Paulussen, S. (2012). Technology and the transformation of news work: Are labor conditions in (online) journalism changing? In E. Siapera & A. Veglis (Eds.), *The handbook of global online journalism* (pp. 192–208). Boston, MA: Wiley-Blackwell.
Paulussen, S., & Harder, R. A. (2014). Social media references in newspapers. Facebook, Twitter and Youtube as sources in newspaper journalism. *Journalism Practice, 8*(1), 1–10.
Paulussen, S., & Ugille, P. (2008). User generated content in the newsroom: Professional and organisational constraints on participatory journalism. *Westminster Papers in Communication & Culture, 5*(2), 24–41.
Philippe, J. (2001). La légitimation de la cause humanitaire: Un discours sans adversaires. In Editions Humanitaire en discours. MOTS. *Les Langages du politique, 6*, 15.
Picard, R. (2013). Killing Journalism. In S. Russ-Mohl, H. W. Nienstedt, & B. Wilczek (Eds.), *Journalism and media convergence* (pp. 19–28). Berlin/Boston: De Gruyter.
Platon, S., & Deuze, M. (2003). Indymedia journalism. A radical way of making, selecting and sharing news? *Journalism, 3*(4), 336–55.
Potts, J. (2011). *Creative industries and economic evolution*. Cheltenham: Edward Elgar.
Quandt, T., & Singer, J. B. (2009). Convergence and cross-platform content production. In K. Wahl-Jorgensen & T. Hanitzsch (Eds.), *The handbook of journalism studies* (pp. 130–143). New York: Routledge.
Reese, S. D. (2001). Understanding the global journalist: A hierarchy-of-influences approach. *Journalism Studies, 2*(2), 173–87.
Rheingold, H. (2002). *Smart Mobs: The next social revolution*. Cambridge, MA: Basic Books.
Riff, D., Lacy, S., & Fico, F. (2014). *Analyzing media messages: Using quantitative content analysis in research*. New York: Routledge.
Robinson, M. (2007, August). Two Decades of American News Preferences. Pew Research Center. Available from http://pewresearch.org/pubs/566/two-decades-of-american-news-preferences
Robinson, P., Goddard, P., Parry, P., & Murray, C. (2009). Testing models of media performance in wartime: UK TV news and the 2003 invasion of Iraq. *Journal of Communication, 59*(3), 534–63.
Robinson, P., Goddard, P., Parry, P., & Murray, C. (2010). *Pockets of resistance: British News media, War and Theory in the 2003 Invasion of Iraq*. Manchester: Manchester University Press.
Russ-Mohl, S., Nienstedt, H. W., & Wilczek, B. (2013). Introduction. In S. Russ-Mohl, H. W. Nienstedt, & B. Wilczek (Eds.), *Journalism and media convergence* (pp. 3–18). Berlin/Boston: De Gruyter.
Sacco, V., & Bossio, D. (2014). Using Storify: Challenges and opportunities for journalists covering conflicts. *Journal of Applied Journalism and Media Studies, 3*(1), 27–45.
Sacco, V., & Bossio, D. (2015). Using social media in the news reportage of War & Conflict: Opportunities and challenges. *The Journal of Media Innovations, 2*(1), 59–76.
Sacco, V., Giardina, M., & Stanoevska, K. (2013). Social media curation and journalistic reporting on the Arab Spring. In J. Hayes, K. Battles, & W. Hilton-Morrow (Eds.), *War of the Worlds to Social Media*. New York: Peter Lang (Chapter 12).
Sambrook, R. (2010). *Are foreign correspondents redundant? The changing face of international news*. Oxford: Reuters Institute for the Study of Journalism.
Scherer, E. (2011). *A-t-on encore besoin des journalistes? Manifeste pour un « journalisme augmenté »*. Paris: puf.

Schwalbe, C. B. (2006). Remembering our shared past: Visually framing the Iraq war on US news websites. *Journal of Computer-Mediated Communication, 12*(1), 264–89.

Shaw, I. (2012). *Human rights journalism: Advances in reporting distant humanitarian interventions*. Basingstoke: Palgrave Macmillan.

Shoemaker, P., & Reese, S. (1996). *Mediating the message*. White Plains, NY: Longman.

Snow, D. A., & Benford, R. D. (1988). Ideology, frame resonance, and participant mobilization. *International Social Movement Research, 1*(1), 197–217.

Snow, D. A., Rochford, E. B., Jr., Worden, S. K., & Benford, R. D. (1986). Frame alignment processes, micromobilization, and movement participation. *American Sociological Review, 51* (4), 464–81.

Sylvester, C. (2011). *Experiencing war*. Oxford: Routledge.

Tester, K. (2001). *Compassion, morality and the media*. Buckingham: Open University Press.

Thussu, D. K. (2007). *News as entertainment: The rise of global infotainment*. London: Sage.

Thussu, D. K., & Freedman, D. (2003). *War and the media: Reporting conflict 24/7*. London: Sage.

Tumber, H. (2009). Covering war and peace. In K. Wahl-Jorgensen & T. Hanitzsch (Eds.), *The handbook of journalism studies* (pp. 386–97). New York: Routledge.

Valkenburg, P. M., Semetko, H. A., & de Vreese, C. H. (1999). The effects of news frames on readers' thoughts and recall. *Communication Research, 26*(5), 550–69.

Volkmer, I. (1999). *News in the global sphere: A study of CNN and its impact on global communication*. Luton: University of Luton Press.

Volkmer, I. (2003). The global network society and the global public sphere. *Development, 46*(1), 9–16.

Wall, M. (2006). Blogging Gulf War II. *Journalism Studies, 7*(1), 111–26.

Williams, R. (1974). *Television: Technology and cultural form*. London and New York: Routledge.

Williams, A., Wahl-Jorgensen, K., & Wardle, C. (2011). More real and less packaged: Audience discourse on amateur news content and its effects on journalism practices. In K. Andén-Papadopulos & M. Pantti (Eds.), *Amateurs images and global news* (pp. 193–209). Bristol: Intellect.

Wu, H. D., & Hamilton, J. M. (2004). US foreign correspondents: Changes and continuity at the turn of the century. *International Communication Gazette, 66*(6), 517–32.

Media Convergence is *NOT* King: The Triadic Phenomenon of Media "Convergence-Divergence-Coexistence" *IS* King

Artur Lugmayr and Cinzia Dal Zotto

1 Introduction

Convergence is a heavily discussed term in the media field today. Where many marketing departments are tagging almost any new digital activity to 'convergence', the underlying meaning of this term in a larger context is rather poorly understood. The multiplicity of uses of the word "convergence" already creates diverging views, conceptualizations and understandings (Arango-Forero, Roncallo-Dow, & Uribe-Jongbloed, 2015). With the introduction of digital TV in Europe, one single standard called DVB was ought to emerge and convergence was to bundle all Internet based services on one single set-top-box. Even radio was ought to be digitalized under a single standard called DAB to provide the consumer with a single platform in enjoying digital services. Reality is far away. Consumers diverge and use multiple platforms such as mobile TV, IP-TV, or Internet TV services to enjoy their interactive experiences. Marketing departments brand this service eco-system as convergence; however, this never really happened as consumers diverge in their use, habits and perceptions.

Discussing convergence or divergence within a media industry context requires a more general and abstract discussion about the evolution of media and its principles. For this purpose we extend our reflection to a larger context, and consider evolution from a biological point of view. In biology evolution is understood as a "process of continuous change from a lower, simpler or worse to a higher, more complex or

A. Lugmayr
Visualisation and Interactive Media, School of Media, Culture, and Creative Arts (MCCA), Curtin University, Bentley/Perth, Western Australia, Australia
e-mail: lartur@acm.org; artur.lugmayr@lugymedia.com

C.D. Zotto
Academy of Journalism and Media, Faculty of Economics and Business, University of Neuchâtel, Neuchâtel, Switzerland
e-mail: cinzia.dalzotto@unine.ch

A. Lugmayr, C. Dal Zotto (eds.), *Media Convergence Handbook - Vol. 1*, Media Business and Innovation, DOI 10.1007/978-3-642-54484-2_21

better state" (Merriam-Webster, http://www.merriam-webster.com/dictionary/). Evolution theory tries to explain all phenomena related to the development of species, and how these species reached their current state. Two main theories attempt to explain the evolution process (Wikipedia, http://en.wikipedia.org/): The first theory (*convergence theory*) assumes, that the present state of a specie was determined, and was bound to develop as it is today. Additional features of individual species developed independently and had to emerge due to natural preconditions (e.g., wings of species had to develop due to the existence of air). The second theory (*contingency theory*) assumes that the present state was determined by randomness and particular contingent events. Thus, new features of species would emerge accidentally, and the present state would not be replicable, even if the same pre-conditions would exist. No matter which theory better describes the phenomenon, sound investigations confirm that the biological evolution led to (Wikipedia, http://en.wikipedia.org/) processes of adaptation, co-evolution, cooperation, specialization, and extinction.

Comparing media evolution with the process of biological evolution seems now to be far-fetched. Nevertheless, we believe it is an excellent starting point for gaining more insights and understanding media industry developments. As following step, we thus link the idea of biological evolution to the world of media, and in particular to the phenomena of convergence, divergence, and coexistence. Indeed, each new form of media has a new set of features that either develops independently or in dependence to other media forms. According to McLuhan's law of retrieval (McLuhan, 1994; McLuhan & Lapham, 1994), each new media retrieves features from an old media. For instance, the World Wide Web inherited features from textual and visual media in form of web pages. Following this argumentation, our main consideration is—can we consider this media evolution as a matter of convergence, divergence, or coexistence? Are we dealing with a parallel development of media, where each media form develops similar additional features independently of other media? Are we dealing with a convergence evolution process, where each new media form is inheriting certain features from other existing media forms? Or is media evolution a matter of contingency, according to which new media forms emerge accidentally? Or is convergence and divergence a phenomenon on markets how new media environments are perceived by customers, following the argumentation in Montpetit (2015), where convergence "is disrupting [and] engendered dislocation and divergence in the way next generation media dissemination and consumption are perceived and marketed"?

As mentioned above, no matter if the root of evolution is convergence or contingency, from a biological perspective the process of evolution leads to adaptation, co-evolution, cooperation, specialization, and extinction (Wikipedia, http://en.wikipedia.org/). We can find numerous examples for these phenomena within the media industry: considering the history of mass-communication and media the biological phenomenon of *specialization* well resembles that of fragmentation, which characterized the development of mass-media towards niche media (see e.g., Kueng, 2008). TV content evolved from fixed program schedules to video on demand options; TV content, if previously delivered through a few channels only to masses, now reaches infinite niche audiences by delivering highly

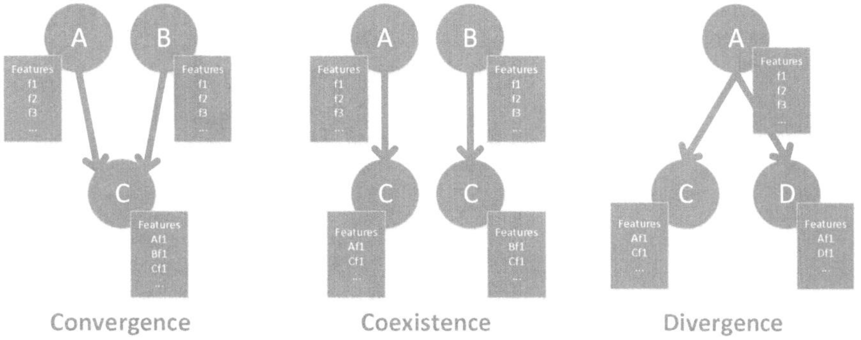

Fig. 1 Convergence, coexistence, divergence: the three processes characterizing the evolution of media

specialized content. The notion of *adaptation* can be compared to the needs of media to adapt to technology developments. Newspapers are finally adapting and moving from print to online publishing, allowing micropayments and online subscription models. *Cooperation* is clearly identifiable with the merger and acquisition activities taking place to allow firms to compete within a changing media landscape or to gain knowledge about how to successfully operate in the digital domain (Lugmayr & Zotto, 2013; Zotto, Galichet, & Lugmayr, 2011). Traditional media houses that are too slow in adapting to the new digital environment might face *extinction*, as Manroland and Schlott in Germany or Border Books in the USA well show. And we could enlist many more examples at this place. Media are currently experiencing a rather complex evolutionary step in their development, a step which was initialized by the advent of digital technologies and the Internet and of which convergence is just one single aspect. Adaptation, specialization, co-evolution and extinction of media are phenomena resulting not only from media convergence but also from divergence and coexistence processes, as depicted in Fig. 1.

The aim of this book chapter is to analyze all contributions to the two volumes of this edited handbook, and extract the trendsetting ideas within the media convergence research field. On the basis of the chapters, we will thus discuss examples of convergence, coexistence and divergence processes at different levels. For doing so we will proceed as follows. First of all, we will describe the method that we used to detect the most frequently discussed topics within the book chapters, and thus to create topic clusters. Then, we will continue with an in depth analysis of the sub-themes that within those clusters mostly reflect phenomena of media evolution and of the interplay between media convergence, divergence and coexistence processes. In particular the tag cloud produced from all book chapters (presented in Fig. 2) demonstrates the one sided approach in research in dealing with the interplay between these three processes. This emphasizes the fact, that scholars need to address more research in all three directions rather than focusing only on investigating one. Finally, an overall discussion will conclude the chapter, raise new questions and open the horizon to further research in the field.

Fig. 2 The triad of media progress: convergence–coexistence–divergence

2 Methodology and Approach

All 38 book chapters of the two edited volumes have been examined within the scope of this study (please see Ala-Fossi, 2015; Arango-Forero et al., 2015; Damasio, Henriques, Teixeira-Botelho, & Dias, 2015; Deliyannis, 2015;Denward, 2015; Dey, 2015; Duh, Meznaric, & Korošak, 2015; Ebner, Schon, & Alimucaj, 2015; Fleury, 2015; Foster & Brostoff, 2015;Gershon, 2015; Georgiades, 2015; Giesecke, 2015; Grüblbauer & Haric, 2015; Hamacher, 2015; Ibrus, 2015; Innocenti, Pescatore, & Rosati, 2015; Immonen & Giesecke, 2015; Karmasin, Diehl, & Koinig, 2015; Leminen, Huhtala, Rajahonka, & Westerlund, 2015; Matteucci, 2015; McKay, 2015; Montpetit, 2015; Mustic & Plenkovic, 2015; Sreekanth et al., 2015; Reichl, 2015; Sabiha, Morrison, Annika, & Obal, 2015; Sacco, 2015; Sousa, de Lemos Martins, & Zagalo, 2015; Spyridou & Veglis, 2015; Svahn, Wahlund, Denward, & Rademaker, 2015; Tuomi, 2015; Veglis, Dimoulas, & Kalliris, 2015; Villi, Matikainen, & Khaldarova, 2015; Vukanovic, 2015; Wellbrock, 2015; Tavares & Schofield, 2015; Zlatanov & Koleva, 2015). Together with the literature review that we have incorporated in the introduction to the edited volume and to this chapter, the corpus of the 38 handbook chapters provided us with sufficient material to identify key discussion clusters within the media convergence research field and to gain deeper understanding of the thematic area to conduct:

- investigation of several contributed book chapters to understand the underlying phenomena, driving forces, and their impact on the transformation of media industry;
- contribution with a new theory to the field of media studies, in particular how convergence-divergence-coexistence implicates in the field of media development;
- understanding traditional perception of media business, and the impact on business emerging from new media ecosystems;
- raising questions about current discussions in research, and pinpointing to areas that are rather scarcely emphasized.

To conduct the study, we aimed at an exploratory process, where we emphasized quantitative methods based on text mining techniques. We adopted the methods presented in Jockers (2014) and Williams, G. Available: http://handsondatascience.com/, resulting into the process illustrated in Fig. 3. Our main tool for conducting the study was the statistical software package R which includes various text mining packages such as tm, NLP, LDA, and worldcloud (R-Project, http://www.r-project.org). We utilized R including its text mining packages in combination with the text mining software package KHCoder (Coder, K. Available: http://khc.sourceforge.net/en/) to simplify the execution of the study. We also experimented with the RQDA package of R, but conducted the qualitative analysis thereafter based on *Keyword in Context Concordance (KWIC)* (see Jockers, 2014; Williams, G. Available: http://handsondatascience.com/) searches to speed up the process. Our basic data material was represented by the 38 book chapters contained in the two edited volumes. The analysis of the book chapters and the review of existing literature in the field of convergence allowed us to synthetize at the end of this final chapter a new viewpoint on this thematic area.

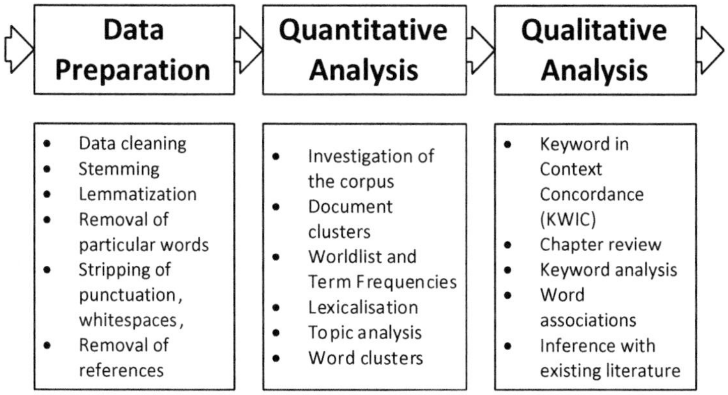

Fig. 3 Methodology and approach

Data Collection and Preparation The corpus of the study is based on the 38 book chapters, which provided us with sufficient material to perform the study and gain a deeper understanding of the thematic area. In the data preparation phase, we performed low level steps such as conversion to machine readable formats, removing special characters and references from each paper. Particular words, that should not be part of the analysis process, have also been removed from the corpus. For the experiments conducted in R, we performed essential corpus pre-preparation tasks such as stripping whitespaces, punctuations, and non-alphanumeric and numeric characters. After some experimentation in R, we decided to avoid stemming, as it led to some misleading results at a later stage of the study. While using KHCoder we relied on the preparation functions of this tool, and the lemmatization function via the Stanford POS tagger in particular became a very handy solution.

Quantitative Analysis This analysis focused on the investigation of the corpus through the generation of word clusters, word frequency lists, the term frequency distribution, the document frequency distribution, TF-DF plot, word association calculations, cluster analysis, and an LDA analysis. The methods applied for the quantitative analysis are based on descriptions in Jockers (2014) and Williams, G. Available: http://handsondatascience.com/. The results of the quantitative analysis are presented in the appendix of this book chapter. These results were analyzed, and led was the basis of the qualitative analysis.

Qualitative Analysis The main tool for the qualitative analysis was the keyword in context (KWIC) technique, according to which we analyzed the context of the keywords *convergence, coexistence,* and *divergence*. In this part of the analysis, we tried to understand the different key-ideas brought forward by the different chapters which we summarized in the following sections of this book chapter.

3 The Dynamics of Media Evolution, or the "Convergence–Divergence–Coexistence" Triad

As a result of our analysis and with reference to the content which the book chapters have been based on, we defined six main topics. Within the scope of the identified topics, we further focused on a few sub-themes more clearly reflecting the media evolution phenomena discussed within the book chapters, as illustrated in Table 1.

In the following subsections the identified sub-themes are extensively discussed through direct references to the related book chapters.

Table 1 Topics and themes across the contributed book chapters

Themes	Discussion topics	Chapter numbers and references
Business, market, and quality	Divergence of traditional media firms' perception of business, and the emergence of convergent media ecosystems	Wellbrock (2015), Ebner et al. (2015), Matteucci (2015), Karmasin et al. (2015), Vukanovic (2015), Leminen et al. (2015), Reichl (2015), Immonen and Giesecke (2015)
Social, interaction and networking	New modes of interaction and media consumption	Arango-Forero et al. (2015), Gershon (2015), Foster and Brostoff (2015), Tavares and Schofield (2015), Sreekanth et al., (2015), McKay (2015), Sabiha et al. (2015)
Content and technology	The "convergence divide"—convergence through divergence, divergence through convergence	Montpetit (2015), Veglis et al. (2015), Deliyannis (2015), Tuomi (2015), Damasio et al. (2015), Fleury (2015)
Media form and audience	Emergence of new media forms: transmedia, social media, ambient media, and the challenges for advertising	Arango-Forero et al. (2015), Tuomi, (2015), Villi et al. (2015), Denward (2015), Sousa et al. (2015), Innocenti et al. (2015), Ibrus, (2015), Sabiha et al. (2015), Svahn et al. (2015), Sacco, (2015)
Information, intelligence and networks	Effects on communication and information processing: data, information, knowledge, and networks of communication	Hamacher (2015), Mustic and Plenkovic (2015), Duh et al. (2015), Grüblbauer and Haric (2015), Giesecke (2015), Immonen and Giesecke (2015)
Media organizations and profession	Driving forces, power, and coexistence of evolution of traditional policies and media	Dey (2015), Spyridou and Veglis (2015), Ala-Fossi (2015), Zlatanov and Koleva (2015), Georgiades (2015)

3.1 Divergence of Traditional Media Firms' Perception of Business, and the Emergence of Convergent Media Ecosystems

Today's media industry requires a rethinking of existing managerial and organizational models. Organizational practices within media organizations seem to diverge and thus not to be appropriate for the reality in which media firms are currently operating. As Karmasin et al. (2015) states, there is a "need [for convergence mastering organizations and collaboration oriented management competences. This requires] that companies [...] not only have to understand [...] changes in society, consumer behavior and business models, but [...] also [...] have to be able to react appropriately and proactively". The divergence between managerial strategies and business reality appears to be a major dilemma for the media industry, engaged since more than a decade in a structural transformation process. This dilemma has been well discussed in Vukanovic (2015). The paper underlines the need for media companies to embrace emerging technologies and recognize that media divergence can "successfully perform as vendor lock-in, top-down corporate process and a bottom-up consumer driven process". On the same path Leminen et al. (2015) points out that the divergence between media strategies and business reality increases if existing business models are not abandoned and new models introduced into the market. According to Leminen et al. (2015) the rigid belief system of traditional media firms prevents them to innovate and develop their business models according to industry and technological changes. New entrants coming from other industries foster cross-industrial convergence as well as industry specific convergence of different business models elements such as technology platforms, revenue logic, content production and value creation as the publishing industry case shows.

A market driven business model convergence can though have a reverse side and lead to journalistic quality divergence of content in traditional and online media (Wellbrock, 2015). The authors of this chapter argue that, because advertising represents their main source of income, online news media are acting as "audience production plants" so that serious biases with regard to high quality content and minority issues might arise. An extension of the activities of public service broadcasters to the online market is suggested as a possible solution to regulate that market, and stimulate news media to engage in fulfilling their public service mission. This is a clear argument driving all news media to converge in ensuring their watchdog and democracy enhancing function.

The case of traditional Television operators seems to minimize the divergence between their perception and the reality of the business situation. TV operators have acknowledged rather early the incoming Internet threat and have been working hard, in several domains to prepare themselves and accommodate convergence while exploiting their TV market incumbency. As Matteucci (2015) concludes, despite a high technological turbulence established TV and telecom operators strive to keep market shares through walled garden solutions and by lobbying for stronger IPR protection. However, audience trends signal already a new storm coming with online TV viewing progressing at high rates and at the expenses of traditional TV consumption. Standardization and regulation have thus been enabling technological and business convergence for a while, but they might not be sufficient to prevent a diverging business trend with the emergence of hybrid media forms in the future.

A media ecosystem where convergence and divergence coexist seems thus an emerging reality within the media industry. Nevertheless, how the industry will cope with the dynamics of such an evolution from a management point of view still remains an open question. A cross-disciplinary approach appears to be the starting point to tackle this issue, and for this purpose the concept of communication ecosystems has been introduced by Reichl (2015). According to the authors that concept represents a multidisciplinary framework to simultaneously address convergence issues from an economic, technical and user perspective. By opening up communication with users and taking into consideration their input, the quality of user experience can be improved and problems of divergence at consumption level might be overcome.

The importance of establishing and nurturing network relations with key stakeholders and in particular with users and customers is underlined also in Immonen and Giesecke (2015). Such relations would increase the possibility to combine new technologies and new meanings into innovative business product ideas. Following a holistic approach the paper claims that applied business concepts have a direct impact on the entire organizational system including the technologies in use, tasks, processes as well as staff roles and competences. Clearly in line with classic organizational development theories, Immonen and Giesecke (2015) points out that losing the customer focus—thus diverging from the customer view—might blur the value proposition picture and bring a company to develop in a wrong if not destructive direction. The impact of customer oriented applications of new technology on business models has been addressed and discussed also in Ebner et al. (2015): the possibility of printing a personalized scientific book is seen as a solution to bridge the gap between open access and printed journals, and thus as a further step towards media convergence. This again requires organizational forms to adapt accordingly and develop into network like converging structures, as stated in Gershon (2015). A cross-disciplinary discussion seems to be appropriate to better understand audiovisual media production, and in particular serial narratives, where planning and practice diverge (Innocenti et al., 2015).

3.2 New Modes of Interaction and Media Consumption

The way how we interact with media and information technologies has been greatly developing and will continue to evolve. If technological convergence is allowing this development, along this path we can already see a divergence process emerging as always more specific and individualized solutions appear with regard to spaces, devices, and modalities in which interaction takes place. One example of divergence in this field has been discussed in Foster and Brostoff (2015), where the authors analyze the application of media interaction in game design and urban planning, two distinct areas which though share the goal of building up a cohesive and user-friendly visual narrative about alternative realities. The dynamics of media evolution extends the impact of interaction modes' development towards other domains, such as urban planning, and considers for instance how people interact with objects in urban spaces. Divergence effects are clearly visible in interaction design and user experience both in urban planning and game design. More and more multisensory devices stimulate human senses, it is thus essential to develop strategies which allow humans to interact with technology in a multimodal way and to "design systems that are sensitive to what the user wants without having been given explicit commands" (Sreekanth et al., 2015). As a consequence, video game simulations might be applied to urban planning and urban planning data might enrich game design. As authors state in Foster and Brostoff (2015), it may only be a matter of time for "real world urban planning and game simulation to converge to the point where both are subsets of some larger field of study". As embedded media are diverging, the development of user interfaces will help to cope with the personal and divergent media stimuli. A solution might be multisensory interaction design, as addressed in Tavares and Schofield (2015). This chapter enhances the need for user interfaces to be designed taking "plurality, adaptability, and cognitive ability aspects" into consideration (Tavares & Schofield, 2015) so to help the consumer cope with a diverging media landscape.

A quite far more complex issue is information system design—in particular the information system design within media industries. The question is to which extent information systems at corporate level can be designed to fulfil the needs of user experience and new media consumption modes, while capturing the tacit intellectual capital within knowledge intensive firms. This issue has been addressed in McKay (2015). If we further consider, as we have already discussed, that for the media industry the key product is still content and convergence leads to the emergence of new forms of media such as transmedia storytelling, then corporate information systems should be designed to support those new media forms as well as the interaction and user-friendly modalities they require as e.g., described in Sabiha et al. (2015).

The change of media consumption behavior is critical issue in the evolution of media, to which chapter Hamacher (2015) devotes particular attention by questioning if traditional media houses are targeting a specific selected audience through their content offerings. Besides presenting an excellent methodology to analyze textual online content, Hamacher (2015) presents a solution for media houses to optimize their portfolios to target particular audience niches.

3.3 "The Convergence Divide": Divergence through Convergence, Convergence Through Divergence

The evolution of media has brought about convergence processes. However, as we have already mentioned above, simultaneous divergence processes can be detected and the issue of a potential "convergence divide" has emerged. Looking at evolution of technology, McLuhan already emphasized (McLuhan, 1994) that technological development has clear social implications. Thus, when discussing media convergence, discussing if technology convergence may or may not lead to social convergence seems to be inevitable. This illustrates the fact that convergence on one level does not essentially lead to convergence on another level. Within the scope of this section, a few examples shall illustrate the "convergence divide", which has been underlined by many chapters in this book. In particular, we recall here the tension arising between "technological convergence" and "social divergence" in mobile Internet discussed in Damasio et al. (2015).

The issue of a convergence divide is especially visible when considering convergence at technology level leading to divergence at cultural and economic level. Such a divergence process is evident in the field of networking, which has been excellently described in Montpetit (2015): the cornerstone of the Internet is network neutrality—each user on the Internet is treated equally, as well as all her/his traffic. Thus, the Internet is a place where traffic is "not managed" and coordinated by a non-central authority. With network providers pushing towards new digital payment models, and the idea of prioritizing consumer traffic related to online paying services, network neutrality seems to be an issue of the past, while the time of managed internet services has come. As stated in Montpetit (2015), Netflix, Instagram, Amazon, Hulu, and Vines are applications that are built on a principle of network neutrality. However, for other services such as IPTV, TV interactive service offerings, and mobile phone TV services the "friction between managed and unmanaged services will continue" (Montpetit, 2015). Even if we know today that on February 26 the US Federal Communication Commission has approved a law enforcing net neutrality, we also know that in the rest of the world the net neutrality issue is still open. Thus, this book chapter nicely demonstrates a good example of convergence divide showing that, while technologies converge, culture and business models are diverging. Thus, convergence on one layer (technology) is leading to divergence on another (social).

An interesting aspect of convergence has been discussed in Deliyannis (2015) where the author states that "interactive TV broadcasting provides [...] backward compatibility that avoids divergence". Following this argumentation, media consumption does not essentially diverge based on interactive content; in fact, content with novel interaction requirements may act as a driver for convergence through the development of interactive television systems featuring advanced interactive and presentation methods. In this example, divergence may thus lead to convergence. A similar line of argumentation has been followed in Tuomi (2015), where this relation has been examined through the example of television and mobile devices acting as second screens and facilitating television watching experiences.

A counter example, convergence as a result of divergence, has been presented in Mustic and Plenkovic (2015). Here the authors discuss the diverging requirements of consumers in terms of visual presentation to increase information credibility, and show how the use of convergent graphic media represents a convergent answer to consumers' diverging perspectives. A discussion around the thematic area of digital TV, which has been conducted in Matteucci (2015), highlights again the interrelation between technological convergence and divergence. The chapter argues that technological standardization led to technical convergence on one hand; however, it might have contributed to the development of different platforms and thus to divergence on the other hand. Technological standardization could thus be considered also as enabler of diverging consumers. This thesis is sustained also in Ala-Fossi (2015) where the convergence divide has been exhaustively discussed with reference to television broadcasting. According to the authors, the development of standardized television platforms did not essentially lead to convergence on other levels. On the contrary, with the digitalization of terrestrial television an increased divergence at content production, distribution and consumption level—rather than convergence—could be observed. Even the number of terrestrial broadcast standards increased from three to five, and with it—following Ala-Fossi (2015)—the consumer divergence.

This phenomenon has also been illustrated in Karmasin et al. (2015), where the authors state that "while content is used on multiple platforms [. . .] and convergent, it takes more devices to display it", meaning that technology is diverging. To cope with these developments, and allow media organizations to apply the latest technology in distributing their content across platforms, a coherent organizational model for cross-media publishing is required. A framework solution for creating such a model has been presented in Veglis et al. (2015).

To conclude the discussion about the interrelationship and mutual influence of convergence and divergence processes, maybe a completely different approach is required to cope with the convergence divide. A user centered approach is proposed in Fleury (2015), in which authors ask themselves if it would not make more sense to find a solution to understand and meet end users' needs to benefit from the evolution of media rather than to statically analyze phenomena of convergence divide. Thus, "how [is it possible] to understand and support end users [. . .] in making sense and benefiting from the evolution [of] media" (Fleury, 2015). Taking the example of convergence between television and mobile technology, the authors show how media consumption is becoming increasingly tight to individual preferences and activities, while media consumers are still bound to content providers as communication channel. Challenges that need to be tackled here are thus mostly of usability nature such as content synchronization across devices which is relevant both for end users and broadcasters. Eventually, a coherent model for cross-media publishing is required to allow media organizations to utilize latest technology in distributing their content across platforms, and a solution framework has been presented in Veglis et al. (2015).

3.4 Emergence of New Media Forms: Transmedia, Social Media, Ambient Media, and the Challenges for Advertising

The history of information processing (see e.g., Available: http://www.historyofinformation.com/index.php) for a complete overview] goes in parallel with the emergence of new media forms. Indeed, the introduction of new technologies brought to the emergence of new forms of media. McLuhan (1994) understood this phenomenon long ago, as also others did when discussing the effects of new media forms. The evolution of media implies the continuous re-invention, testing of new ideas, adapting these to the current state, and the development of new theories in the field of media studies to describe the phenomenon. New media features are added, other features mutate and some others get completely removed within processes of both convergence and divergence. A lengthy discussion illustrating the multidimensional nature of convergence, as well as the coexistence of convergence and divergence processes within the dynamic evolution of media is featured in Ibrus (2015). Convergence as part of an historical media evolution process has been underlined also in the introduction section of this book chapter and in Arango-Forero et al. (2015), where the multi-faceted use of the word convergence has been addressed.

A consequence of technological convergence and media convergence processes in terms of new media forms is *transmedia storytelling*, i.e., the possibility to tell stories though and across different media. As stated in Sousa et al. (2015), transmedia storytelling requires the addition of new information pieces via each of the different communication channels engaged in telling the story. This is what makes the difference from traditional storytelling where one complete content piece is then published across different media forms. Alice in Wonderland published as film, novel, videogame, website, and illustrated book is an example of traditional storytelling (Sousa et al., 2015). The emergence of new cross-media technology enabling new ways to develop content and tell stories across media is discussed also in Ibrus (2015) and Innocenti et al. (2015), as well as in Denward (2015) where a practical example of transmedia storytelling has been presented: *Sanningen om Marika* (*The Truth about Marika*), a program which has been produced by the Swedish public broadcaster SVT in collaboration with a pervasive games startup company.

Such examples emphasize the fact that media evolve also at a narrative and semiotic level and media convergence affects not only the industry as such but also culture and society (Sabiha et al., 2015). This is why a cross-disciplinary approach is necessary to understand nature of media convergence, the impact of which is evident also in the development of notions such as narrative and/or story as discussed in Innocenti et al. (2015). The need for clearer understanding and definition of narrative as we know it today, and in particular of transmedia storytelling as a result of media convergence, is highlighted also in Sabiha et al. (2015) by focusing on the importance of considering user experience design early in the story writing process. According to the authors of this chapter transmedia storytelling "discards linearity, provides several perspectives to a story, promotes a participatory culture, and often harnesses the added value of collective intelligence". Diversity and creativity, the two main characteristics of "all innate learning beings" (Sabiha et al., 2015), are thus at the very base of media convergence and divergence processes: learning beings participate, share, thus converge; however, by adding new perspectives to a story they stimulate divergence as well. A contemporary example of converging different perspectives into one channel is represented by today's social networks such as Twitter. The convergence of news distribution in such networks has been discussed in Villi et al. (2015), and authors conclude that delegating news diffusion through social networks enables the news corporations to connect to a diverging audience on diverging devices.

Social media have a further effect that is to be included in the dynamics of media evolution. By illustrating the complex relationship existing between professional journalists and social media, Sacco (2015) points out how journalists can act as moderators within social networks so that multiple functions such as those of producer, mediator and curator of news converge within the role of a professional journalist.

Another example well representing another side of the media evolution process is *ambient media* (also known as ubiquitous media or pervasive media). This term describes media embedded throughout our environment, integrating sensor data such as location based services, and adapting to the context of the consumer (Lugmayr, Serral, Scherp, Pogorelc, & Mustaquim, 2013; Lugmayr, Zou, Stockleben, Lindfors, & Melakoski, 2013). One very basic example of ambient media is Foursquare, or Facebook displaying location based information. User interfaces enabling interaction with content are becoming part of our daily living environment and constitute a base for further developments of interaction forms and towards "ambient media convergence" (Immonen & Giesecke, 2015). Understanding the dynamics of media convergence is or utmost importance in order for media evolution processes to be sustainable. Thus, the discussion about the evaluation and development of advertising within a converging media world, as well as the inclusion of converging media solutions in media and advertising planning, as problematized in Svahn et al. (2015), becomes fundamental.

3.5 Effects on Communication and Information Processing: Data, Information, Knowledge, and Networks of Communication

To further elaborate the argumentation of Zlatanov and Koleva (2015), technology convergence leads to wider and easier accessibility to information, and thus to information overload. In order to cope with information overload and to interpret the information stream of individuals new social structures emerge. Technology savvy brokers of information and influencers alter the structure of existing social networks within the process of information diffusion (Zlatanov & Koleva, 2015). As a consequence, we can say that technology convergence leads to social divergence. Nevertheless, the divergence of information overload into different social network information streams requires a minimum of visual convergence in representing information and knowledge. As suggested in Mustic and Plenkovic (2015), visual communication allows for acting as carrier, and for easy interpretability of media messages. Thus, if information overload is one aspect of convergence, the way how information spreads through networks and peers is both an aspect of visual convergence and of consumption divergence. To make it more predictable new models are required and one possible solution has been presented in Duh et al. (2015). This paper defines a predictive model describing how information can spread between nodes within a communication network. From a communication theory point of view, measurements and in particular quantitative metrics for building theoretical models of information are a necessity. An excellent overview and introduction into the basic metrics that can be used in gaining quantitative results from empirical data have been presented in Hamacher (2015).

So far theoretical aspects related to information processing and communication have been discussed. However, to a larger extent, information processing and communication have a direct impact on organizational practice. For instance, convergence through company mergers most probably leads to divergence in content handling—due to different ways of processing information and different company cultures—and thus to a higher content management complexity (Grüblbauer & Haric, 2015). This is a rather tricky issue, as content and information typically represent the main product of any media firms. In Grüblbauer and Haric (2015) authors discuss this aspect of convergence by problematizing the complexity of business intelligence and corporate information systems. In Gershon (2015) this discussion is extended to transnational corporations. The author highlights the importance of having an intelligent network system in place to allow international business communication, as well as the fact that intelligent network systems directly affect the spatial design and operations of transnational organizations. In other words, convergent communication networks allow organizational structure divergence. This diverging organizational process, which can be defined as an ecosystem of which media firms are part, is discussed in Giesecke (2015) where diversification processes through the example of live event ticket selling are presented. The emergence of ecosystems where convergence and divergence processes coexist have been also discussed in Immonen and Giesecke (2015).

3.6 Driving Forces, Power, and Evolution of Traditional Policies and Media

Where does convergence come from, and who is the driving force behind convergence? Is convergence a phenomenon that media and IT firms are pushing forward in order to expand their business and reach a dominant position in the digital world? According to Dey (2015) "The inherently disaggregated, divergent structure of the Internet has spawned a democratic culture [where] it is unlikely that even media conglomerates will find it easy to dominate the media". This perspective is shared also in Immonen and Giesecke (2015), where the authors argue that convergence is the driving force allowing large media companies to stay in power, and divergence empowers small and innovative newcomers to flourish in niche markets. This might explain the current dilemma of mid-sized media companies attempting to compete with emerging digital companies such as music services as e.g., Spotify, which are pushing audiences to diverge (https://www.spotify.com). This last viewpoint leads to another argumentation, which has been presented in Ala-Fossi (2015) citing the case of the integrated digital broadcasting system in Japan: "convergence [must] be pursued and initiated by people in power [as] it [does not happen by itself] as an automatic process".

The case of information spread discussed in Zlatanov and Koleva (2015) represents a good example of how technological convergence may lead to divergence. The chapter points out that while the quantity of available information rises and converges into certain social media networks, technology savvy actors' ability to process information leads to their superiority in influencing information spread and stream direction. Information spread, which is fundamental to form public opinion, depends on how information is altered and interpreted by individuals within communication processes. When people are confronted with an information overload, information brokers interpreting the flood of information become necessary. Furthermore, media channels' convergence together with the dominance of social media as information diffusion tools leads to divergence at social level, as the structure of existing information networks is altered in the information diffusion process and technology illiterate actors are unconsciously segregated. As a practical case, we would like to refer to Zlatanov & Koleva (2015), where this effect has been illustrated on the example of the introduction of Blue-Ray and HD DVD.

The superiority of technology literate actors is well described also in Veglis et al. (2015) where the transformative power and effects of convergence processes within news media firms are discussed. If on one hand technological convergence seems to open up new opportunities for established media to consolidate and new media to emerge, on the other hand its effects on journalistic content quality, newsroom structures and working conditions are diverging from traditional standards and raise serious doubts. Organizational issues have also been discussed in Georgiades (2015), where the author problematizes the media managers' understanding of information provision/communication. He argues that media managers provide information in order to achieve employees' trust and focus on ensuring employees' independence in their job instead of trying to build up a convergence enhancing dialogue. This leads not only to diverging understandings but also to diverging organizational behaviors. Especially the pressure to transform towards digital media houses and work more efficiently causes divergence on organizational level. As pointed out in Spyridou and Veglis (2015) on the example for journalism—the pressure to create a super-journalist, thus a journalist literate in all new technologies, causes a divergence between journalistic quality and efficiency. A similar issue is raised in Georgiades (2015), which emphasizes employee involvement, and communication to achieve convergence rather than pure efficiency.

4 Discussion

Browsing and analyzing the content of all chapters within the two volumes of this edited handbook on media convergence we cannot conclude that convergence is the only process actually characterizing and shaping the evolution of media. Media evolution appears to be more the result of convergence, divergence and coexistence processes at different levels: technological, organizational, and social, to name the ones that have been mostly approached in this handbook. There is a strong interaction between the processes taking place at those levels, so that convergence processes are not independent from divergence processes and vice versa. More appropriately we can speak of an evolution where media convergence and divergence processes happen simultaneously, coexist and often influence each other building up on and developing existing features, forms and structures. Furthermore, sometimes media convergence hinders divergence in terms of business diversification as for instance discussed in Deliyannis (2015). Here the authors point out that interactive television should offer new features of television experience but always allow the consumer to jump back to an old passive non-interactive content mode at the same time. Such an offer is based on modular software components enabling the support of new interactive features by simplifying the "tuning" process. In this case convergence is basically hindering business diversification—such as the creation of a new device—through the requirement of e.g., backward compatibility within an existing device, i.e., the television. This would lead to the conclusion, that convergence might be a hurdle in the development of innovations in the world of media.

Most contributions to this edited handbook focus on the thematic area of convergence, only a few touch upon divergence, and almost none on the notion of co-existence. This clearly demonstrates a general lack of research in wider and more diverse directions for gaining understanding on the phenomenon of media evolution. Media divergence processes characterized by the development of additional features, and thus of new media forms emerging from similar forms, in each evolutionary step are still poorly investigated. Coexistence processes, in which new media rather develop in parallel to similar media forms, are poorly investigated, too. Chapters' contributions further highlight the notion of media ecosystem opposed to the one of "single" media entities; they underline the complexity of a dynamic media context, the evolution of which is characterized by the interplay of media convergence, divergence and coexistence processes.

We believe that focusing research on the sole phenomenon of convergence, thus neglecting other perspectives explaining media evolution, will hinder the building of theory in media studies. On the contrary, considering media convergence, coexistence, as well as divergence processes as part of a larger context such as the wide and dynamic evolution of media—where new features emerge, disappear, or are combined with one another as a matter of convergence or contingency—is the real key to understand them as well as their future development. Biology evolution docent and so do the chapters included in this edited handbook. We can thus conclude that future research within media divergence and coexistence processes as well as combining media convergence, divergence and coexistence phenomena is therefore very much needed.

Appendix A: Quantitative Analysis

Topic Analysis (LDA)

The topic analysis has been performed based on R, (R-Project, http://www.r-project.org), and consisted on the building of a lexicon, whose vocabulary was used as input for the *Latent Dirichlet Allocation* (*LDA*) analysis. We used a Gibbs sampler and built six topic clusters, which have been examined for topic clusters to identify higher level phenomenon based on the techniques described in Jockers (2014) (Table 2, Fig. 4).

Table 2 Topic analysis

	Business, market, and quality	Social, interaction, and networking	Content and technology	Media form and audience	Information, intelligence, and networks	Media organizations, and profession
1	Business	Interaction	Mobile	Media	Information	Journalism
2	Market	Design	Content	Transmedia	Data	Broadcasting
3	Model	User	Social	Audience	Event	Information
4	Media	Multimodal	Interactive	Story	Media	Organizational
5	Quality	Human	Television	News	Network	European
6	Services	Information	Users	Game	News	Radio
7	Models	Tools	Devices	Games	Roles	Employees
8	Open	Speech	Internet	Storytelling	Content	Policy
9	Markets	Knowledge	Services	Production	Ecosystem	Digital
10	Industries	System	Access	Medium	Social	Employee
11	Company	Experience	Networks	Jenkins	Role	Network
12	Product	Intelligent	Video	Narrative	Model	Regulation
13	Companies	Virtual	Media	Social	Business	Standard
14	Industry	Interface	Phone	Series	Article	Managers
15	Service	Systems	Screen	Audiences	Thus	Ownership
16	Publishing	Gesture	Convergence	Twitter	Actors	Regulatory
17	Billion	Learning	Network	Brand	Results	Management
18	Book	Game	Interaction	Facebook	Provider	Competition
19	Content	Computer	Publishing	Stories	Table	Speech
20	Economics	Video	Activities	Participation	Sources	Involvement

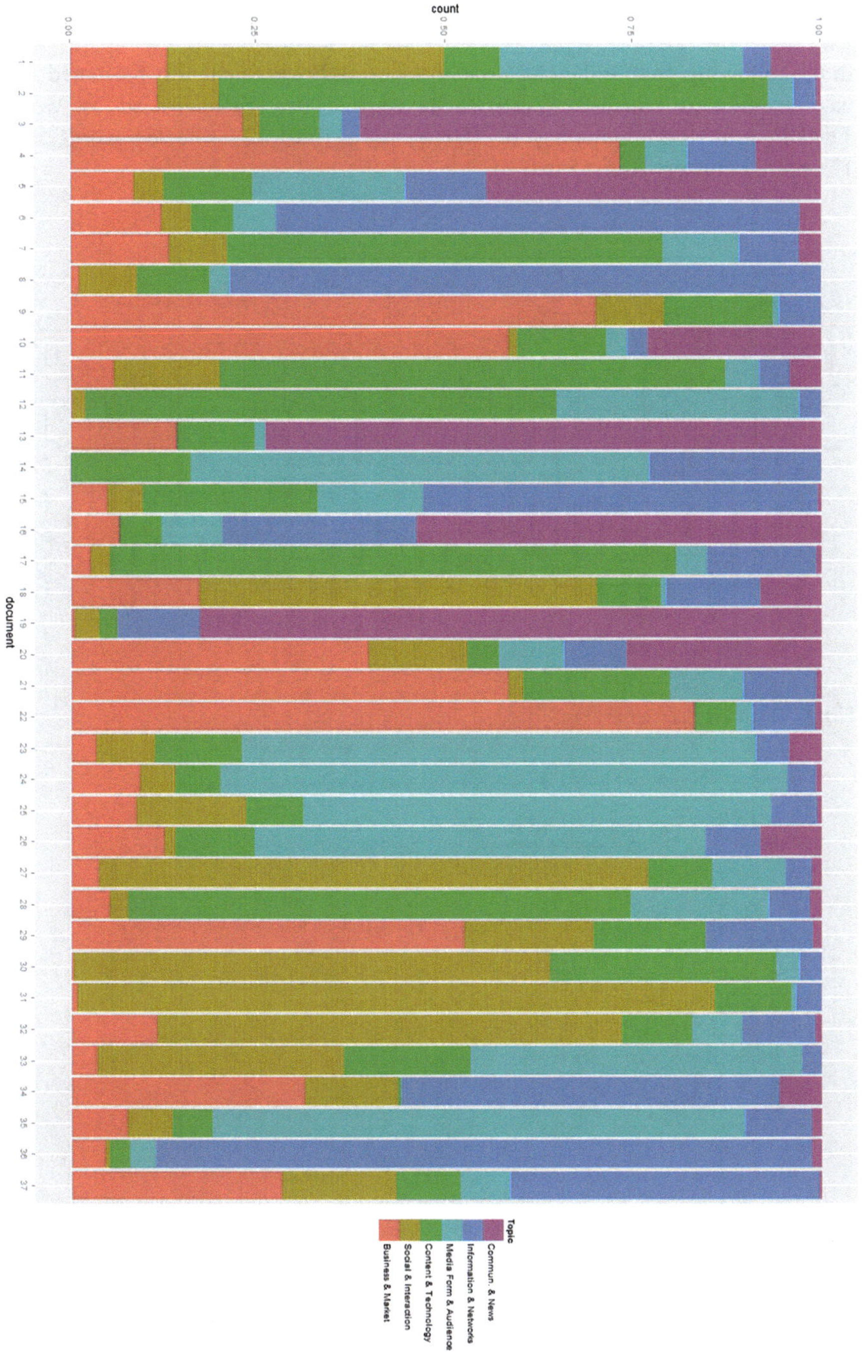

Fig. 4 Dominant topics within the 37 contributed book chapters

Document Cluster Analysis (Incl. Agglomeration Stages)

In the following picture the cluster analysis of all book chapters is presented. The parameters were: method (ward), distance (Cosine), Standardized by words, TF-IDF, min. term frequency of 150 (resulting into 127 selected words), and 13 optimal clusters (Figs. 5 and 6).

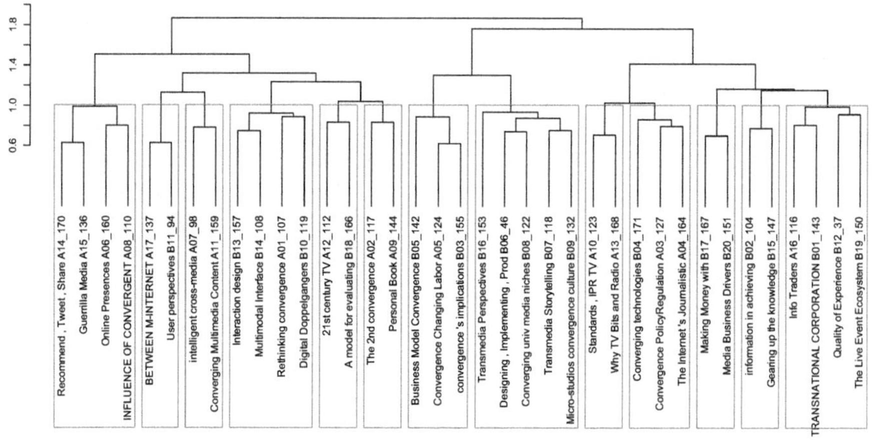

Fig. 5 Document cluster analysis

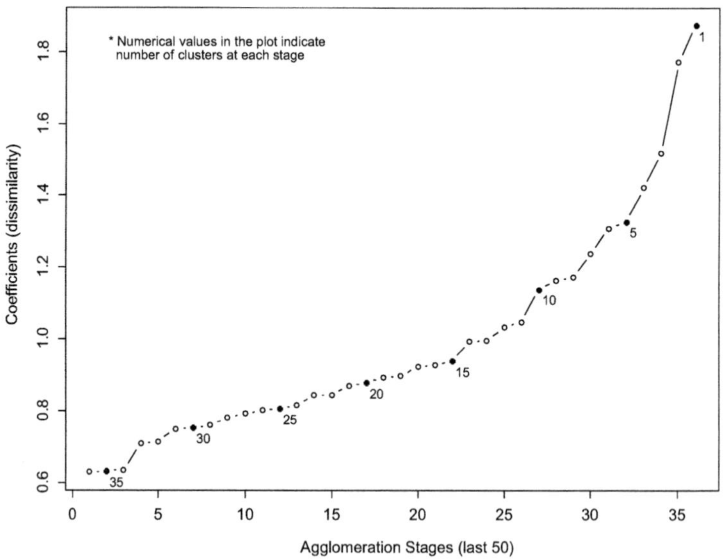

Fig. 6 Agglomeration stages of the document cluster analysis

Term Frequencies (TF) (Fig. 7)

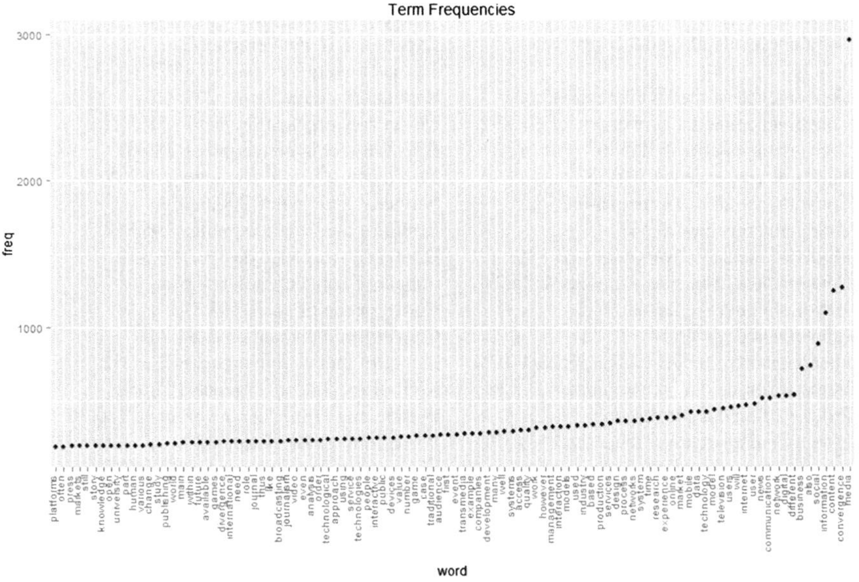

Fig. 7 Term frequencies of the most 40 used terms

Word Clusters

Word Clusters: Convergence

Media convergence/41459.593	Effects of convergence/263.659
Convergence of media/5276.675	Technical convergence/257.737
Technological convergence/4830.857	Digital broadcast media convergence/255.824
Organizational convergence/3734.403	Future of convergence/255.563
Technology convergence/2502.125	Form of convergence/244.492
Digital convergence/2411.864	Media convergence results/244.399
Market convergence/2034.277	Industrial convergence/243.142
Convergence process/1986.008	Concept of convergence/242.388
Network convergence/1985.531	Convergence trends/242.388
Convergence era/1833.161	Economic convergence/240.419
Process of convergence/1390.205	Impact of media convergence/240.212
Device convergence/1250.46	Specific convergence/238.91
Convergence of content/1247.204	Issue of convergence/235.675
Industry convergence/1115.899	Abstract media convergence/232.006
Convergence culture/1091.031	2nd convergence/231.612
Media convergence/960.024	Age of media convergence/229.792
Digital media convergence/898.199	Effects of media convergence/229.116
Business model convergence/886.534	Convergence of media industries/226.303
Content convergence/831.469	Data convergence/224.521

(continued)

Tv convergence/817.11	Network convergence success/223.393
Cultural convergence/807.233	Interactive tv convergence/222.39
Global convergence/796.413	Convergence problem/214.309
Media convergence/753.811	Convergence experience/211.342
Different levels of media convergence/753.34	Experience of convergence/211.342
Convergence of media content/645.631	Media convergence issues/209.786
Convergence media convergence/591.892	Regulatory convergence/209.321
Convergence policy/552.556	Media convergence predictor/208.712
Model of convergence/507.953	Effect of media convergence/205.61
Convergence of news media/489.921	Media convergence phenomenon/205.02
First convergence/475.537	Service convergence/200.684
Model of media convergence/464.839	Result of media convergence/196.813
Convergence of business models/462.427	Digital convergence design/196.399
Power of convergence/440.672	Impact of convergence/188.695
Convergence research/417.651	Convergence of users/185.315
Phenomenon of media convergence/410.04	Convergence phenomena/184.748
Introduction media convergence/403.226	Driver of media convergence/183.952
Process of media convergence/394.543	Kind of media convergence/183.952
Effective media convergence/380.561	Phenomena of media convergence/180.75
Era of convergence/366.632	Word convergence/180.665
Management of media convergence/356.183	Media convergence of today/177.237
Use of media convergence/354.155	Convergence age/176.552
Case of convergence of media/352.602	Convergence effects/175.773
Models of convergence/351.545	Debate of media convergence/174.685
Divergent convergence/346.535	Different types of convergence/173.845
Technological convergence of media/343.065	Convergence of industries/172.546
Media convergence analysis/331.954	Forces of media convergence/171.932
Successful media convergence/307.611	Convergence management/170.353
Global convergence of media/301.6	Other media convergence dynamics/170.02
Digital tv convergence/298.539	Convergence of networks/169.961
Aspects of media convergence/298.344	Use convergence/168.9
Term convergence/296.863	Use of convergence/168.9
Digital convergence issues/294.413	Case of convergence/167.79
Area of media convergence/289.837	Media research the concept of convergence/164.647
High convergence media content/288.358	Literature review media convergence/163.93
Concept of media convergence/283.853	Modern media convergence/161.03
Evolution of media convergence/283.766	Technological convergence/161.029
Social convergence/280.315	Structural convergence/160.957
Idea of convergence/280.267	Problem discussion media convergence/158
First convergence media/280.236	Viewpoint of media convergence/157.9
Cultural media convergence/274.605	Cross-industrial convergence/155.362
Age of convergence/264.828	

World Clusters: Divergence

Media divergence/1347.552	Issue of divergence/28.087
Social divergence/901.992	Divergence issues/27.531
Divergence of media business models/890.854	Effect divergence/26.713
Convergent divergence/331.053	Various divergence issues/26.137

(continued)

Media convergent divergence/303.159	Divergence of broadcast media/25.901
User divergence/186.746	Further divergence processes/24.827
Divergence of media forms/167.753	Possibilities of divergence/24.367
Convergent divergence media/151.579	Jensen-shannon divergence/21.379
Content divergence/148.638	Specific convergence-divergence aspects of qoe/20.385
Divergence of content/148.638	Audience convergence/divergence/19.554
Divergence of media business drivers/148.027	Divergence operating/19.242
Divergence models/125.688	Situation of divergence/18.6
6.5 convergent divergence/94.587	Terminal divergence/17.387
Technological advantage of media divergence/92.315	Content-based divergence/12.712
Market divergence/90.915	Intended divergence/11.83
Complementary process of media divergence/88.307	Divergence cross-technology issues/11.487
Divergence trends/86.661	Behavioural divergence/11.486
Divergence processes/75.066	So-called Kullback–Leibler divergence/9.491
Processes of divergence/75.066	Convergence/ divergence issue/9.196
Technological impact of media convergence/divergence/60.609	Strong divergence ion/9.071
Case of divergence/59.99	Outlined divergence/8.989
Ecosystem divergence/53.987	Convergence/ divergence effect/8.746
Kullback–leibler divergence/50.848	Phenomenon of convergence/ divergence/8.708
Divergence projects media disruption/49.475	Conditioned divergence/8.122
Post divergence/44.437	Divergence transpire/8.122
Divergence phenomena/44.035	Supports divergence/8.122
Degree of divergence/42.978	Main convergence/ divergence trade-offs/7.985
Cultural divergence/41.23	Convergence-divergence dilemma of qoe/6.685
One source of divergence/39.845	Convergence/ divergence trough/6.223
Divergence paradigm/39.721	Jensen-shannon divergence of eq./5.447
Need of user divergence/39.236	Convergence-divergence mechanism/3.798
Various divergence processes/32.136	Pro-divergence incentives/3.13
Convergent divergence framework/31.626	Space/ time divergences/1.414
Additional divergence/29.328	
Divergence dynamics/28.612	

Full World Clusters

Social media/69484.384	Value of information/1260.623
Media convergence/41459.593	Several media/1255.266
Media content/35209.445	Device convergence/1250.46
News media/24409.342	Convergence of content/1247.204
Media companies/18968.177	Digital broadcasting/1223.402
Media industry/17979.539	Journalistic quality/1220.407
New media/16171.599	New business model/1208.572
Digital media/13949.863	Online content/1159.214
Different media/13385.665	4.2 social media/1158.073
Business model/11168.862	Video games/1149.24
Traditional media/10582.376	Social networking/1145.295
Business models/8281.9	Media owners/1142.52
Social networks/7963.842	Media business development/1138.263

(continued)

Media business/7103.481	Social media features/1137.947
TV content/6516.154	Social interaction/1126.77
Social tv/6276.593	Cross-media content/1122.023
Converged media content/5565.936	Industry convergence/1115.899
Convergence of media/5276.675	New ways/1097.223
Digital tv/4860.404	Media landscape/1093.938
Technological convergence/4830.857	Convergence culture/1091.031
Interactive tv/4340.214	Media studies/1090.474
Convergent media/4098.611	Information stream/1087.57
Mass media/3941.827	Media producers/1070.026
Media experience/3929.058	Media market/1050.532
Social network/3921.883	User interaction/1049.775
Organizational convergence/3734.403	TV broadcasting/1036.183
Online media/3363.022	Belief systems/1008.26
TV series/3360.4	Content types/1000.741
Same time/3314.651	Business model change/997.96
Other media/3250.895	Digital information/985.2
Other hand/3192.07	Convergent media content/984.679
2.1 media/3114.236	Information architecture/975.07
New business models/3025.946	Current business model/973.303
Mobile tv/2971.688	Content management/970.359
Media space/2833.878	Content consumption/966.376
User experience/2722.473	Media convergence/960.024
Transmedia storytelling/2631.679	Media forms/954.362
Mobile phone/2629.286	Cross-media publishing/951.904
Media types/2540.364	Business ecosystem/948.63
News content/2504.56	Media content provider/944.804
Technology convergence/2502.125	User-generated content/943.604
Media managers/2497.945	New model/939.412
Digital convergence/2411.864	New content/922.635
Experience design/2398.807	TV set/920.299
Media markets/2391.198	Digital radio/919.156
Media industries/2316.741	New media corporations/917.241
Content production/2278.271	Traditional tv/912.466
Event ecosystem/2272.082	Multiple media/910.394
New media business models/2242.509	Publishing industries/905.305
New information/2205.451	Social divergence/901.992
New technologies/2195.902	Media design/900.632
Media models/2178.52	Digital media convergence/898.199
Media product/2157.365	Social media networks/894.864
Mobile devices/2123.649	Divergence of media business models/890.854
Media consumption/2044.274	Business model convergence/886.534
Market convergence/2034.277	Revenue model/874.517
New technology/2005.228	Content distribution/873.892
Convergence process/1986.008	Media tools/871.117
Network convergence/1985.531	Information management/869.823
TV experience/1892.884	Publishing industry/862.818
Media sector/1892.244	Provision of information/861.371
Business system/1890.668	Video content/860.841
Media production/1835.985	Media platforms/859.422
Convergence era/1833.161	TV market/854.06
TV screen/1804.645	Business concepts/852.939
News media industry/1764.942	Information technology/851.725

(continued)

Mobile phones/1750.489	New business/847.976
Media technology/1722.848	Different types of media/845.909
Media planning/1719.023	Internet media/839.119
Content of media/1717.534	Social media role/836.478
Media company/1712.974	Content convergence/831.469
Public media/1708.607	Other roles/818.495
Information content/1698.196	Tv convergence/817.11
Digital content/1648.606	Selected media/809.551
Social media content/1624.557	Advertising market/808.469
Media user/1618.407	Cultural convergence/807.233
2.0 media/1557.118	New services/806.275
Media users/1531.194	Open access/800.774
Media consumers/1511.248	Global convergence/796.413
Interaction design/1502.207	Social value/790.198
Social media platforms/1482.328	Intelligent network/783.903
Distribution of media content/1449.303	Multimedia experience/783.543
Online market/1418.07	Mobile media/783.28
Process of convergence/1390.205	New models/780.18
Media divergence/1347.552	Employee involvement/774.3
Content providers/1339.509	Same content/767.689
New service/1336.13	(the rest of the list is suppressed)
Media organizations/1329.929	
Mobile access/1322.896	
Interactive media/1281.277	
Quality of experience/1279.314	
Business model evolution/1275.07	

References

Ala-Fossi, M. (2015). Why TV bits and radio bits did not fit together? digitalization and divergence of broadcast media. In A. Lugmayr & C. Dal Zotto (Eds.), *Convergent divergence—Crossdisciplinary viewpoint on media convergence*. Germany: Springer.

Arango-Forero, G., Roncallo-Dow, S., & Uribe-Jongbloed, E. (2015). Rethinking convergence: A new word to describe an old idea. In A. Lugmayr & C. Dal Zotto (Eds.), *Convergent divergence—Crossdisciplinary viewpoint on media convergence*. Germany: Springer.

Damasio, M., Henriques, S., Teixeira-Botelho, I., & Dias, P. (2015). Between m-internet users and stakeholders: Convergence or divergence? In A. Lugmayr & C. Dal Zotto (Eds.), *Convergent divergence—Crossdisciplinary viewpoint on media convergence*. Germany: Springer.

Deliyannis, I. (2015). Converging multimedia content presentation requirements for interactive television. In A. Lugmayr & C. Dal Zotto (Eds.), *Convergent divergence—Crossdisciplinary viewpoint on media convergence*. Germany: Springer.

Denward, M. (2015). Designing, implementing and producing for participation: Media convergence practice. In A. Lugmayr & C. Dal Zotto (Eds.), *Convergent divergence—Crossdisciplinary viewpoint on media convergence*. Germany: Springer.

Dey, S. (2015). Convergence policy and regulation—A free speech perspective. In A. Lugmayr & C. Dal Zotto (Eds.), *Convergent divergence—Crossdisciplinary viewpoint on media convergence*. Germany: Springer.

Duh, A., Meznaric, S., & Korošak, D. (2015). Guerrilla media: Interactive social media. In A. Lugmayr & C. Dal Zotto (Eds.), *Convergent divergence—Crossdisciplinary viewpoint on media convergence*. Germany: Springer.

Ebner, M., Schon, S., & Alimucaj, A. (2015). Print your personal book. In A. Lugmayr & C. Dal Zotto (Eds.), *Convergent divergence—Crossdisciplinary viewpoint on media convergence*. Germany: Springer.

Fleury, A. (2015). User perspectives on media convergence: Results from quantitative and qualitative studies. In A. Lugmayr & C. Dal Zotto (Eds.), *Convergent divergence—Crossdisciplinary viewpoint on media convergence*. Germany: Springer.

Foster, S., & Brostoff, J. (2015). Digital doppelgangers: Converging technologies and techniques 3D world modeling, video game design and urban design. In A. Lugmayr & C. Dal Zotto (Eds.), *Convergent divergence—Crossdisciplinary viewpoint on media convergence*. Germany: Springer.

Georgiades, S. (2015). The effects of information in achieving employee involvement and thus organizational convergence in a media organization. In A. Lugmayr & C. Dal Zotto (Eds.), *Convergent divergence—Crossdisciplinary viewpoint on media convergence*. Germany: Springer.

Gershon, R. (2015). Redefining business space: Intelligent networking and the transnational organization. In A. Lugmayr & C. Dal Zotto (Eds.), *Convergent divergence—Crossdisciplinary viewpoint on media convergence*. Germany: Springer.

Giesecke, R. (2015). The live event ecosystem? local media against global ticket sellers? In A. Lugmayr & C. Dal Zotto (Eds.), *Convergent divergence—Crossdisciplinary viewpoint on media convergence*. Germany: Springer.

Grüblbauer, J., & Haric, P. (2015). Making money with & through information and knowledge? competitive intelligence, trend analysis and revenue models of classic media. In A. Lugmayr & C. Dal Zotto (Eds.), *Convergent divergence—Crossdisciplinary viewpoint on media convergence*. Germany: Springer.

Hamacher, K. (2015). Online presences of traditional media vs. blogs—Redundancy and unique news coverage. In A. Lugmayr & C. Dal Zotto (Eds.), *Convergent divergence—Crossdisciplinary viewpoint on media convergence*. Germany: Springer.

Ibrus, I. (2015). Micro-studios meet convergence culture: Clustering, dialogues, innovation. In A. Lugmayr & C. Dal Zotto (Eds.), *Convergent divergence—Crossdisciplinary viewpoint on media convergence*. Germany: Springer.

Immonen, S., & Giesecke, R. (2015). Media business drivers 2020—The case of David and Goliath revisited. In A. Lugmayr & C. Dal Zotto (Eds.), *Convergent divergence—Crossdisciplinary viewpoint on media convergence*. Germany: Springer.

Innocenti, V., Pescatore, G., & Rosati, L. (2015). Converging universes and media niches in serial narratives. An approach through information architecture. In A. Lugmayr & C. Dal Zotto (Eds.), *Convergent divergence—Crossdisciplinary viewpoint on media convergence*. Germany: Springer.

Jockers, M. (2014). *Text analysis with R for students of literature*. New York: Springer.

Karmasin, M., Diehl, S., & Koinig, I. (2015). Media convergence? implications for businesses: Is there a demand for new business models, organizational structures and management/employee competencies? In A. Lugmayr & C. Dal Zotto (Eds.), *Convergent divergence—Crossdisciplinary viewpoint on media convergence*. Germany: Springer.

Kueng, L. (2008). *Strategic management in the media—From theory to practice*. London: Sage.

Leminen, S., Huhtala, J.-P., Rajahonka, M., & Westerlund, M. (2015). Business model convergence and divergence publishing industries. In A. Lugmayr & C. Dal Zotto (Eds.), *Convergent divergence—Crossdisciplinary viewpoint on media convergence*. Germany: Springer.

Lugmayr, A., Serral, E., Scherp, A., Pogorelc, B., & Mustaquim, M. (2013). Ambient media today and tomorrow. *Multimedia Tools and Applications,* 1–31 Available: 10.1007/s11042-012-1346-z

Lugmayr, A., & Dal Zotto, C. (2013). Convergent divergence—Cross-disciplinary viewpoint on media convergence. In *European Media Management Education Association Conference 2013—Digital Transformations and Transactions in Media Industries*. Bournemouth.

Lugmayr, A., Zou, Y., Stockleben, B., Lindfors, K., & Melakoski, C. (2013). Categorization of ambient media projects on their business models, innovativeness, and characteristics evaluation of Nokia Ubimedia MindTrek Award Projects of 2010. *Multimedia Tools and Applications, 66*, 33–57. doi:10.1007/s11042-012-1143-8.

Matteucci, N. (2015). Standards, IPR strategies and digital TV convergence: The EU experience. Part I: Theories. Part: Empirical evidence. In A. Lugmayr & C. Dal Zotto (Eds.), *Convergent divergence—Crossdisciplinary viewpoint on media convergence*. Germany: Springer.

McKay, E. (2015). Gearing up the knowledge engineers: Effective experience design through effective human-computer interaction (HCI). In A. Lugmayr & C. Dal Zotto (Eds.), *Convergent divergence—Crossdisciplinary viewpoint on media convergence*. Germany: Springer.

McLuhan, M. (1994). *Understanding media: The extensions of man*. Cambridge, MA: MIT Press.

McLuhan, M., & Lapham, L. H. (1994). *Understanding media—The extentions of man*. Cambridge, MA: MIT Press.

Merriam-Webster. *Merriam-Webster online dictionary*. Available: http://www.merriam-webster.com/dictionary/

Montpetit, M.-J. (2015). The 2nd convergence: A technology viewpoint. In A. Lugmayr & C. Dal Zotto (Eds.), *Convergent divergence—Crossdisciplinary viewpoint on media convergence*. Germany: Springer.

Mustic, D., & Plenkovic, M. (2015). Influence of convergent media on the perception of information credibility. In A. Lugmayr & C. Dal Zotto (Eds.), *Convergent divergence—Crossdisciplinary viewpoint on media convergence*. Germany: Springer.

Reichl, P. (2015). Quality of experience convergent communication ecosystems. In A. Lugmayr & C. Dal Zotto (Eds.), *Convergent divergence—Crossdisciplinary viewpoint on media convergence*. Germany: Springer.

Sabiha, G., Morrison, P. A., Annika, W.-E., & Obal, D. (2015). Transmedia user experience design. In A. Lugmayr & C. Dal Zotto (Eds.), *Convergent divergence—Crossdisciplinary viewpoint on media convergence*. Germany: Springer.

Sacco, V. (2015). How does social media shape media convergence? The case of journalists covering war and conflict. In A. Lugmayr & C. Dal Zotto (Eds.), *Convergent divergence—Crossdisciplinary viewpoint on media convergence*. Germany: Springer.

Sousa, M. N., de Lemos Martins, M., & Zagalo, N. (2015). Transmedia storytelling: The roles and stakes of the different participants in the process of a convergent story, in divergent media and artifacts. In A. Lugmayr & C. Dal Zotto (Eds.), *Convergent divergence—Crossdisciplinary viewpoint on media convergence*. Germany: Springer.

Spyridou, L.-P., & Veglis, A. (2015). Convergence and the changing labor of journalism: Towards the 'super Journalist' Utopia. In A. Lugmayr & C. Dal Zotto (Eds.), *Convergent divergence—Crossdisciplinary viewpoint on media convergence*. Germany: Springer.

Sreekanth, N. S., Varghese, N., Pradeepkumar, C., Pal, V., Prasad, G., Pal, S., et al. (2015). Multimodal interface for effective man machine interaction. In A. Lugmayr & C. Dal Zotto (Eds.), *In Convergent divergence—Crossdisciplinary viewpoint on media convergence*. Germany: Springer.

Svahn, M., Wahlund, R., Denward, M., & Rademaker, C. (2015). A model for evaluating converged media for advertising purposes. In A. Lugmayr & C. Dal Zotto (Eds.), *Convergent divergence—Crossdisciplinary viewpoint on media convergence*. Germany: Springer.

Tavares, T., & Schofield, D. (2015). Interaction design for convergence medias and devices: A multisensory challenge. In A. Lugmayr & C. Dal Zotto (Eds.), *Convergent divergence—Crossdisciplinary viewpoint on media convergence*. Germany: Springer.

Tuomi, P. (2015). 21st century television—Gathering around second screens. In A. Lugmayr & C. Dal Zotto (Eds.), *Convergent divergence—Crossdisciplinary viewpoint on media convergence*. Germany: Springer.

Veglis, A., Dimoulas, C., & Kalliris, G. (2015). Towards intelligent cross-media publishing: Media practices and technology convergence perspectives. In A. Lugmayr & C. Dal Zotto

(Eds.), *Convergent divergence—Crossdisciplinary viewpoint on media convergence*. Germany: Springer.

Villi, M., Matikainen, J., & Khaldarova, I. (2015). Recommend, tweet, share: User-distributed content (UDC) and the convergence of news media and social networks. In A. Lugmayr & C. Dal Zotto (Eds.), *Convergent divergence—Crossdisciplinary viewpoint on media convergence*. Germany: Springer.

Vukanovic, Z. (2015). Converging technologies and diverging market trends of internet/web and Traditional media. In A. Lugmayr & C. Dal Zotto (Eds.), *Convergent divergence—Crossdisciplinary viewpoint on media convergence*. Germany: Springer.

Wellbrock, C. M. (2015). The internet's impact on journalistic quality. In A. Lugmayr & C. Dal Zotto (Eds.), *Convergent divergence—Crossdisciplinary viewpoint on media convergence*. Germany: Springer.

Wikipedia. *Wikipedia, the free encyclopedia*. Available: http://en.wikipedia.org/

Zlatanov, B., & Koleva, M. (2015). Info traders, innovation and information spread. In A. Lugmayr & C. Dal Zotto (Eds.), *Convergent divergence—Crossdisciplinary viewpoint on media convergence*. Germany: Springer.

Dal Zotto, C., Galichet, G., & Lugmayr, A. (2011). Managing media convergence: An economic and editorial project, In *European Media Management Conference 2011*, Moscow: European Media Management Association (EMMA).

About the Authors

Marko Ala-Fossi, University lecturer (Radio), Adjunct Professor, D.Soc.Sc., University of Tampere. Member of a pan-European research team on digital radio, (DRACE) (2004–). Former news journalist, producer and market analyst at the Finnish Broadcasting Company (YLE) (1989–2004). Research interests include media policy, political economy and social shaping of the new media delivery technologies.

Aldi Alimucaj holds an Master Degree on Technical Science. He studied computer science at Graz University of Technology and wrote his master thesis about "Individualization possibilities for Books with "Print On Demand" Technique".

A. Lugmayr, C. Dal Zotto (eds.), *Media Convergence Handbook - Vol. 1*, Media Business and Innovation, DOI 10.1007/978-3-642-54484-2

Germán Arango-Forero earned his doctorate degree in Communication at Universidad Austral, Buenos Aires, Argentina. He also attended the Media and Entertainment Management Program at the Institute for Media and Entertainment in New York. He is associate professor at the Faculty of Communication at Universidad de La Sabana, Colombia. Media Economics, Audiences Analysis and Entertainment Industry are his main areas for teaching. Audiences and media markets are his main topic for research. Board member of the International Journal on Media Management. He has been also working as a sports journalist since 1986 in print, radio, television and Internet. Currently he serves as TV sports anchor of Noticias Uno, Channel One in Colombia.

Prof. Dr. Cinzia Dal Zotto is Professor of Media Management at the Faculty of Economics and Business at the University of Neuchâtel, Switzerland. Before that and until December 2008 she was Research Director at the Media Management and Transformation Centre and Associate Professor at the Jönköping International Business School in Sweden. Dal Zotto received her Ph.D. in Organizational and Human Resource Development at the University of Regensburg in Germany. Between 2001 and 2003 she had a post doc scholarship funded by the German Ministry of Education and her research dealt with new ventures' growth in the media sector. In that program she supervised doctoral students and was a consultant for start-up firms. Dal Zotto has published three books, various book chapters, and papers in the fields of organization, human resources, entrepreneurship and strategy. She has taught at the universities of Regensburg and Passau in Germany, at the University of Westminster in London, at Trento and Bolzano University, as well as at the ESC Toulouse in France. She further held seminars for the Joint Research Centre in Ispra, Italy. Dal Zotto was a visiting scholar at UC Berkeley in 2002. She also worked as Research Analyst at International Data Corporation in London during 1995 and at Reuters Venture Capital in Munich in 2001.

Manuel José Damásio is the head of the film and multimedia department at Universidade Lusófona de Humanidades e Tecnologias (ULHT) in Lisbon, Portugal. He holds a Ph.D. from Universidade Nova de Lisboa and a M.Sc. from Napier University in the UK. He has worked both as a trainer, consultant and producer in the audiovisual industry at an European and international level and also has a strong past experience as head researcher and project manager in several national and European research and development projects.

Dr. Ioannis Deliyannis is an appointed lecturer at the Audio & Visual Arts department, Ionian University, Corfu, Greece. He holds a teaching post in the field of Interactive Multimedia. His research interests include the application of Interactive and Networked Multimedia Technologies in the area of Television and Audiovisual Arts. He is the author of a series of journal and conference publications, followed by a series of books targeting the experimental and creative aspects of the technologies involved, all available. At his free time he heads the Ionian University Sailing Team competing at regattas all over the Mediterranean sea with the sailing yacht Burgundy GRE-1364.

Shantanu Dey has been part of the Communication faculty of the '*Business Ethics & Communication Group*' of the Indian Institute of Management Calcutta during 2008–2013.

Dey holds a Ph.D. (Inter-disciplinary) degree from the National Law School of India University Bangalore, India; a Masters degree in Communication from the University of Massachusetts, Amherst, USA; and a Masters degree in Economics from the Delhi School of Economics, India.

Earlier, he has worked as Copy Editor, INDIA TODAY magazine; Executive Director, EXPRESS TELEVISION; Director-Programming, TV TODAY (India Today Group), Head of Programming, TIMES TELEVISION and Business Editor, TEHELKA. He has also worked as a Consultant to Star TV, Indian Broadcasting Foundation & Broadcast Worldwide.

Dey has written, produced and directed around 45 documentaries/non-fiction television programmes some of which have been screened in other countries.

Patrícia Dias is the author of "The Mobile Phone and Daily Life", a book about the social impact of the mobile phone. She is a lecturer and a researcher at the Catholic University of Lisbon (UCP) and at Lusófona University (ULHT), specialized in social media and digital marketing. She coordinates the Advanced Training in Social Media at UCP. She holds a Ph.D. in Communication Studies from Catholic University of Portugal with *Summa* Cum *Laude* distinction.

Charalampos A. Dimoulas was born in Munich, Germany on August 14, 1974. He received his diploma and Ph.D. from the Department of Electrical and Computer Engineering of Aristotle University of Thessaloniki (AUTh) in 1997 and 2006, respectively. In 2008, he received scholarship on post-doctoral research at the Laboratory of Electronic Media of the Department of Journalism and Mass Communication of AUTh. Both his doctoral dissertation and his post-doc research deal with advanced audio-visual processing and content management techniques for intelligent analysis of prolonged multi-channel recordings. He currently serves as Lecturer of Electronic Media in the Department of Journalism and Mass Media Communication, AUTh. His current scientific interests include media technologies, signal processing, machine learning, multimodal intelligent content analysis including multimedia semantics, audiovisual content description and management automation. Dr. Dimoulas is member of IEEE, EURASIP and AES.

Andrej Duh is a senior researcher at the University of Maribor. His current interests are distributed systems, Big Data analytics and mobile communications. Andrej has 20 years of experience in industrial production and research work at universities. He is the founder of several start-up companies in the field of ICT. Andrej was director of Solutions in a large international telecommunication company. He holds a Ph.D. in physics from the University of Ljubljana.

Martin Ebner is currently head of the Department for Social Learning at Graz University of Technology and therefore responsible for all university wide e-learning activities. He holds an Assoc. Prof. on media informatics and works also at the Institute for Information System Computer Media as senior researcher. His research focuses strongly on e-learning, mobile learning, learning analytics, social media and the usage of Web 2.0 technologies for teaching and learning. Martin gives a number of lectures in this area as well as workshops and talks at international conferences. For publications as well as further research activities please visit the website: http://martinebner.at

Kay Hamacher studied physics at Dortmund University, Germany. He obtained a Ph.D. in theoretical physics focusing on novel computational techniques to understand quantum mechanical systems in greater detail. After a deroute through business development and IT-security, he was awarded a Liebig-Fellowship to pursue studies in theoretical chemistry at the University of California, San Diego in J.A. McCammon's group. Following his passion to develop computational methods to understand complex systems he accepted a position as a Distinguished Fellow at the Max-Planck-Institute für Physics of Complex Systems in Dresden, Germany. From there he moved on to a faculty position in bioinformatics and computational biology at the Technical University Darmstadt, where he is now affiliated with the departments of biology, of physics, and of computer science. Accordingly, his research his multi-disciplinary and he is eager to reach out to other areas of science and transfer knowledge and insight back an forth.

Sara Henriques is a researcher in Communication Sciences at the Centre for Research in Applied Communication, Culture and New Technologies and a Ph.D. student in Communication Sciences at Lusófona University. She has developed academic research in the developmental and cognitive psychology field and more recently in the communication sciences field. Sara holds a M.Sc. in Educational and Developmental Psychology from ISPA—University Institute with *Summa* Cum *Laude* distinction.

Juho-Petteri Huhtala is currently a doctoral student in Marketing at the Department of Marketing, Aalto University School of Business, Finland. He holds a master's degree in marketing from the Aalto University School of Economics. His research interests include market orientation, product and service innovations, organizational learning, new product/service development and business models.

George Kalliris was born (1964) in Nicosia, Cyprus. In 1989 he received his Diploma in Electrical Engineering with Telecommunications from the Aristotle University of Thessaloniki. In 1995 he received his Ph.D. from the same University. His doctoral research was carried out at the Laboratory of Electroacoustics and TV Systems. During and after completing his doctoral studies he worked in research projects and as a part-time teacher. His current position is associate professor and director of the Electronic Media Lab at the Department of Journalism and Mass Communication. He has also taught in two Master degree programs, to the Film Studies Department and as a visiting professor of the Frederick University of Cyprus. His current research interests include audiovisual technologies for the new media, radio and television studio design, digital audio-video processing–production–broadcasting-webcasting, multimedia content, restoration, management and retrieval.

Irina Khaldarova, originally from Kyrgyzstan, is a Ph.D. candidate at the University of Helsinki, Media and Communication Studies. Her academic interests focus on how an idea or a news item travels through a population of the blogosphere and what role the web influencers, power bloggers in particular, play in this process. Before starting the Ph.D. course in Helsinki she obtained her master's degree from the School of Journalism, Mass Media and Cultural Studies at Cardiff University, UK. The thesis project was based on the examination of British political bloggers, especially their role in media agenda-setting from the perspective of two-step flow of information.

Maya Koleva is a Ph.D. candidate in the Faculty of Journalism and Mass Communication at Sofia University. Her research focuses on the application of network analysis to media content analysis in order to identify and measure the influence of key stakeholders featured in the media coverage of a certain issue. Maya holds a B.A. in Journalism from Sofia University and an M.A. degree in Public Policy with focus in media policy and regulation from Central European University in Budapest. She has worked as a foreign news editor in one of Bulgaria's largest dailies. Currently, simultaneously with her Ph.D. research, Maya is also working as a project manager in a specialised media research and evaluation company.

Dean Korošak is a physicist, Professor at University of Maribor and Head of Chair for Applied Physics. His current interests are network theory and its applications to complex systems, as well as Big Data analytics in particular in connection with social media. He is also the director of RAZ: UM—"Research and Arts Zone at University of Maribor"—established with a mission to help create, coordinate and advance innovative, transdisciplinary thinking at the interface between arts and science.

Dr. Seppo Leminen holds positions as Principal Lecturer at the Laurea University of Applied Sciences and Adjunct Professor in the School of Business at Aalto University in Finland. He holds a doctoral degree in Marketing from the Hanken School of Economics and a licentiate degree in Information Technology from the Helsinki University of Technology, currently School of Electrical Engineering at Aalto University. His research and consulting interests include living labs, value co-creation and capture with users, relationships, services and business models in marketing as well as management models in high-tech and service-intensive industries. Results from his research have been reported e.g. in Management Decision, International Journal of Product Development and Technology Innovation Management Review, among many others.

Artur Lugmayr is A/Prof. at Curtin University, Australia, where he teaches and supervises students in visualization technologies, interactive media, media management, and digital humanities. Artur was Professor for digital media management at the Tampere Univ. of Technology (TUT), Finland 2009–2014 establishing the Entertainment and Media Management Lab. and the New Ambient Multimedia Lab. 2004–2009. Artur holds a Dr.-Techn. degree (Information Technology, TUT), and is pursuing his Dr.-Arts studies at Aalto Univ., Helsinki, Finland in motion pictures. He was visiting scientist in Singapore, Brisbane, Austria, Ghana; since 2000 raised/involved in 1.7+ MEUR funding (excl. 2014/2015 applications); 170+ publications; 24+ invited keynotes; and 27+ invited guest lectures. He founded and chairs the Association for Information Systems (AIS) Special Interest Group "AIS SIG eMedia" and the International Ambient Media Association (iAMEA); and is active member of the ACM TVX steering board, IFIP TC 14 for Entertainment Computing, MindTrek Association, Academic MindTrek steering board, and IEEE IG MENIG. Previously he was international advisor for the NeoTV Lab/Buenos Aires/Argentina, founder/chair of the ISO/IEC SC29 WG 11 MPEG "MPEG-21 in Broadcasting". He won the Nokia Award of 2003 with the text-book "Digital Interactive TV and Metadata—Future Broadcast Multimedia" published by Springer-Verlag in 2004; ELISA award for the text-book "Ambient Media and Beyond"; a special mention for his NAMU research group of the ELISA Ubimedia Award at MindTrek in 2007 for the open source Portable Personality (P2) software. Artur has founded the Academic MindTrek Conference series, the Semantic Ambient Media (SAME) workshop series, the Nokia Ubimedia MindTrek Award series, the EuroITV Grand Challenge series, and the Interactive Digital TV in Emerging Economies workshop series. He organized the Academic MindTrek Conference for 7 years, as well as EuroITV conference in Tampere in 2010. He also is publisher and founder of the open access journal and series entitled International Journal on Information Systems and Management in Media Industry (ISMMI). He acted as review/committee member and editorial board member of several publishers and journals, as e.g. SERC Press, Wiley&Sons, IGI-Global, Springer-Verlag, IEEE, Inderscience, Int. Journal on Media Management; and as part of the review/committees of several conferences, as e.g. ACM-MM, AmI, ICEC, MobileCHI, ICME, EuroITV, NordiCHI, UIC, MMEDIA, Interact, uxTV, ACM-CHI, EBRF, IEEE-IE, VC, PDP. However, in the future he will only accept these "positions" when knowing the organizers. He organized 8+ int. conferences, 20+ int. workshops at conferences, held 13+ conference tutorials, organized 6+ media awards, edited 6+ books; and edited 12+ special issues.

Janne Matikainen, D.Soc.Sc., is a Researcher in Media and Communication Studies at the Department of Social Research, University of Helsinki. His research focuses on the social aspects of the Internet, especially the interaction in online communication and online participation. In the recent years, he has studied the relation between traditional media and social media. He has edited many books in Finnish on online communication, new media and research methods.

Nicola Matteucci (B.Sc., M.Sc. and Ph.D. in Economics) is Assistant Professor in Applied Economics at the Faculty of Economics of Marche Polytechnic University (Ancona, Italy), where he teaches Industrial Economics, Networks Economics and Game Theory, at the undergraduate and postgraduate levels. He is also a Research Fellow at UNU-MERIT, Netherland. He worked as researcher and consultant in national and international projects (including EU Framework Projects and EU Commission studies) dealing with ICT (Information and Communication Technologies), R&D and productivity issues. His main research interests are the economics and management of ICT and media industries, intellectual property rights in creative industries and antitrust and regulation profiles. His most recent work focuses on the economics and policy for broadband and digital TV. He regularly serves as referee for a number of scientific journals, and sits on various examination committees.

Sebastian Meznaric has recently has completed a D.Phil. in quantum information theory at University of Oxford, where he studied the ability of quantum mechanical systems to communicate and process information more efficiently than their classical counterparts. He developed quantitative techniques to identify those quantum systems and operations that posses uniquely quantum properties, making them into potential candidates for future applications in quantum technologies. After the completion of D.Phil., Sebastian has moved into the corporate world and currently works as an analyst at Goldman Sachs, developing systematic trading strategies at the asset management part of the firm.

Dr. Marie-José Montpetit is an internationally recognized Social TV, video and wireless innovator with over 15 years experience in architecture, and technologies. She is a lecturer on the future of TV MIT Media Lab and an advisor to local Boston startups. Her cross-platform IPTV project won the Motorola Innovation Award in 2007 and her work on Social TV was rewarded with a MIT Tech Review TR'10 in 2010. In 2012, Appmarket.tv named her one of the 20 top visionaries in second screen. Dr. Montpetit lives in Boston, MA with her daughter Mathilde.

Daria Mustić graduated in journalism at Faculty of Political Science, University of Zagreb in 2004. After graduation, she worked as a media analyst. Since 2007. she is working at the Faculty of Graphic Arts, where she finished her postgraduate study. Her area of interest are visual communication and media convergence. She is the editorial board member of scientific journals «Informatologia» (2007–2013) and «Media, Public and Public Relations» (2012–2013).

Mario Plenković, Ph.D., was elected as a full professor in Social Science scientific-educational field; scientific field—Information Science in 1998 (University of Zagreb, Zagreb). He is the President of Croatian Communication Association (1993–2013), President of International Federation of Communication Associations (2012–2014); leader of international scientific projects, director and leader of 20 international scientific conferences at «Inter University Center Dubrovnik» and co-organizer of 20 international scientific gatherings, on subject «Society and Technology». He is Editor-in-Chief of scientific journals «Informatologia» (1980–2013) and «Media, Public and Public Relations» (2001–2013).

Mervi Rajahonka is a post-doctoral researcher in Logistics at the Department of Information and Service Economy, Aalto University School of Business, Finland. She holds a doctor of science degree in economics (logistics) from the Aalto University School of Business, a master's degree in technology (industrial management) from the Helsinki University of Technology and a master's degree in law from Helsinki University. Her research interests include supply chain management, business models, modularity, processes and service innovations.

Sergio Roncallo-Dow, Ph.D. (Philosophy) and M.A. (Communications) from the Pontificia Universidad Javeriana, Bogotá Colombia. He has been professor and researcher at several universities in Colombia and is now associate professor at the Faculty of Communication at Universidad de La Sabana. He is a board member of Palabra Clave journal. His research interests are the philosophy of technology and communication as well as contemporary aesthetics and, in particular, the relationship between art and technology. In recent years he has dedicated his research to the thought of Marshall McLuhan.

Vittoria Sacco, Ph.D., is a post-doctoral assistant at the Academy of Journalism and Media of the University of Neuchâtel in Switzerland. She has publications on topics covering social media, new media technologies, the changes and values on journalism in times of crisis, and changes in journalism practices and identity. She teaches quantitative methods for social sciences. She participated in the data collection for the international research project Worlds of Journalism. She won a best paper award at IAMCR 2013 and in 2014 she got a scholarship from the Swiss National Science Foundation to finish her Ph.D. researches at Swinburne University of Technology, Melbourne, Australia.

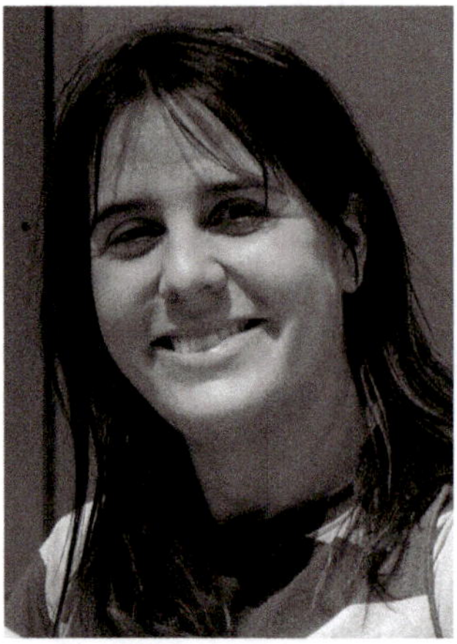

Dr. Sandra Schoen (born 1976) studied educational science, psychology and computer science at the Ludwig-Maximilians-University in Munich and received a master and a Ph.D. degree (Dr. phil.). Since 2006 she works at Salzburg Research Forschungsgesellschaft (Austria) in the InnovationLab. Her focus is on ICT innovations for learning, online communities and future studies. More here: http://sandra-schoen.de

Lia-Paschalia Spyridou is Visiting Lecturer at the Department of Social and Political Sciences at the University of Cyprus. She holds an M.A. in Communication from the University of Westminster and a Ph.D. in New Media and Journalism from the Aristotle University. For almost a decade she worked as a journalist in television and magazines. In 2012 she received an academic excellence grant for her postdoctoral research on participatory journalism from the Aristotle University. Dr. Spyridou has been among the main research members of the www.choose4greece.com voting aid application and www.wereportcyprus.org, the first participatory journalism platform in Cyprus. Her research work has been published in *International Communication Gazette*, *Journalism Practice* and *Journalism*. Her research interests lie in the fields of the 'technologisation' of journalism and news work, participatory journalism, social media and meaning construction in the public sphere, media and democracy, protest movements, e-campaigning.

Inês Teixeira-Botelho is the author of "Extreme Generation", a Vodafone's sponsored book about the impact of new technologies both in digital natives and emigrants. She has worked both as author for several publications, books and screenwriters, as speaker, consultant and advisor on communication, media, technology and entrepreneurship and also as researcher and project manager in Lusófona University (ULHT). She holds a M.Sc. from Catholic University of Portugal with *Summa* Cum *Laude* distinction.

Pauliina Tuomi, Master of Arts, is a Ph.D. student in Digital Culture at the University of Turku and works as a researcher at the Tampere University of Technology, in Advanced Multimedia Center (AMC), Pori Unit. Her doctoral thesis is set to define the different characteristic elements of participatory and intermedial TV culture in the twenty-first century. She is currently finishing her thesis on interactive, participatory and social television. Basically focusing and the areas of expertise include interactive SMS-based TV-entertainment, participatory features of convergence, both technical and thematical, between television and internet (web 2.0) + linkages between television and the use of social media (Twitter, Facebook).Tuomi has been working in various research projects on TUT concentrating for example on mobile learning, game-based learning and exergaming. Tuomi has published scientific papers and articles both on her work projects as well as a part of her theme of the thesis. In addition, Tuomi has been organizing international conferences and acting as a committee member and reviewer as well.

Enrique Uribe-Jongbloed holds a Ph.D. (Theater, Film and TV Studies) from Aberystwyth University, Wales. He is currently Assistant Professor and Director of the research group in Audiovisual Culture at the Faculty of Communication, Universidad de La Sabana, where he has lectured since 2007. His interests rest at the crossroads of identity, culture and media production, with a focus on minority media and cultural adaptations of audiovisual products.

Andreas Veglis is Professor, head of the Media Informatics Lab at the School of Journalism and Mass Communication at the Aristotle University of Thessaloniki. He received his B.Sc. in Physics, M.Sc. in Electronics and Communications, and Ph.D. in Computer Science, all from Aristotle University. In November of 2010 he was elected head of the postgraduate programme of the Department of Journalism and Mass Media Communication, Aristotle University of Thessaloniki. In March of 2012 he was elected Deputy Chairman of the Department of Journalism Mass Media Communication. His research interests include information technology in journalism, new media, course support environments, data journalism, open data and distance learning. He is the author or co-author of 12 books, he has published 50 papers on scientific journals and he has presented 74 papers in international and national Conferences. Professor Veglis has been involved in 18 national and international research projects.

Mikko Villi, D.A., M.Soc.Sc., works as a University Lecturer in Media and Communication Studies at the Department of Social Research, University of Helsinki. He acts also as the director of the Communication Research Centre CRC at the University of Helsinki. Villi has gathered knowledge especially on themes related to new communication technology and forms of communication, in particular on mobile, visual and digital media. Lately, he has focused his attention on social media, the media industry and the media ecosystem, where different media, devices and actors converge. His recent publications include articles in Interactions: Studies in Communication and Culture, Visual Studies, Participations: The International Journal of Audience and Reception Studies and several Finnish journals.

Christian-Mathias Wellbrock is an Assistant Professor for media management at the University of Hamburg's School of Business, Economics and Social Sciences since October 2012.

He has conducted research employing quantitative statistics as well as formal modeling. His research interests are in the fields of media industry regulation, media sponsorships, journalistic quality, and sports economics.

Christian studied economics at the Université Paris 1 Panthéon-Sorbonne, France and the University of Hamburg, Germany, where he received his Master's degree. He received his Ph.D. from the Department of Economics, Business and Social Sciences at the University of Hamburg for his work on economic analyses of journalistic quality.

His teaching experience includes courses on media economics, media management and microeconomics in both undergraduate and graduate education.

Christian was a visiting doctoral fellow at the Media Management and Transformation Center in Jönköping, Sweden, in 2009 and a visiting assistant professor at Michigan State University. His research has been published in several peer-reviewed journals.

Dr. Mika Westerlund is an Assistant Professor at Carleton University's Sprott School of Business. He previously held positions as a Postdoctoral Scholar in the Haas School of Business at the University of California Berkeley and in the School of Economics at Aalto University. His research interests include open innovation, business models, and management models in high-tech and service-intensive industries. Results from his research have been reported in numerous scholarly journals including California Management Review, Industrial Marketing Management, and European Journal of Marketing.

Biser Zlatanov (Asst. Prof. Dr.) teaches Media Marketing, Media Management and Business Communication. He studied Journalism, Financial Management, Business and Finance in Sofia. He earned his doctorate in 2009. His main research interests are in the fields of Media Innovation, Investment Valuation and Strategic Marketing. His latest research has focused on social network analysis and emerging value networks in the media industry. Zlatanov explores the shifts in growth opportunities in sustainable and disruptive economic environment. He regularly publishes articles in different journals and is a co-author of books and scientific research projects.